Sharlissa Moore's highly readable account of renewable energy transformations at the intersection of Europe and North Africa provides new insights into this rapidly changing and resource-rich region. Equally important, it synthesizes the very latest in multidisciplinary approaches to energy transitions within the complex sociotechnical systems that surround and enable our lives.

— Adam Reed, *Education Director, Renewable and Sustainable Energy Institute, University of Colorado Boulder, USA*

Many turn to renewable energy because of climate change and other reasons. Energy, however, is more than clever engineering. Politics, economics, ethics, and justice interweave with technology—locally, nationally, and internationally. Concentrating solar power projects to produce electricity in Morocco illustrate these challenges in this pathbreaking analysis, a must-read.

— John H. Perkins, *Member of the Faculty Emeritus, The Evergreen State College, USA*

In an engaging narrative style, Moore has created an important sociotechnical study of Morocco's push for renewable energy, dispelling for once the notion that energy is just about science and technology. This rich case study provides insights into issues of power and justice relevant for other regions of the world.

— Mary Jane Parmentier, *Clinical Associate Professor, School for the Future of Innovation in Society, Arizona State University, USA*

Sharlissa Moore skillfully shows how the imbalances of power in the formation of a collective vision of energy systems stubbornly replicate themselves as imagination turns into design, implementation and social outcomes. An important contribution to the scholarship of Science and Technology Studies and essential reading for everyone involved in managing and shaping energy systems. This study, empirically centered in Morocco, offers a more general lesson: that sustainability is not merely determined by technological possibility or the economics of energy production, but also by more subtle forces such as those determining whose voice is amplified or muted when we imagine our shared energy future.

— Walter D. Valdivia, *George Mason University, USA*

Sustainable Energy Transformations, Power, and Politics

This book analyzes energy transitions and the opportunities and challenges for building sustainable energy systems to improve human capabilities while protecting the environment.

Sufficient and secure energy supply is critical to human thriving and socioeconomic development. Yet energy systems are also implicated in the most pressing socioenvironmental challenges of our time—climate change, air pollution, and water and land use. This book examines what is arguably the most ambitious vision for a renewable energy–based system worldwide. This vision, often called Desertec, is for a regional electricity system supplying North Africa, Europe, and the Middle East with sustainable and affordable power. The behemoth plan would entail building dozens of large-scale solar and wind power plants, mostly in North Africa; interconnecting the fragmented transmission infrastructure of 38 Mediterranean countries; and linking North Africa to the European Union through undersea transmission cables. Within the Mediterranean region, the book focuses on Morocco, which is one of the most advanced developing countries in renewable energy scale-up, to understand its motivations for building renewable energy infrastructure and the effects of this scale-up on sustainable development. The book therefore takes a unique multi-scalar approach to understanding the social and political aspects of energy transitions, weaving together the views of villagers living near Morocco's first solar energy zone with the perspectives of national decision makers in Morocco and with the views of European policymakers and major transnational energy companies in the Mediterranean region.

This book will be of great interest to students, scholars and policymakers interested in energy transitions, sustainable and renewable energy, Mediterranean politics, sustainable development, and the environment and sustainability more generally.

Sharlissa Moore is an Assistant Professor of International Energy Policy, jointly appointed between the International Relations fields in James Madison College and the Civil and Environmental Engineering department at Michigan State University, USA.

Routledge Studies in Energy Transitions

Series Editor: Dr. Kathleen Araújo

Stony Brook University, USA

Considerable interest exists today in energy transitions. Whether one looks at diverse efforts to decarbonize, or strategies to improve the access levels, security and innovation in energy systems, one finds that change in energy systems is a prime priority.

Routledge Studies in Energy Transitions aims to advance the thinking which underlies these efforts. The series connects distinct lines of inquiry from planning and policy, engineering and the natural sciences, history of technology, STS, and management. In doing so, it provides primary references that function like a set of international, technical meetings. Single and co-authored monographs are welcome, as well as edited volumes relating to themes, like resilience and system risk.

Titles in this series include:

How Power Shapes Energy Transitions in Southeast Asia
A Complex Governance Challenge
Jens Marquardt

Energy and Economic Growth
Why We Need a New Pathway to Prosperity
Timothy J Foxon

Accelerating Sustainable Energy Transition(s) in Developing Countries
The Challenges of Climate Change and Sustainable Development
Laurence L. Delina

Petroleum Industry Transformations
Lessons from Norway and Beyond
Edited by Taran Thune, Ole Andreas Engen and Olav Wicken

Sustainable Energy Transformations, Power, and Politics
Morocco and the Mediterranean
Sharlissa Moore

For more information about this series, please visit: www.routledge.com/Routledge-Studies-in-Energy-Transitions/book-series/RSENT

Sustainable Energy Transformations, Power, and Politics

Morocco and the Mediterranean

Sharlissa Moore

Routledge
Taylor & Francis Group
LONDON AND NEW YORK

earthscan
from Routledge

First published 2019
by Routledge
2 Park Square, Milton Park, Abingdon, Oxon OX14 4RN

and by Routledge
52 Vanderbilt Avenue, New York, NY 10017, USA

First issued in paperback 2020

Routledge is an imprint of the Taylor & Francis Group, an informa business

British Library Cataloguing-in-Publication Data
A catalog record for this book is available from the British Library

Library of Congress Cataloging-in-Publication Data
Names: Moore, Sharlissa, author.
Title: Sustainable energy transformations, power, and politics : Morocco and the Mediterranean / Sharlissa Moore.
Description: 1 Edition. | New York : Routledge, [2018] | Series: Routledge studies in energy transitions | Includes index.
Identifiers: LCCN 2018017286 | ISBN 9781138579460 (hardback) | ISBN 9780429508011 (ebook)
Subjects: LCSH: Power resources—Environmental aspects—Morocco. | Renewable energy sources—Morocco.
Classification: LCC TD195.E5 M66 2018 | DDC 333.790964—dc23
LC record available at https://lccn.loc.gov/2018017286

ISBN 13: 978-0-367-58678-2 (pbk)
ISBN 13: 978-1-138-57946-0 (hbk)

Typeset in Times New Roman
by Apex CoVantage, LLC

Contents

Figures

Tables

Acknowledgments

A research project that crosses continents is invariably indebted to a network of supportive people, all instrumental in the realization of my dream to conduct fieldwork on a regional sociotechnical vision. I am deeply grateful to Ed Hackett for his countless hours of advice, feedback, and support that shaped the intellectual contributions of this book. I aspire to both his standards of scholarship and his principled mentorship. Many thanks to Mary Jane Parmentier, who has been a supportive colleague and friend who facilitated my work in Morocco and my participation in study-abroad programs in Morocco and Spain. Jamey Wetmore's detailed feedback over many years has been invaluable. Ben Minteer's class on environmental ethics, and later his participation on my dissertation committee, were of inspiration to the normative parts of this book. I appreciate the feedback and support I received from Monamie Bhadra and Jen Fuller, along with the support of many colleagues at the School for the Future of Innovation in Society at Arizona State University (ASU).

I deeply appreciate the support administrators and colleagues at James Madison College provided to me as I transitioned this work from a dissertation to a book. James Madison College students Iza Graham and Courtney Bourgoin and Michigan State University electrical engineering student Emily Fluent provided essential assistance with research on transnational grid projects worldwide.

I am thankful to the numerous people who generously shared contact lists with me, particularly Georgeta Vidican and Jonathan Walters. I appreciate the time of a number of interviewees associated with Dii, the Desertec Foundation, and German and Spanish governments and NGOs.

I am grateful to the ASU Study Abroad Office for supporting me in four trips to Morocco and Spain. I am humbled by the student cohorts from these trips, many of whom left the United States for the first time and embraced Morocco with compassion, excitement, and an eagerness to learn. I learned much from seeing Morocco through their eyes. My thanks go to Parker Helble and Mary Munoz Encinas, from the 2014 trip, for their assistance with energy justice research. Many thanks to Hakim Akki for his generous support. *Shukran bizef* to Iman Essamati from International Studies Abroad (ISA) in Meknes, Morocco for valuing my research when I was discouraged and to her family for welcoming me into their home and feeding me the best meals I have had in my life. If everyone shared Iman's empathy, tolerance, and inquisitiveness, the world would be a better place. Many thanks go to Dani Ostad at ISA for his undying enthusiasm and upbeat spirit.

I owe a deep debt of gratitude to the generous El Hilali family in Morocco for their hospitality, which, in part, ensured I did not sleep on the streets during my fieldwork. Meriem El Hilali was instrumental in helping me develop collaborations between experts in Morocco and the United States, and I am delighted to have collaborated with the MENA Policy Hub, which she headed. Youssef El Hilali's gracious assistance was essential in navigating daily life and language in Morocco. Also, thanks to Mehdi Ounjema for his insights and assistance. Without Mehdi I likely would have become irreparably lost navigating the city of Rabat. Many Moroccan government employees, researchers, and businesspeople showed me great hospitality, welcoming me into their offices for hours at a time, always with mint tea.

I also want to thank the Egyptians I met for welcoming me to their country during a precarious time. Thanks to Angeline Cione for her support and for exploring Cairo with me in the midst of a revolution. None of the Moroccans listed here directly contributed to or reviewed this book; any errors or misperceptions about Morocco are my own.

I am fortunate to have forged a serendipitous research collaboration with Boris Schinke of Germanwatch, important to Chapter 7. Thank you to Boris and the collaborators associated with the *Social CSP* report, including our skilled Moroccan research assistants: Fatima Ahouli, Souaad Ouzahim, Ghizlane Atik, Mohamed Ait El Haj, and Zakariya Ait El Bouhali. I am deeply grateful to the NGOs Association Tichka and The Human Touch for welcoming me, along with a group of American students, to Ouarzazate with heartfelt hospitality.

My thanks for financial support go to the US National Science Foundation's Doctoral Dissertation Research Improvement Grant, # DNS0307; ASU's Quantum Energy and Sustainable Solar Technologies Center; and ASU's Consortium for Science, Policy and Outcomes.

Last, but certainly not least, none of this would have been possible without my partner Aubrey Wigner's support, for which I am deeply grateful.

Abbreviations

COP: Conference of the Parties to the United Nations Framework Convention on Climate Change
CSP: Concentrating Solar Power (or Concentrated Solar Power)
Dii: Desertec Industrial Initiative/ Desert Power Industry Initiative
DLR: German Aerospace Center
ESMAP: Energy Sector Management Assistance Program (World Bank)
EU: European Union
EU-MENA: European Union, Middle East, North Africa
FDI: Foreign direct investment
FIT: Feed-in tariff
GEIDCO: Global Energy Interconnection Development and Cooperation Organization
GIZ: German Agency for International Cooperation
GDP: Gross domestic product
GW: Gigawatt
HVDC: High-Voltage Direct Current transmission (lines/links/cables/interconnections)
IEA: International Energy Agency
IR: international relations
ISO: Independent System Operator
ISCCS: Integrated Solar Combined Cycle System
kV: kilovolt
LCOE: levelized cost of electricity
MAD: Moroccan dirham
MASEN: Moroccan Agency for Solar (or Sustainable) Energy
MAScIR: Moroccan Foundation for Advanced Science, Innovation and Research
MCINet: Ministère de l'Industrie, de l'Investissement, du Commerce, et de l'Economie Numérique (Ministry of Industry, Investment, Commerce and the Digital Economy)
MEMEE/MEM: Ministère de l'Energie, des Mines, et du Développement Durable (Ministry of Energy, Mines, and Sustainable Development)
MENA: Middle East and North Africa
MOU: Memorandum of Understanding

MW: Megawatt
NGO: Nongovernmental organization
ONEE: Office National de l'Electricité et de L'eau Potable (National Office for Electricity and Potable Water, or Drinking Water)
IRESEN: l'Institut de Recherche en Energie Solaire et Energies Nouvelles (Institute for Research on Solar Energy and New Energy Technologies)
KfW: German-government owned development bank, Kreditanstalt für Wiederaufbau
kWh: kilowatt-hour
PSA: Plataforma Solar de Almería (Solar Research Platform in Almeria, Spain)
Solar PV: Solar Photovoltaic cell
RCREEE: Regional Center for Renewable Energy and Energy Efficiency
R&D: Research and development
RWE: German electric utility, Rheinisch-Westfälisches Elektrizitätswerk AG
SIE: Société d'investissements énergétiques (Energy Investment Company)
SMADER: La Société Marocaine de Développement des Energies Renouvelables (Moroccan Society for the Development of Renewable Energy)
STEP: *Station de transfert d'énergie par pompage* (pumped hydroelectric storage)
STS: science and technology studies
TREC: Trans-Mediterranean Renewable Energy Cooperation
TSO: Transmission System Operator

Part I

Large-scale energy system transformations

1 Introduction

Visions for sustainable energy transformations

Just after sunset, I am sitting outdoors at a café on the Friedrichstraße in Berlin waiting for dinner. As I look around me, I see all of the trappings of modernity on this commercial, cosmopolitan street. Nearby there is an H&M outlet, a Starbucks café, a Hugo Boss store, and the Westin Grand hotel. I hear bustling traffic and crowds of people speaking various languages, and the U-Bahn electric train rumbles below. As the sunlight fades, signs and streetlights begin to illuminate this modern "Great White Way," a term that described the world's first electrified streets. I imagine how much the Friedrichstraße has changed throughout its storied history. The new Kaiser Wilhelm Memorial Church is a half-mile from here, standing amid the ruins of the original 1890 church that was mostly destroyed in intense bombing during World War II. As recently as 1989, the Berlin Wall divided the street separating East from West Berlin, and the memory of the wall remains in the brick line that bisects the city. Checkpoint Charlie was active just down the street. As my thoughts return from the Berlin of the past to the present, I begin thinking of my recent interview with a consultant who is part of a group of investors seeking to build a Mediterranean-wide electricity grid and to export solar energy from the Sahara Desert to Europe. As I look around at the lights forming constellations in the city, I find myself wondering what this would all look like if such a vision were to come to pass. In fact, what would it look like if the entire Mediterranean region were powered with Saharan power? And what effects would an energy vision that stiches together areas encompassing broad power and wealth disparities have in other regions of the globe?

While it is dark now here in Berlin, the late-afternoon sun is still shining on the site of a solar power plant under construction in rural Morocco. Some imagine that the electrons generated at that site could be transmitted underneath the Mediterranean and across Europe to light the streetlights in front of me on the Friedrichstraße. It occurs to me that very little would look different here, yet much would change as a behemoth, complex system would be constructed to make Berlin's Great White Way come to life at night with renewable energy from the desert sun rather than from the burning of coal and the fissioning of uranium. I envision a sea of mirrors glimmering against an imagined backdrop of empty sand dunes in the Sahara Desert, making the desert blossom with green electrons. This dream is not new, but has been dreamt for over a century.

Six months later, I wander through the village of Tasselmante just south of the High Atlas Mountains of Morocco. My colleague remarks that this is an idyllic place. The mountain air smells fresh, and there is not a cloud in the sky. Children play in the street. Palm trees glisten in the oasis, and the village is nearly silent. There are ruins of a kasbah in the middle of the village, a testimony to the 3,000-year history of the Amazigh (Berber) dynasties in this area. Despite this peaceful scene, these are not empty sand dunes—Ouarzazate, a city of over 100,000 people, is only about 10 miles away, and the broader Province of Ouarzazate is home to a fast-growing population of almost 300,000.

A woman in a pink traditional *djellaba* (robe) pulling brackish water up from a rudimentary well in the village center invites us to talk with her. Clean drinking water is a long walk from here. I climb up the berm on the side of the village and see the beginnings of construction on the world's largest solar power plant just a stone's throw away. I wondered, what would the outcomes be for social and ecological justice if the power system were globalized to bring together disparate places, such as the commercial center of East Berlin with its storied past and a small village at the door of the Sahara that lacks clean drinking water? And what are the broader implications for sustainable development for large-scale solar power?

The context: visions for transnational renewable energy systems

Electrical grids are "critical infrastructure" in the industrialized world, driving economies supporting modern life (Farrell, Zerriffi, & Dowlatabadi, 2004). Electricity powers the production of materials and goods, the operation of the Internet, the provision of water, household appliances, and the production and transportation of food. At the nation-state scale, energy is a strategic resource relating to geopolitics and power, but also a source of socioeconomic development. At the local scale, access to electricity improves human capabilities through services such as lighting, refrigeration, water filtration, and heating and cooling.

At the same time, climate change mitigation is largely an energy problem, and greenhouse gas emissions from dirty electricity generation contribute substantially to increasing global temperature, as well as local air and water pollution and public health effects. This requires countries to shift their electrical power systems from fossil fuel sources to renewable electricity sources, such as sun and wind. These are not matters of mere technological substitution, but instead have significant human and social dimensions, relating to the ways in which nation-states and citizens imagine what an ideal future would be. Although it is often assumed that renewable energy will de facto obviate the social, political, and justice problems fossil fuels cause and therefore will be intrinsically sustainable, this study illustrates that this is not the case. Rather, sustainability must be intentionally planned in energy system design. A transition has the potential to negatively affect people's lives, displacing workers from the fossil fuel industry, rewriting geographies from large central station power plants to sprawling solar and wind facilities, bypassing marginalized communities, and inequitably distributing costs through subsidies that disproportionately benefit upper-income communities. Sustainable energy policy requires a higher standard than generating electrons for the sake of electrons or than climate change mitigation alone. Instead, it requires an improvement in the capacity of people to secure their human rights, to sustain what they value, and to chart a path to human thriving, fostering what Karlsson-Vinkhuyzen and Jollands (2013) call "deep energy security."

Interdependent renewable electricity systems are being envisioned that require partnership, integration, and negotiation among developed and developing countries, pairing the challenge of building sustainable energy systems with the complexity of North-South cooperation. This book explores how actors separated by broad power and wealth disparities and heterogeneous energy needs envision regional sociotechnical systems and the technologies embedded within them. It asks how actors and nation-states envision what a sustainable and a "good" energy system would look like. An exploration of visions for electricity systems illustrates the values and goals underpinning competing dreams for energy futures and their anticipated outcomes.

After the Chernobyl nuclear reactor accident in 1986, physicist Gerhard Knies dreamed of safer large-scale sources of energy. He made a back-of-the-envelope calculation finding that "within six hours deserts receive more energy from the sun than humankind consumes within a year" (http://www.desertec.org). Because 90% of the world lives within 1,800 miles of a desert, and since high-voltage direct current (HVDC) transmission lines can transmit electricity with losses of only 3% over 1,000 kilometers, he calculated that the world could be powered by renewable energy from deserts.

Building on this idea, from 2008 to 2015 an ambitious vision for a renewable electricity system unfolded in the Mediterranean region. This vision would entail building dozens of large-scale solar and wind power plants, mostly in North Africa; interconnecting the somewhat fragmented transmission infrastructure of 38 Mediterranean countries; and linking North Africa to the European Union (EU) through undersea transmission cables to develop a power system fueled almost exclusively by renewable energy. As detailed in this book, multiple versions of the Desertec vision were developed. One of several organizations known for promoting this vision, the Desertec Industrial Initiative (Dii), is a limited-liability company and a conglomerate of companies that at its peak reached 56 shareholder and partner companies. The novelty and complexity of Dii's organizational structure and network rivaled the novelty and complexity of its vision, and this is in addition to the numerous actors promoting similar regional electricity systems worldwide.

Morocco is also an important country in this vision, but—more importantly—it is a strategic site for understanding renewable energy scale-up and energy security in developing countries. Morocco is committed to developing 10 gigawatts (GW) of solar and wind power by 2030 and is interested in regional renewable energy integration. Morocco has charted an ambitious path for renewable energy that illustrates the reasons that developing countries might pursue clean energy as part of a socioeconomic development strategy. Construction is well underway on the Noor-Ouarzazate solar power facility near the High Atlas Mountains of Morocco, which, when complete, will be the largest concentrating solar power plant worldwide. Noor provides important lessons for understanding the effects of large-scale solar power on local villages and for understanding how power plant siting would affect the energy justice dynamics of the regional system.

The intended audience of this book includes sustainability and sustainable development scholars, international relations scholars interested in energy, scholars interested in the societal and ethical dimensions of science and technology, energy policy and sustainable development practitioners, and students studying climate change and renewable energy. My aim is to write accessibly about complex and wide-reaching topics.

The challenge of envisioning and designing sustainable energy systems

This book contributes to two themes. The first is the role of the future in electricity systems: How actors imagine the future and integrate it with the past, and how people envision a transition to future sustainable electricity systems, particularly across nation-state lines. This visioning process plays an important role in social and technological innovation and in shaping the electrical power systems of the future. The second theme is the relationship between energy systems and sustainable development, particularly socially sustainable development. As the goals for electrical power systems expand from powering cities to empowering human development, this book explores the energy justice aspects of changes in energy systems.

Socio-energy systems: Crucial to understanding the societal and policy significance of energy transitions is conceiving of electrical power systems not as technological systems but as sociotechnical systems. Sociotechnical systems theory examines the co-shaping of the hardware of the system, (e.g., power plants, transmission lines, transformers), as well as the political, social, and institutional aspects of systems, (e.g., regulatory institutions, political frameworks, activist groups, citizens, engineers, and financiers). Analyzing sociotechnical systems requires understanding both the material techno-economic aspects of electricity and the less tangible socioinstitutional aspects, including actors, norms, visions, and values (Sovacool & Geels, 2016).

Historians who studied the electrification of Western countries laid the groundwork for this approach (see, for example, Glaser, 2009; Hughes, 1983; Nye, 1990; Rose, 1995; Tobey, 1996). Hughes (1983) first argued that electrical power systems are better characterized as sociotechnical systems, as they are shaped by economics, politics and political institutions, and geography. These systems develop within exercises of power to reflect the interests, values, and politics of powerful actors. Therefore, technological standards—such as the frequency of the transmission network—are influenced by political and social factors; they are not established for technical reasons alone. Hughes viewed electricity systems as groupings of interacting parts arranged in a network configuration, parts which have historically been centrally controlled and hierarchically arranged (Hughes, 1983). The system exerts an influence on its environment (e.g., on communities, landscapes, institutions) and is also shaped by its environment (e.g., by geography, major events, policies) (ibid.). Hughes's comparison among three countries illustrates that the electricity system is a "cultural artifact" influenced by the technological style of each individual country. Hughes describes "system builders" as critical actors in the construction of electrical power systems who possessed both the technological acumen to invent new technologies and the business and networking prowess to develop a system into which these new technologies would be integrated. I refer to the actors working today to reimagine and reshape electricity systems as system imaginers.

As the system advances in this exercise of social and mechanical power, it achieves substantial momentum with mass, speed, and direction (Hughes, 1983). Once systems have achieved momentum, also termed path dependence or technological lock-in, they are remarkably difficult to change (Arthur, 1994; Hughes, 1994). Historically, a major event, such as a world war, has been required to transform such systems (Hughes, 1983). However, an intentionally governed energy transition could perhaps occur without such a momentum-breaking event because of today's explicit policy and social goals governing transitions (Kern & Rogge, 2016).

Transforming these systems is a central policy challenge of our time. Geels and collaborators developed the "multi-level perspective" on sociotechnical transitions to sustainability, which focuses on innovation-driven change in systems. Geels (2004) conceives of "niches" as incubators in which radical innovations develop and potentially carve out a place in the system. However, it is difficult for new technologies to break into a path-dependent system (Verbong & Geels, 2007). Another body of work focuses on systems transitions enacted on the basis of the choices that policymakers and citizens make among various energy technologies (Meadowcroft, 2009; Perkins, 2017). Miller, Richter, and O'Leary (2015) borrow from sociotechnical systems theory to define "socio-energy systems" as "a way of recognizing that energy systems involve the work, behavior, and choices of many different kinds of people" (p. 30). A third body of work focuses on change evoked by a visioning process that ultimately results in material reconfigurations, as discussed later.

Imaginaries and visions for sociotechnical systems: In today's knowledge societies, large-scale sociotechnical systems' evolution is often guided by a distinct and systematic visioning phase for specific technologies and systems, as well as broader imaginaries for how technological progress relates to social progress. Jasanoff and Kim (2015) define sociotechnical imaginaries as "collectively held, institutionally stabilized, and publicly performed visions of desirable futures animated by shared understandings of forms of social life and social order attainable through, and supportive of, advances in science and technology" (p. 4). An imaginary relates to how meaning is made for material objects and how nation-states imagine what a "good life" would be (ibid.). This theory places greater emphasis on the state and on policy than the multi-level perspective does, which focuses on how technological innovations break through a system. I draw on sociotechnical imaginaries theory to discuss Morocco's national imaginary for renewable energy in Chapter 5.

While a nationally-held imaginary is by definition promulgated by a nation-state, I define a sociotechnical vision as a specific plan for a system developed and promoted by a network of actors. I use the concept of a sociotechnical vision to discuss concrete plans for a regional sustainable electricity system, and I map out the constituent parts of a sociotechnical vision in Chapter 4. Studying an energy transformation in its early—visioning—stage makes visible the sociopolitical conflicts surrounding changes to modern infrastructure that are often only visible when it fails (Edwards, 2003). The invisible infrastructure includes, for example, the power poles that blend into the urban landscape, the electricity generation that is out of sight, and the gas stations people are accustomed to seeing at major intersections.

The visioning process plays a crucial role in guiding innovation and has a lasting influence on a system's shape and societal and justice outcomes. It encompasses formal processes, such as developing targets, legislation, and technological roadmaps, as well as conducting studies and making models to prove that the vision is feasible. The visioning process also includes informal processes—for example, conversations and decisions made behind the closed doors of corporate conference rooms and government offices. It plays a crucial role in proffering, jettisoning, and cementing aspects of the system's design. Decisions made in the visioning process steer innovation and have a lasting influence on the system's shape if it is built. The visions I discuss in this book go beyond what Sovacool (2014) calls the "rhetoric of energy transitions." They are powerful and do work in the world. They shape what is materially constructed in direct and indirect ways. They are also resilient, reemerging in different time periods and contexts over more than a century.

Understanding the future from a sociological perspective is essential for understanding major changes in energy systems. Theories on the "sociology of expectations," or the "sociology of anticipation," explain the importance of the normative work actors perform in the visioning stage to influence future materiality (Borup, Brown, Konrad, & Van Lente, 2006). While the future does not yet exist, it gains an ontological status by being constructed in the present, as the "immaterial real" (Adam, 2004). Actors make representations of the future, which are "objectified on paper" (Brown, Rappert, & Webster, 2000). Positioning the future as proximal or distant does work in the present (Selin, 2008). As technologists and politicians make promises about the future benefit of technologies, they become requirements that can lead to the failure of those technologies when expectations are disappointed (Borup et al., 2006).

It is important not just to understand but also to improve the visioning phase, as energy system builders grapple with long time horizons and uncertainty. An investment cycle in the power sector is between 25 and 50 years long, encompassing planning, permitting, constructing, operating, and decommissioning.

Poorly planned dreams for electricity systems could later turn into nightmares. For example, the complex energy market and management institutions created in California in the 1990s disrupted the complex power system—which also was socially manipulated by Enron—resulting in rolling blackouts and bankruptcies (Hirsh, 1999). The pressure can be great to maintain a new system, even if it perpetuates the social inequities of the current system. Better planning in the visioning phase could help system imaginers to anticipate future problems and avoid building technological juggernauts. Thus, anticipatory governance of socio-energy systems is needed, to anticipate problems and inequities and to plan for them early in the design phase using methodologies of foresight, public engagement, and the integration of social scientists into laboratories (Barben, Fisher, Selin, & Guston, 2008).

Sustainable development and energy systems change: It is important to layer onto sustainable development a more complex understanding of the future as an ontological object and an examination of the upfront stages of innovation. Sustainability is intricately connected to temporality (Held, 2001). The concept relates to how something can be propped up over time and how the current generation's choices affect future generations. The Brundtland Report (1987) defined sustainable development loosely in terms of intergenerational justice, as "development that meets the needs of the present without compromising the ability of future generations to meet their own needs" (p. 41). The definition of sustainable development is contested and pluralistic. It is culturally specific, and ought to be since it relates closely to what people value (Lélé, 1991). It is important to study visions and imaginaries related to sustainability, as they contain important indicators of how a social group, community, or nation defines its values on the basis of what members cherish about the past and what they wish to build for the future. Visions elucidate the goals of particular actors in current generations, and also allow for reflection on what will be inherited by future generations.

Just as it is clear that sustainable development is a fundamentally temporal concept, it is also clear that it is a moral concept (Thompson, 2010). Ethical principles could productively influence the sociotechnical imaginary that organically emerges from the state, market, and society. This is important because popular discourses on sustainability can be linguistically impoverished (Norton, 2005) and serve to narrow and close down debates about sustainable energy futures (Leach, Scoones, & Stirling, 2010).

Some environmental ethicists emphasize the importance of the non-anthropocentric value of nature, or that each organism has a "good of its kind," or an intrinsic value, separate from its human-attributed instrumental value (Rolston, 1998, p. 131). However, environmental pragmatists emphasize that developing non-anthropocentric value is difficult, or even impossible, in practice. Norton's "convergence hypothesis" speculates that environmental policy goals based on long-sighted anthropocentric views would converge in the long term with those based on non-anthropocentric values (Minteer, 2009). In contrast, weak, econometric conceptions of sustainability focus on maintaining a fair savings rate of natural capital among generations to avoid a decline in welfare in the short term

(Norton, 2005). Strong sustainability conceptions could aid in achieving convergence, protecting "things" that cannot be valued merely economically and assuming resources are not fungible. Valuing multiple, noneconomic instrumental qualities in nature, including recreational, aesthetic, and scientific values, can increase the likelihood that stakeholders will protect the environment (Minteer & Manning, 2009). I illustrate in this book that this is the case in Morocco, which is focused on social rather than environmental goals in its climate change mitigation policy (see Chapter 5).

Norton (2005) argues that just pluralistic participatory processes must be developed to select what should be valued and sustained. This selection process is not entirely abstract, nor must it occur only through an intentionally planned public participatory process. Instead, I argue that it is a challenge of building sociotechnical systems that enable human thriving as well as the protection of nature. Because technology is often seen as a vehicle for social progress, the sociotechnical realm is useful for understanding values and shared priorities, to conceive of what a socially and environmentally sustainable socio-energy system might look like early in the system's design.

Often, sustainability is defined as achieving optimal outcomes across the triple bottom line of people, planet, and profits, or the pillars of society, environment, and economy, which the UN Sustainable Development Goals seek to balance. Partly because of its complexity, the social pillar has been insufficiently conceptualized. The concept of energy justice could aid in understanding social sustainability for energy systems design. This avenue reflects the "ethical turn" in sustainable development research toward "moral imperatives" (Holden, Linnerud, & Banister, 2017). It requires a broader, more holistic and qualitative understanding of social sustainability than the approach of developing social sustainability indicators (for example, Strezov, Evans, & Evans, 2017), which can end up requiring analysts to weigh absurdly incongruent categories against one another in a model.

Sovacool and Dworkin (2015) view energy justice as a "conceptual tool" for evaluating different categories of justice in one framework, an "analytical tool" for unmasking values embedded in energy systems design, and a "decision-making tool" for planning. Jenkins, McCauley, Heffron, Stephan, and Rehner (2016) argue for a whole systems approach that accounts for "three prongs" of energy justice: distributive, procedural, and recognition justice. Distributive justice examines the fairness of outcomes, accounting for who benefits and who loses from energy systems. Procedural justice refers to the fairness and inclusivity of the stakeholder process for energy decision making. Recognition justice argues that marginalized groups, based on race, ethnicity, class, and gender should be recognized and given full political rights. Finally, it is important to understand the human capabilities energy systems provide for people to meet their basic needs, as a precursor to human thriving and imagination (Nussbaum, 2003).

Energy justice in a globalized world is fundamentally multi-scalar. Injustice must be diagnosed in complex energy systems that affect people differently at local, regional, national, and international scales. Complex energy supply chains and energy systems interweave justice issues in disparate locations. The win-win solutions discourse popular in sustainable development overlooks the complexity of trade-offs and distributive injustice at different scales (McShane et al., 2011). Some attention has gone toward issues of cross-border emissions, including who is responsible for which emissions (Sovacool, Burke, Baker, Kotikalapudi, & Wlokas, 2017). It is also important to examine systems being built across scales, at what scales people win and lose, and the history of inequity layered onto new projects, including structural injustice relating to North-South socioeconomic disparities and colonial history.

Economic justice among developing and developed countries is sometimes conceived of in terms of access to technology and technological systems. Graham (2000) argued that an individual's power in society relates partly to his or her level of access to high-reliability networks. Graham calls areas (e.g., neighborhoods, countries, communities) with preferential access to water, mobility, electricity, and communications infrastructure "customized precisely to the needs of powerful users and spaces, whilst bypassing less powerful users and spaces" "premium networked spaces" (ibid., p. 185). Premium networked spaces are constructed by "*strategies* of [powerful] coalitions of interests within the contested and highly complex geopolitical and governance context of their respective cities" (ibid., p. 186). I use this term throughout this book. I found that many Moroccan officials see Moroccans as being unjustly severed from premium networked spaces owing to European politics.

Access to technology is far from the only North-South issue at stake. The South risks being coerced by more powerful countries to meet environmental sustainability goals in a way that does not match the interests of countries in the Global South. Western countries often focus on the environmental pillar of sustainability and on changing culture to encourage citizens to make environmentally responsible decisions. Developing countries often focus on socioeconomically sustainable development, as well as on the question of whether economic development should occur before environmental protection or whether the two are "mutually reinforcing and prerequisites for each other" as the Brundtland Report envisioned (Hjerpe & Linnér, 2009, p. 238). Capacity building is needed to situate countries on a level playing field, to begin to chart out new futures for current generations of social groups experiencing distributive injustice stemming from energy systems.

Dreaming the future: scale and energy system transformation pathways

Building a sustainable energy system includes a refurbishing and retrofitting agenda, as system builders cannot start from scratch but must graft new infrastructure onto existing infrastructure that has achieved momentum, as well as onto a sociopolitical landscape with disparate energy policies and goals. Moreover, electrical power systems are no longer called on only to provide light and heat, as they were in the early 1900s, but rather to improve the environmental and social sustainability of complex, interconnected infrastructure. The relevant issue areas are wide-ranging, including climate change, development, population growth, poverty, water, and food. Rather than comparing one technology to another, energy policy for the twenty-first century must view energy as an integrated-systems-level challenge to provide even greater human services and capabilities while avoiding environmental harm.

Energy systems are continually in transition to a certain extent, as power plants are retrofitted or decommissioned and fuel sources shift. "Techno-economic rationales" relating to sunk costs in infrastructure, economic logic, and steep learning curves for new technologies have contributed to slow, incremental change in energy systems (Sovacool & Geels, 2016). Energy historian Vaclav Smil argues that it will take generations to achieve a global energy transition, although some national energy transitions could be achieved in several decades (Smil, 2016). While some transitions historically have taken a long time, Sovacool (2016) offers some national-scale examples in which transitions have occurred more rapidly. Intentional policy goals could allow for more expeditious changes than in the past (Kern & Rogge, 2016).

Laird (2013) argues that the term "energy transition" has been used so narrowly to describe techno-economic fuel substitutions that it masks the important social and political stakes of energy transitions. I use Miller, Iles, and Jones's (2013) term "energy transformation" to highlight the extent of the systems-level changes that this book considers. This is not merely an issue of semantics but rather emphasizes the need to improve on frameworks until they are sufficient to the task of understanding *both* the techno-economic and the socioinstitutional aspects of major changes in energy systems (Sovacool & Geels, 2016).

Power systems overview: Today, electrical power systems, which are mostly centralized in configuration, encompass grids that include generation, transformation, transmission, and distribution (see Figure 1.1). First, electricity is generated mainly in large-scale power plants, which are often sited in remote areas. Then the voltage is ramped up for efficient transmission along transmission lines to cities. Once the power reaches cities, the voltage is ramped down for safe distribution along power lines to consumers. Utility companies sell electricity to consumers to use for heating and cooling, refrigeration, entertainment, and other end uses.

Electricity demand, or load, changes with respect to the time of day and season. Baseload demand refers to the minimum power demand throughout the day. Peak hours of demand (peak load) vary across countries and regions, on the basis of the type of industry, the climate and season, and the patterns of household demand. Supply and demand must be balanced on the grid at every second, which is a technological feat. Blackouts occur when demand exceeds supply. Load shedding occurs when supply cannot meet demand and some consumers are deprived of electricity to prevent a full blackout.

Currently, different types of generation facilities are used to supply electricity at different times of day. Baseload power plants, traditionally coal and nuclear, provide power 24-7 and ensure the system operates at a constant voltage. However, they are inflexible and cannot be easily or cheaply turned on and off or ramped up and down. Power plants that can be easily controlled are called "dispatchable" sources of electricity. Previously, natural gas power plants, which are dispatchable, were used to meet peak demand, but natural gas plants are increasingly being used

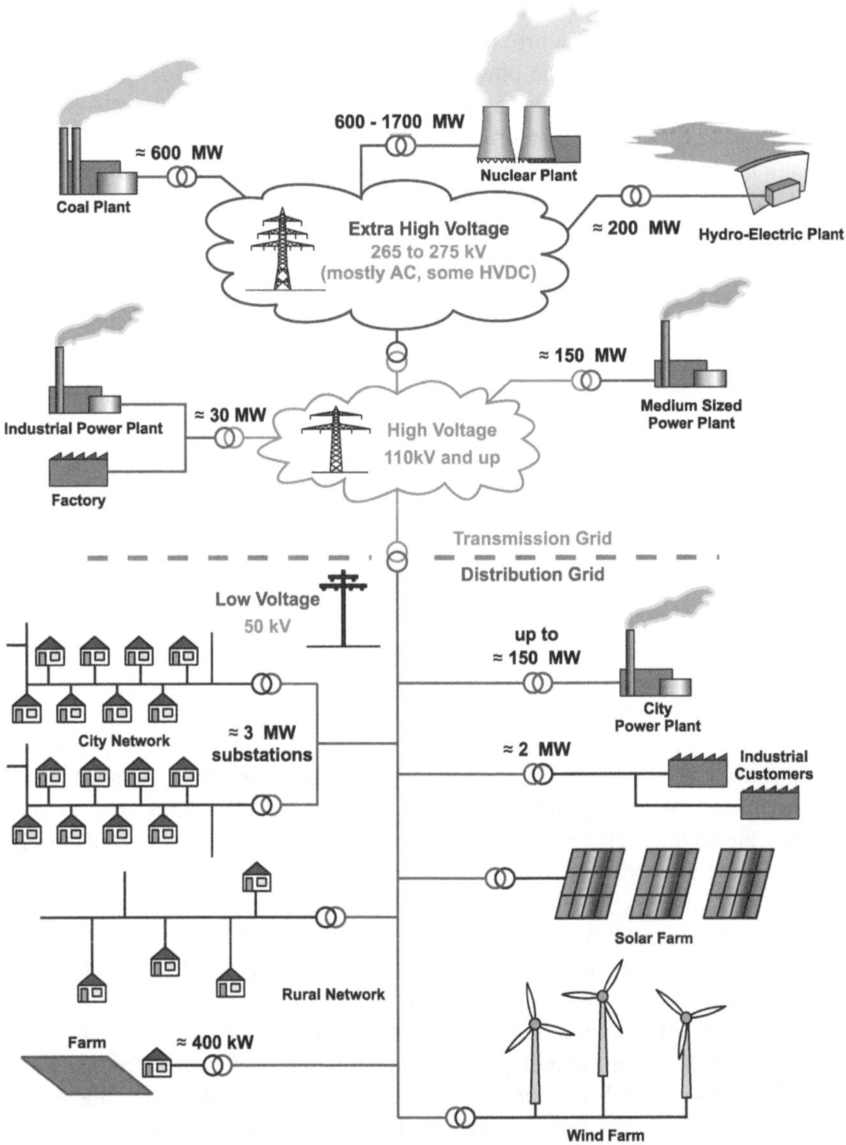

Figure 1.1 The grid

From Wikimedia Commons by MBizon

to meet baseload demand. While there is often no direct market for the service of dispatchability, it is valued indirectly for its role in maintaining reliable supply.

Because electricity supply and demand are balanced within a system, the market for electricity differs from markets for commodities such as crude oil or oranges. The discourse on supply and demand is limiting, but is also crucial to understanding electricity systems. Özden-Schilling (2015) observes that the process of maintaining supply and demand in electrical power systems is so unique that what is typically a discourse used by economists has been reified as a priority for a variety of actors in the power system. In the early 1900s, system builders treated baseload power plants as "load-seeking turbines," constantly creating new demands for electricity in order to utilize surplus supply, rather than designing storage technologies or more flexible power plants (Nye, 1990; Rose, 1995). Unlike coal and nuclear, renewable energy sources are intermittent because the sun obviously does not always shine, nor does the wind always blow; renewable resources vary by season as well. "Curtailment" of renewable electricity occurs when renewable facilities are shut down because baseload power plants cannot easily be ramped down to prioritize supply from green electrons. Thus, issues of timing relate not only to how fast an energy transformation could occur, but also to how temporality is managed within the electrical power system, as more asynchronous renewable energy, subject to the natural temporal cycles of wind, sun, and seasons, is added to the electricity system. Useful here is the concept of "timescapes" developed by Adam (1998) to stress "rhythmicities, timings and tempos, changes and contingencies" (p. 10). The timescape of the grid relates to the temporalities of how people consume electricity, and how new technologies, such as distributed and renewable supply or electric vehicle charging, are changing the temporal management of the power system.

New technologies: Three main renewable energy technologies are included in the Desertec vision—solar photovoltaics (PV), concentrating solar power (CSP), and wind. First, solar PV cells convert light from the sun directly to electricity by photons exciting electrons through the photoelectric effect. Large PV panels can be sited en masse in the desert, consuming large land footprints. Smaller panels can be placed on rooftops for point-of-use generation. A system in which onsite small-scale technologies provide power for homes or businesses is called a distributed generation system.

Second, CSP uses the sun's heat, rather than its light, to generate electricity. For example, a CSP power tower plant consists of a field of heliostats, or flat, garage-door-sized mirrors affixed to posts, which focus sunlight on a central power tower receiver, heating a heat-transfer fluid used to turn water into steam to produce electricity in a turbine generator. Thus, a CSP plant is similar to a conventional steam/thermal power plant, but the heat comes from the sun rather than from fossil fuel combustion. Heat storage can be used to make CSP a dispatchable source of electricity for 20 or more hours per day. CSP can be coupled with natural gas backup for morning startup and passing clouds. Generally, CSP is a centralized, large-scale technology that consumes thousands of contiguous acres. Some smaller-scale options exist, including solar powered Dish Stirling engines, which could be sited in a distributed fashion or en masse in the desert, solar cookers, and community-scale CSP–natural–gas systems. CSP and PV are distinct in their technological attributes, history, and sociopolitical implications.

Third, wind turbines use the kinetic energy from wind to turn the turbine's blades around a rotor. Wind power has become cheaper than both PV and CSP, reaching prices that make it competitive with fossil fuels in many markets. Wind power plants also consume thousands of acres, although the land sometimes can be simultaneously used in other ways, such as for grazing cattle.

The reliable provision of electricity and low upfront costs are often valued over other priorities, such as affordability, sustainability, and justice. Some utility companies even view themselves as selling the service of reliability, rather than the commodity of electrons. The concept of resilience has been added to that of reliability, referring to maintaining reliability even when the system undergoes a shock or perturbation. Regulatory bodies may require 10%–25% reserve generation capacity that immediately can be turned on in case of an unexpected spike in demand or a failure. Lack of reliability is costly because it disrupts economic productivity. A reliable power system is often seen as a prerequisite to growing a robust industrial sector in a developing country. Utility companies have traditionally viewed intermittent renewable energy as a challenge to reliability, presuming that a certain percentage of renewable capacity would "break" the grid. While reliability is essential, traditional views on reliability will need to shift in order for a renewable energy transformation to occur.

Low cost is also highly valued. Electrical power systems are capital intensive; for example, the consolidated holding company–level assets of investor-owned utilities in the United States total $1.5 trillion (EEI, 2016). A number of electricity-related companies are included in the FT Global 500 on the basis of their high market values, including Siemens (Germany), ABB (Switzerland), NextEra Energy (US), EDF (France), Enel (Italy), Iberdrola (Spain), and Dominion Resources (US). Cost encompasses not only construction costs but also associated infrastructure costs, fuel costs, interest rate on capital, the amount of electricity generated by a power plant each year, and operation and maintenance costs. Sunk costs make transformations difficult. Even though a transformation to renewables likely would be cost-effective in the long term if well planned, it requires large upfront investment. It is commonly assumed that renewable energy has not become a significant part of the energy mix because it is too expensive. This book will illustrate that the transformation challenge is more complex than this. Even the powerful companies involved in Dii struggled to make progress on transformations.

Pathways: Today, multiple studies illustrate that a fully renewable electricity system, or at least very close to it, is technologically feasible (NREL, 2012). However, system transformations will be required, making the pathway to achieving a low-carbon and renewable system more complex, with various societal, geographic, and temporal implications. Below I describe four overarching transition or transformation pathway scenarios. These scenarios are simplified and are not mutually exclusive. They are intended to aid in considering possible scalar arrangements for energy futures. These pathways and related trends also help illustrate why system imaginers would consider the extreme measure of sourcing electricity generation from a different continent.

Scenario 1: extreme fossil fuel transition

In this transition, rather than transformation, scenario, the system would remain in the same configuration as today, substituting extreme methods of extracting primary fossil fuels, such as hydraulic fracturing (fracking) to exploit shale oil and gas, ultra-deepwater drilling, extraction of bitumen from oil sands, and exploitation of methane hydrate deposits. In this scenario, low prices, reliability, and maintaining the traditional fossil fuel sector would be prioritized over environmental protection and social justice, precluding a renewable energy transformation at enormous detriment to society and environment. Utilities in the United States are shifting toward using natural gas generation, which is dispatchable but not renewable, for meeting baseload demand rather than peak. The wide distribution of shale and oil gas resources across the globe, including in countries that do not have major domestic conventional hydrocarbon reserves, such as Argentina, Colombia, and China, suggests that natural gas could problematically compose a significant portion of the electricity mix worldwide for decades to come. Additionally, coal remains the most abundant fossil fuel reserve worldwide. This scenario is more likely in the United States than in the EU, which is politically committed to ending business as usual.

Scenario 2: distributed generation and storage transformation

In this scenario, the centralized grid would be dismantled and replaced with a completely distributed, small-scale, point-of-use electricity system as advocated by the appropriate technology movement, originally grounded in E.F. Schumacher's assertion that small is beautiful (Schumacher, 1973). This scenario includes small-scale energy technologies, such as rooftop solar panels and backyard wind turbines, community solar panel installations, and distributed storage options, such as the Tesla home battery. Energy "prosumers"—consumers who are energy producers—would control the generation of electricity (Ruotsalainen, Karjalainen, Child, & Heinonen, 2017).

This scenario reflects the most profound transformation and likely would require a breakthrough in distributed storage technologies coupled with a "death spiral" for utility companies. Social movements often associate this transformation with broader societal and lifestyle change, as citizens would be in control of generation and more aware of their consumption, leading to conservation and efficiency. It would require large upfront capital investment directly from citizens, or new models of providing capital for household generation. It likely would necessitate conservation in order for household demand to be met without grid backup in high-energy-consuming countries (Hanser, Lueken, Gorman, & Mashal, 2017).

Activists advocating for this scenario come from diverse social groups, including underrepresented people of color, libertarian groups, and environmental organizations, which share a deep distrust of utility companies. For example, the German environmental NGO Eurosolar advocates for a distributed renewable energy transition. Activists from marginalized communities that have shouldered a disproportionate share of the environmental and social harms of energy generation believe that the grid as a sociotechnical regime is beyond social repair and incapable of legislating equitable outcomes. Lennon (2017) describes Black Lives Matter activists in the United States seeking to liberate themselves from a "colonial grid system" that has caused public health impacts that have disproportionately affected communities of color. Their activism constitutes an important step toward energy justice and community empowerment.

For some, the grid is colonizing and hierarchical, or altogether absent, and for others, it provides a public service of reliable and affordable electricity. Many consumers who have purchased rooftop PV panels are wealthy, as it can cost $15,000 to invest in rooftop solar panels and even more for batteries. Controversies have played out in regulatory arenas about the amount rooftop PV system owners should be compensated for electricity fed back into the grid, and how much they should pay for grid maintenance, since consumers without batteries still rely on the grid for reliability, including at night and during times of cloud cover. Several US state regulatory commissions controversially have instituted monthly fees requiring PV system owners to invest back into the grid (Federal Trade Commission, 2016).

The scalability of a fully distributed system is questionable, as not every disadvantaged community in countries around the world has access to an NGO that can raise the capital and coordinate an exit from the grid. Moreover, large-scale industrial uses may not be able to be met with small-scale energy supply. At high levels of partial grid defection among wealthy consumers, low-income communities could be left with aging and unreliable grid infrastructure. Moreover, the locked-in techno-economic aspects of electricity systems make a fully distributed transformation unlikely. These are difficult issues that electricity decision makers will grapple with for decades.

Scenario 3: blended distributed and centralized generation at the national scale

It is likely that future electricity systems will be an amalgam of centralized and distributed generation, rather than purely one or the other. In 1977, Amory Lovins developed the term "soft path," in which he called for electricity generation that is appropriately scaled to match its use. In a mixed scenario most closely resembling the soft path, household demand could be most efficiently met with distributed PV panels and battery storage, whereas a factory would be supplied with large-scale power sources. Centralized generation for large-scale use could come from dispatchable renewable generation, such as CSP, or from natural gas or advanced nuclear power plants. More likely, some households would adopt rooftop PV panels, with or without batteries and without dramatic conservation, while feeding surplus power into the grid during the day and purchasing power from the grid at night. This distributed generation could be subsumed under the overall centralized system, with utility companies and transmission system operators accounting for it in integrated resource planning. Centralized grid storage could be added, and price signals could encourage consumers to shift some of their electricity usage to off-peak hours.

Scenario 4: regional energy interconnection and integration

One potential solution to renewable energy intermittency is long-distance, cross-border trade of electricity (Cottier & Espa, 2017). In this scenario, centralized grids could sprawl across various regions to balance intermittency across multiple time zones and peak hours of electricity usage. Additionally, or alternatively, electricity from dispatchable CSP plants in sunny areas could be exported to cloudy areas. As in Scenario 3, distributed and centralized renewables could coexist, but the need for large amounts of storage and dirty baseload or dispatchable power sources could be reduced or eliminated. This pathway could be viewed as a technological fix (techno-fix), substituting centralized renewables for fossil fuels while retaining the centralized configuration. However, this book illustrates the complexity, geopolitics, number of actors, and societal dimensions that make this techno-fix far from straightforward.

Major trends

The direction of energy transformations remains uncertain. Three major trends underway in power systems across the globe will affect which of the end scenarios might be reached. The Desertec vision is situated at the nexus of these three trends.

The first trend affecting energy transformations is the *nation-state and international policy goals to scale up renewable energy* being implemented for reasons including climate change mitigation, energy security improvement, socioeconomic development, and local pollution reduction. In 2015, the Paris Agreement on climate change instituted the most comprehensive country commitments to ending climate change. Denmark, Spain, Germany, Italy, and the UK have renewable shares above 20% (REN21, 2017). Renewable energy targets are not limited to wealthy countries; in 2016, 48 developing countries had committed to 100% renewable energy targets, with high renewable energy growth rates in China, India, and Brazil (ibid.). While the percentage of renewable electricity capacity is small, at 9% including hydroelectricity, the growth rate of renewables is high, with 62% of new capacity additions in 2016 coming from renewables (ibid.). In 2017, $249 billion was invested worldwide in renewables, including solar PV (47%), wind (34%), and hydropower (15%) (ibid.).

Second, there are longstanding *liberalization, or deregulation, trends in power systems*. Vertically integrated utility monopolies control generation, transmission, and distribution. Market liberalization efforts aim to break up these monopolies, separating generation, transmission, and distribution, and to eliminate state-run utility companies. Most power markets are neither fully liberalized nor fully monopolistic.

Liberalization efforts have been shaping the power sector in Europe since the 1960s (Verbong & Geels, 2007) and in California since the 1970s (Hirsh, 1999). During the 1980s and 1990s, multilateral development banks promulgated a global deregulation trend that shaped the power sectors of at least 70 countries, including Morocco (Dubash & Williams, 2006). In many cases, the World Bank and the

International Monetary Fund mandated deregulation as a prerequisite to receiving energy loans (Dubash, 2003). These efforts were partially derailed by blackouts in California, Argentina, and elsewhere and by the exposure of the depth of Enron's global malefactions in the energy sector (Dubash & Singh, 2005). Early in the twenty-first century, the World Bank acknowledged that the needs of each country's electrical power system are unique (Byrne & Mun, 2003) and shifted from a deregulation model to a public-private partnerships model (Dubash & Williams, 2006), which balances state and private control and is used today in the Moroccan renewable energy sector. However, liberalization is still often seen as more efficient and desirable. It is easier to integrate liberalized electricity markets across nation-state borders, but not all states buy into this model, as I discuss in Chapter 6.

The third, and least explored, trend, relates to *electrical power systems crossing nation-state borders*. This adds further complexity to the national scope Cherp, Vinichenko, Jewell, Brutschin, and Sovacool (2018) take in their meta-theoretical framework on energy transitions. Dubash and Williams (2006) argue that the internationalization of electricity will cause many of the same sociopolitical problems as the globalization of oil. They speculate that proposed interconnections across nation-state lines in Asia could result in an inequitable "electricity resource colony" relationship (ibid.). Additionally, the internationalization of electricity could lead oil-poor states that export significant quantities of electricity to become "rentier states," which are less accountable to their citizenry because they are reliant on electricity rents rather than tax revenue (see Ross, 2001). However, unlike oil and gas pipelines, regional electricity integration requires concurrent and highly temporal integration of technological, social, political, and regulatory institutions, to an extent that oil markets do not.

Countries' motivations for pursuing transnational interconnection, and the type of interconnection pursued, vary. In some cases, "deep integration" is pursued, with a single market for electricity in the region and a fully synchronized grid (ESMAP, 2010). In other cases, "shallow integration" is pursued, with limited interconnection providing for backup or limited export (ibid.). Some countries have built cross-border transmission because of a shared waterway with hydroelectricity capacity, as in the cases of the Itaipú Dam between Brazil and Paraguay and the Mekong River dams in Southeast Asia. Some countries with low rates of energy access, such as Ethiopia, seek to export electricity to bolster macroeconomic growth. In many cases, the existence of a regional trade union led to political integration followed by a regional electricity plan. A more recent impetus for integration is to address renewable intermittency (Cottier & Espa, 2017). Put differently, renewable energy, often promoted by environmentalists and energy security advocates as part of an energy independence strategy, is leading some powerful stakeholders to pursue electricity *inter*dependence. For example, an expert from the large power technologies manufacturer ABB stated, "The only solution for balancing seasonal mismatches of load and [renewable] generation that we see with our knowledge of available technologies today is large, interconnected power systems."[1]

Dozens of examples of transnational and regional electricity integration projects exist worldwide, suggesting a growing trend that requires attention from social scientists to understand the implications for international relations, energy security, and energy justice. Thus, the conclusions from this book about transnational grid integration are not limited to the Mediterranean region but rather apply to other grid integration processes. The following is a summary of the state of play in grid integration worldwide.

- **Europe:** The EU has demonstrated a strong commitment to electricity integration. The Nordic electricity market, or Nord Pool, is arguably the world's most successful integrated electricity market. The European Network of Transmission System Operators for Electricity (ENTSO-E) includes 36 countries with a total of 423 cross-border transmission lines as of 2016.
- **Africa:** There are five regional power pools on the African continent. The most deeply integrated is the Southern African Power Pool, which connects 12 countries. The additional pools include the Central Africa Power Pool, the Eastern Africa Power Pool, the West African Power Pool, and the Comité Maghrébin de l'Electricité, which is pursuing further connection among Tunisia, Algeria, Morocco, Libya, and Mauritania. Morocco is connected to Spain through two lines and is discussing integration with Portugal. Interest in electricity integration is growing among the African economic communities. For example, Gosaye Mengistie Abayneh, the CEO of the Ethiopian Electric Utility, was quoted in January 2016 in *AllAfrica* stating, "Our vision is to interconnect the entire network in the African continent and beyond" (Sambo, 2016).
- **Middle East:** In the Middle East, the Gulf Cooperation Council Interconnection Authority began trading electricity in 2010, linking Saudi Arabia, Kuwait, Bahrain, Qatar, Oman, and the United Arab Emirates for reliability purposes, with deeper integration under consideration. The Eight Countries Interconnection Project includes limited interconnection among Egypt, Syria, Jordan, Iraq, Turkey, Lebanon, Jordan, and Palestine.
- **Asia:** There are two emerging power pools in southern Asia: The Association of Southeast Asian Nations (ASEAN) Power Grid and the South Asian Association for Regional Cooperation power grid. The Asia Super Grid and Gobitec visions seek to develop massive export of coal, wind, and solar electricity from the Gobi Desert to surrounding countries. In Central Asia, the grid formed under the Soviet Union has mostly disintegrated, while a project called CASA-1000 seeks to connect Central Asian countries with Afghanistan and Pakistan. China also has bilateral lines with approximately nine countries and is pursuing further integration.
- **North America:** Thirty electricity interconnections exist between the United States and Canada, focused mostly on Canadian export of hydroelectricity. Nine cross-border transmission lines exist between the United States and Mexico.
- **Central and South America:** The Central American Electrical Interconnection System (SIEPAC) connects grids across Panama, Costa Rica, Nicaragua,

Honduras, El Salvador, and Guatemala. A link between Guatemala and Mexico was completed in 2009. In 2012, the US State Department launched the initiative "Connecting the Americas 2022" to further integrate the grids in Central and South America. Electricity integration among Mercosur countries (Mercado Común del Sur, including Argentina, Brazil, Paraguay, and Uruguay) is largely limited to cross-border dams. The Andean Community decided to integrate in 2002, but limited action has been taken. Finally, feasibility studies have been conducted for Caribbean island integration.

- **Global Grid:** The Global Energy Interconnection Development and Cooperation Organization (GEIDCO), initiated by the former head of the State Grid Corporation of China (or State Grid), is pursuing the development of a global grid, which is further discussed in Chapter 4.

Dii envisioned a supergrid and a single electricity market in the Mediterranean. This would link all three of the trends—liberalization, renewable energy, and transnationalization—because integrating intermittent renewable energy into the centralized system is the main impetus for regional electricity integration, and in order to construct this regional market the power sectors of each country would likely have to be liberalized.

Multi-scalar methodology on energy transformations

This book uses multi-scalar analysis to understand energy transformations. Edwards (2003) argued that multi-scalar analysis should be used as a method of examining modern technological systems because infrastructures look different on different scales of force, time, and social organization. Each scale yields potentially true but incomplete stories (ibid.). Smith and High (2017) discuss how ethnographers of energy study different scales to understand how the local and global and micro and macro are "inseparable." To understand the effects of a transnational energy system, it is important to examine the scale of local communities that will be affected by power plant siting, national governments whose imaginaries and policies affect energy transformations, and regional processes and politics. I explore the vision for a sociotechnical system across multiple scales including: (1) the global history of CSP technology (2) regional visions for an integrated power system that includes CSP (3) a national Moroccan sociotechnical imaginary for renewable energy and its relationship to regional integration (4) local villages adjacent to the CSP plants under construction in Morocco. Multi-scalar analysis must capture complexity and plurality at different scales, a challenge with which analysts of sociotechnical systems must grapple through both theoretical and methodological triangulation (see Sovacool & Hess, 2017).

The Mediterranean is a strategic site for conducting research on regional and transnational energy systems. Its relevance extends beyond a single case study, as it draws out multiple facets of the transformation trends and scenarios. The EU's roots in regional energy integration trace back to the European Coal and Steel Community, which was not merely an economic liberalization project, but rather a political initiative and fulcrum for spurring European integration and trust (Spierenburg & Poidevin, 1994). EU policymakers have provided some incentive for North-South integration, including (through Article 9) allowing member states to meet renewable energy requirements with imported electricity from non-member states. Spain and Germany have already taken ambitious steps toward an energy transformation. The German *Energiewende*, or energy turnaround, has sought to eliminate nuclear power and decarbonize the electricity sector through subsidies for renewable generation. Additionally, Morocco is an important case study within this vision because of its sociotechnical imaginary for renewable energy scale-up and its interest in North-South integration.

I studied the vision for an integrated Mediterranean electricity system and the Moroccan Solar Plan from 2009 to 2017. My objectives were analytic and normative: (1) to describe and interpret the role of visions for future sociotechnical systems and energy transformations and (2) to evaluate the anticipated justice and sustainable development outcomes of transforming sociotechnical systems in a complex world.

To understand visions for cross-border clean energy transitions in the Mediterranean, I conducted discourse analysis of interviews, news analysis, reports, and project-related documents, as well as visual analysis of advertisements, images, and online videos with interviews of officials. Discourse analysis examines the language that energy stakeholders use in order to understand their interpretation of the world (Sovacool & Hess, 2017). It also concerns how actions and performance affect the material world and the constitution of new power systems (ibid.). Documents are useful in examining the formal processes of shaping a vision, while interviews offer insight into some of the informal processes that are difficult to observe.

In order to understand the global history of CSP, I drew on French texts from the late 1800s, magazine articles describing efforts to build CSP in Egypt in the early 1900s, 1970s documents from Spain and California, as well as secondary literature. I studied the history of large-scale engineering projects planned for the Sahara Desert in order to situate the Desertec vision in context. For Chapter 2, I also drew on a qualitative study of controversies surrounding CSP siting in California from 2009 to 2014.

As an event ethnographer, I attended the Dii conference and the 10th World Wind Energy conference held in Cairo in 2011, and the 2013 Dii conference in Rabat. I also watched the webstream of the 2012 Dii conference in Berlin. I conducted 32 interviews at the European scale, 23 in person and 9 by video call. The interviews lasted at least one hour, with some in-person interviews extending into two or even three hours. I recorded almost all of the interviews and transcribed them verbatim in NVivo. In July 2013, I conducted interviews in various locations in Germany and Spain. The interview sample included four employees of the Dii plus a Dii consultant and the two Desertec Foundation directors. It also included nine Dii shareholder companies: one interviewee each at SCHOTT Solar, Cevital, Munich Re, HSH Nordbank, ABB, Deutsche Bank, and two (simultaneous) interviewees each at M+W Group, Red Eléctrica de España, and RWE. I also spoke informally with several employees from the Dii shareholder company, First Solar, in Arizona. To understand the initial Desertec vision, I interviewed two scientists from the German Aerospace Center (DLR) involved in the 2006 Desertec feasibility studies, as well as the creator of the Desertec vision. Outside the Dii network, I interviewed the head of the Mediterranean Solar Plan (MSP) and a consultant from the German Agency for International Cooperation (GIZ) who worked on the MSP, the head of the World Bank's MENA CSP program, an official from the International Renewable Energy Agency (IRENA), and two German researchers who studied the Desertec vision. I interviewed German government officials from the Federal Ministry for Economic Affairs and Energy (BMWi), the Federal Foreign Office, the Federal Ministry for Economic Cooperation and Development (BMZ), the Federal Environment Agency (BMU), and the German government-owned development bank (KfW). To gain an oil industry perspective on renewable electricity integration, I interviewed a BP employee. In 2017, I conducted a video call interview with the new leadership of Dii-MENA.

At the nation-state scale, I drew on fieldwork, interviews, and document and news analysis to understand national Moroccan views on solar power and regional power integration. I first visited Morocco in June 2012 for a scoping trip. I conducted 20 interviews in Morocco from August to December of 2013 while living in Salé, the sister city of Morocco's capital, Rabat. I sampled for interviews on the basis of representation from the three main energy agencies plus the state-run utility company, major industry actors, state-run investment funds, and energy researchers and research funding agencies. The sample included two leading solar researchers; two officials from the Moroccan Agency for Solar/Sustainable Energy (MASEN); the head of the state-run energy investment company, Société d'investissements énergétiques (SIE); three researchers from the country's leading research center, Centre National pour le Recherche Scientifique et Technique; two officials from the water and energy utility company, Office National de l'Electricité et de L'eau Potable (ONEE); the head of the funding agency for solar energy and new technologies, or l'Institut de Recherche en Energie Solaire et Energies Nouvelles (IRESEN); one official from the Energy Ministry, Ministère de l'Energie, des Mines, et du Développement Durable (MEMEE/ MEM); the head of Sahara Wind; the head of the Moroccan Foundation for Advanced Science, Innovation and Research (MAScIR); and one official from the World Bank in Rabat. I also gained the perspectives of four Germans working on energy in Morocco at: the Germany embassy, the Agence Nationale pour le Développement des Energies Renouvelables et de l'Efficacité Energétique, and MASEN. I filled in the gaps in my data by attending professional events, such as the November 2013 Renewable Energy Exposition in Casablanca, Morocco.

I also helped to lead three study-abroad trips for Arizona State University students to Morocco and Spain from 2013 to 2015 and one to Morocco in 2016, which allowed me to update my research.[2] The programs provided a mobile classroom through which we studied multi-scalar energy development from the upper echelons of the Moroccan government to NGOs that work at the village level. The trips to Spain included meetings with government officials and researchers, two visits to the solar research platform in Almería, and two visits to the solar Solucar Complex in Spain built by (now former) Dii shareholder Abengoa Solar.

At the local scale, I used interviews and fieldwork to understand how communities would be affected by CSP siting and the construction of a regional power system. I made seven trips to Ouarzazate, Morocco, including three visits to the first CSP plant—one before construction began, one during construction, and one after the first phase was completed. In February 2014, I joined part of the fieldwork led by the NGO Germanwatch to evaluate the socioeconomic impacts of the CSP facility. We conducted interviews in the villages of Tasselmante, Idelsane, and Agdz. On the study-abroad trips, we learned more about life in the Ouarzazate area by staying at the *Dar Taliba*, a girls' boarding school.

Chapter overviews

Chapter 2 describes how actors have constructed CSP as feasible and advantageous, on the basis of varying ideological goals, as CSP emerged in three time periods from 1850 to today. Longitudinal analysis can help energy policy analysts understand why a technology framed as emerging has not successfully been taken up in the past and how it has related to social inequity. This chapter compares contemporary visions for CSP in the Sahara Desert to colonial-era megaprojects for the Sahara. It then describes and interprets the history of CSP from 1850 to today. Proponents of CSP from 1860 to 1914 thought it would provide a necessary substitute once coal ran out. They also focused on its immediate feasibility in areas with abundant sunshine, expensive coal, and a need for irrigation, which marketed it toward use in colonies where coal was expensive and sunlight abundant. Following the oil embargo, states gained interest in CSP to meet energy independence and national security goals. CSP re-emerged after 2000 because of policy interest in climate change mitigation. Today, CSP is also intertwined with complicated questions about what counts as a sustainable technology across the triple bottom line, as well as questions about the role of renewable energy in sustainable development.

Chapter 3 discusses the constructed geographical framing of the Mediterranean region to examine energy integration from the perspectives of energy geopolitics and energy justice. Scholars theorize energy geopolitics as the combative quest by nations for scarce resources, or as a concept unnecessarily fixated on zero-sum games while overlooking institutions' potential to legislate win-win outcomes, or as a concept constructed by actor perspectives (i.e., critical geopolitics). A critical geopolitics approach illustrates that advocates of constructing CSP plants for exporting power portrayed North Africa as possessing a geographical advantage over Europe for the generation of solar electricity on specific parcels of land—or renewable energy terroir—while downplaying the geopolitical risk of European energy dependence on North Africa through an Orientalist framing of the land in North Africa as abundant, barren, and devoid of people. These geographical framings relate closely to energy injustice, including a lack of recognition and spatial justice for citizens living in desert landscapes, distributive injustice in energy access, and human insecurity relating to the politics of immigration. Energy security challenges in the twenty-first century go beyond addressing the insecurities stemming from unevenly distributed resources and relate instead to complex debates about sustainable development and energy justice.

Chapter 4 explores the challenges of retrofitting path-dependent, entrenched electricity systems to improve their sustainability. Significant sociopolitical challenges are entailed even in energy transformations that leave the centralized electricity system configuration largely intact. This chapter tells a theoretically inflected story about the evolution of the vision for a Mediterranean-wide electricity system, contributing to the academic literature on Desertec, which is extensive but lacks an overarching narrative and theoretical framework. It maps out the components of an energy vision and the processes and forces that shape it over time. Actors developed at least three different designs for a Mediterranean electricity system over eight years. The characteristics of the visions, including technological feasibility, societal benefits, and sustainability, evolved and were reframed in response to sociopolitical pressures and controversy. This chapter discusses generalizable issues that ought to be taken into account in the early stages of similar visions for regional, and even global, grid integration, including policy and institutional innovation in regulatory and grid management organizations, energy security and cybersecurity, and new price mechanisms.

The next section of this book investigates sustainable energy transformations in Morocco. At the national scale, Chapter 5 analyzes Morocco's energy system and its renewable energy policies. Morocco's sociotechnical imaginary for energy treats renewable energy as part of national pride and socially sustainable development and reflects a dialectic between Morocco's past and its vision for a desirable future. The Moroccan Solar Plan is couched in terms of Morocco's quest for a charismatic centralized technology that will bring it national pride by exploiting the country's renewable energy wealth, and also will compensate for its lack of fossil fuel resources. Morocco is pursuing CSP over cheaper renewable energy technologies because of CSP's relationship to Morocco's specific development needs. The state frames the Moroccan Solar Plan as addressing various socioeconomic challenges, for instance by providing energy for industrialization, achieving energy security, securing foreign direct investment, encouraging job growth, avoiding poor educational outcomes, and developing an R&D policy. Recently, some social movements have called into question the inclusivity of Morocco's sustainable development goals and the importance of the participation of citizens in equitable and socially sustainable development. Chapter 5 argues that without recognition and distributive justice for present generations, the natural capital left to future generations will continue to be unevenly distributed among social groups, even if equivalent amounts are sustained between generations.

Chapter 6 explores Moroccan perspectives on building transnational clean energy grids. It considers the challenges associated with North-South electricity integration projects that cross different political systems and deep energy and wealth disparities by examining energy integration between Morocco and the EU. Electricity integration is not an automatic win-win, as advocates suggest. Rather, a win-win at the nation-state scale could be achieved only by offering value-added for North African national innovation systems, rather than focusing only on the often-discussed, mutual priority of security of energy supply. Claims of win-win outcomes that focus narrowly on security of supply leave out the broader policy goals in the South for renewable energy transitions to achieve sociopolitical benefit and sustainable development. Different perspectives on the role for state versus market between North and South, and a lack of procedural justice in the development of the vision, suggest that deep electricity integration in the Mediterranean region would be difficult to achieve. These findings are applicable to other electricity integration projects that are in the planning phases worldwide, such as the CASA-1000 project to export electricity from Central Asia to Afghanistan, US-Mexico electricity integration, African regional power pools, and the vision to develop a global grid.

Focusing on the local scale, Chapter 7 analyzes the social sustainability of CSP by examining its relationship to existing social inequities in the region where CSP plants are under construction, investigating public participation in the siting process, and recognizing connections between CSP and scarce water resources. The construction of large-scale renewable energy infrastructure has obvious environmental benefits of reducing greenhouse gas emissions and other forms of pollution. However, poorly planned clean energy development risks perpetuating the inequities of the existing electrical power system, as well as producing new configurations of winners and losers. CSP, if well planned, can provide vast improvements over dam building, but problems need to be solved in terms of conservation of scarce water resources and lack of procedural justice in siting. This chapter concludes by comparing the benefits and drawbacks of CSP plants to those of local development projects initiated by NGOs, arguing that both are needed simultaneously to achieve socially sustainable development. The best practices for siting CSP plants should be paired with local development to meet immediate citizen-defined needs. Long-term efforts must be made to address local-scale inequities in order for local citizens to benefit from the development of CSP and electricity exports from CSP.

Chapter 8 synthesizes the book's main arguments using justice theory, fulfilling the normative goals outlined here in Chapter 1. It discusses the theory of energy justice and the challenge of assessing energy justice in multi-scalar systems. It then describes and characterizes the energy injustices revealed in this empirical study of a vision for a multi-scalar energy system, and how they intersect with issues of global energy security. It analyzes issues of justice at the transnational scale in North-South collaboration, assessing the neocolonial critique of Desertec using categories of justice. It explores whether a global framing of energy security could be adopted by nation-states without infringing on energy sovereignty or reproducing neocolonial relations, and whether an imaginary for an interconnected energy system would need to be shared at the regional level in order for secure energy integration to occur. It recommends the principle of freedom from coercion in North-South energy integration, rather than energy sovereignty, and recommends a just regional process for achieving integration. This chapter then discusses the citizen scale and recommends an approach that focuses on cosmopolitan justice and "deep energy security" to avoid building energy systems as if some lives do not matter.

Notes

1 The quotation is from an interview conducted for this book.
2 For more information on the study-abroad trips, see Parmentier and Moore (2016).

Works cited

Adam, B. (1998). *Timescapes of modernity: The environment and invisible hazards*. London: Routledge.

Adam, B. (2004). Memory of futures. *KronoScope*, *4*(2), 297–315. Retrieved from https://doi.org/10.1163/1568524042801392

Arthur, W. B. (1994). *Increasing returns and path dependence in the economy*. Ann Arbor, MI: University of Michigan Press.

Barben, D., Fisher, E., Selin, C., & Guston, D. (2008). Anticipatory governance of nanotechnology: Foresight, engagement, and integration. In E. Hackett, O. Amsterdamska, M. Lynch, J. Wajcman, & W. Biker (Eds.), *The handbook of science and technology studies*. Cambridge, MA: MIT Press.

Borup, M., Brown, N., Konrad, K., & Van Lente, H. (2006). The sociology of expectations in science and technology. *Technology Analysis & Strategic Management*, *18*(3–4), 285–298. Retrieved from https://doi.org/10.1080/09537320600777002

Brown, N., Rappert, B., & Webster, A. (Eds.). (2000). *Contested futures: A sociology of prospective techno-science*. Burlington, VT: Ashgate.

Brundtland, G. H. (1987). *Report of the World Commission on Environment and Development: Our common future*. Oslo: United Nations.

Byrne, J., & Mun, Y. M. (2003). Rethinking reform in the electricity sector: Power liberalisation or energy transformation? In Wamukonya, N. (Ed.), *Electricity reform: Social and environmental challenges* (pp. 48–76). Roskilde, Denmark: UNEP/RISO Centre.

Cherp, A., Vinichenko, V., Jewell, J., Brutschin, E., & Sovacool, B. (2018). Integrating techno-economic, socio-technical and political perspectives on national energy transitions: A meta-theoretical framework. *Energy Research & Social Science*, *37*, 175–190. Retrieved from https://doi.org/10.1016/j.erss.2017.09.015

Cottier, T., & Espa, I. (2017). *International trade in sustainable electricity: Regulatory challenges in international economic law*. Cambridge: Cambridge University Press.

Dubash, N. K. (2003). Revisiting electricity reform: The case for a sustainable development approach. *Utilities Policy*, *11*(3), 143–154. Retrieved from https://doi.org/10.1016/S0957-1787(03)00044-4

Dubash, N., & Singh, D. (2005). Alternating currents: Introduction to an international review of electricity restructuring. *Economic and Political Weekly*, *40*(5), 5242–5248.

Dubash, N., & Williams, J. (2006). The political economy of electricity liberalization. In J. Byrne, N. Toly, & L. Glover (Eds.), *Transforming power: Energy, environment, and society in conflict* (pp. 155–190). Piscataway, NJ: Transaction Publishers.

Edwards, E. 2003. Infrastructure and modernity: Force, time, and social organization in the history of sociotechnical systems. In T. Misa, P. Brey, & A. Feenberg (Eds.), *Modernity and technology* (pp. 185–225). Cambridge, MA: MIT Press.

EEI Edison Electric Institute. (2016). *Industry data*. Retrieved February 14, 2018, from www.eei.org/resourcesandmedia/industrydataanalysis/industrydata/Pages/default.aspx

ESMAP. (2010). *Regional power integration: Lessons from global case studies and literature review* (No. Briefing Note 004/10). Washington, DC: Energy Sector Management Assistance Program, The World Bank. Retrieved from www.esmap.org/node/353

Farrell, A. E., Zerriffi, H., & Dowlatabadi, H. (2004). Energy infrastructure and security. *Annual Review of Environment and Resources*, *29*(1), 421–469. Retrieved from https://doi.org/10.1146/annurev.energy.29.062403.102238

Federal Trade Commission. (2016). *Something new under the sun: Competition & consumer protection issues in solar energy workshop transcript*. Washington DC: Federal Trade Commission. Retrieved March 18, 2017, from https://www.ftc.gov/system/files/documents/public_events/943943/solar_workshop_transcript.pdf

Geels, F. W. (2004). From sectoral systems of innovation to socio-technical systems: Insights about dynamics and change from sociology and institutional theory. *Research Policy*, *33*(6–7), 897–920. Retrieved from https://doi.org/10.1016/j.respol.2004.01.015

Glaser, L. (2009). *Electrifying the rural American West: Stories of power, people & place*. Lincoln, NE: University of Nebraska Press.

Graham, S. (2000). Constructing premium network spaces: Reflections on infrastructure networks and contemporary urban development. *International Journal of Urban & Regional Research*, *24*(1), 183–200.

Hanser, P., Lueken, R., Gorman, W., & Mashal, J. (2017). The practicality of distributed PV-battery systems to reduce household grid reliance. *Utilities Policy*, *46*, 22–32. Retrieved from https://doi.org/10.1016/j.jup.2017.03.004

Held, M. (2001). Sustainable development from a temporal perspective. *Time & Society*, *10*(2–3), 351–366.

Hirsh, R. (1999). *Power loss: The origins of deregulation and restructuring in the American electric utility system*. Cambridge, MA: MIT Press.

Hjerpe, M., & Linnér, B-O. (2009). Utopian and dystopian thought in climate change science and policy. *Futures*, *41*(4), 234–245. Retrieved from https://doi.org/10.1016/j.futures.2008.09.007

Holden, E., Linnerud, K., & Banister, D. (2017). The imperatives of sustainable development. *Sustainable Development*, *25*(3), 213–226. Retrieved from https://doi.org/10.1002/sd.1647

Hughes, T. (1983). *Networks of power*. Baltimore, MD: Johns Hopkins University Press.

Hughes, T. (1994). Technological momentum. In M. Smith & L. Marx (Eds.), *Does technology drive history?* (pp. 101–114). Cambridge, MA: MIT Press.

Jasanoff, S., & Kim, S-H. (2015). *Dreamscapes of modernity: Sociotechnical imaginaries and the fabrication of power*. Chicago, IL: University of Chicago Press.

Jenkins, K., McCauley, D., Heffron, R., Stephan, H., & Rehner, R. (2016). Energy justice: A conceptual review. *Energy Research & Social Science*, *11*, 174–182. Retrieved from https://doi.org/10.1016/j.erss.2015.10.004

Karlsson-Vinkhuyzen, S., & Jollands, N. (2013). Human security and energy security: A sustainable energy system as a public good. In M. J. Trombetta & H. Dyer (Eds.), *International handbook of energy security* (pp. 507–525). Cheltenham, UK: Edward Elgar Publishing.

Kern, F., & Rogge, K. S. (2016). The pace of governed energy transitions: Agency, international dynamics and the global Paris agreement accelerating decarbonisation processes? *Energy Research & Social Science*, *22*, 13–17. Retrieved from https://doi.org/10.1016/j.erss.2016.08.016

Laird, F. N. (2013). Against transitions? Uncovering conflicts in changing energy systems. *Science as Culture*, *22*(2), 149–156. Retrieved from https://doi.org/10.1080/09505431.2013.786992

Leach, M., Scoones, I., & Stirling, A. (2010). *Dynamic sustainabilities: Technology, environment, social justice*. London: Earthscan.

Lélé, S. (1991). Sustainable development: A critical review. *World Development*, *19*(6), 607–621.

Lennon, M. (2017). Decolonizing energy: Black Lives Matter and technoscientific expertise amid solar transitions. *Energy Research & Social Science*, *30*, 18–27. Retrieved from https://doi.org/10.1016/j.erss.2017.06.002

McShane, T. O., Hirsch, P. D., Trung, T. C., Songorwa, A. N., Kinzig, A., Monteferri, B., . . . O'Connor, S. (2011). Hard choices: Making trade-offs between biodiversity conservation and human well-being. *Biological Conservation*, *144*(3), 966–972. Retrieved from https://doi.org/16/j.biocon.2010.04.038

Meadowcroft, J. (2009). What about the politics? Sustainable development, transition management, and long term energy transitions. *Policy Sciences*, *42*(4), 323–340.

Miller, C. A., Iles, A., & Jones, C. (2013). The social dimensions of energy transitions. *Science as Culture*, *22*(2), 135–148. Retrieved from https://doi.org/10.1080/09505431.2013.786989

Miller, C. A., Richter, J., & O'Leary, J. (2015). Socio-energy systems design: A policy framework for energy transitions. *Energy Research & Social Science*, *6*, 29–40. Retrieved from https://doi.org/10.1016/j.erss.2014.11.004

Minteer, B. (Ed.) (2009). *Nature in common? Environmental ethics and the contested foundations of environmental policy*. Philadelphia, PA: Templeton University Press.

Minteer, B., & Manning, R. E. (2009). Convergence in environmental values: An empirical and conceptual defense. In B. Minteer (Ed.), *Nature in common?: Environmental ethics and the contested foundations of environmental policy*. Philadelphia, PA: Templeton University Press.

Norton, B. G. (2005). *Sustainability: A philosophy of adaptive ecosystem management*. Chicago, IL: University of Chicago Press.

NREL. (2012). *Renewable electricity futures study* (No. NREL/TP-6A20–52409). Golden, CO: National Renewable Energy Laboratory. Retrieved from www.nrel.gov/analysis/re_futures/

Nussbaum, M. (2003). Capabilities as fundamental entitlements: Sen and social justice. *Feminist Economics*, *9*(2–3), 33–59. Retrieved from https://doi.org/10.1080/1354570022000077926

Nye, D. (1990). *Electrifying America : Social meanings of a new technology, 1880–1940*. Cambridge, MA: MIT Press.

Özden-Schilling, C. (2015). Economy electric. *Cultural Anthropology*, *30*(4), 578–588. Retrieved from https://doi.org/10.14506/ca30.4.06

Parmentier, M. J., & Moore, S. (2016). "The camels are unsustainable": Using study abroad as a pedagogical tool for teaching ethics and sustainable development. *Teaching Ethics*, *16*(2), 207–221. Retrieved from https://doi.org/10.5840/tej2016113038

Perkins, J. (2017). *Changing energy: The transition to a sustainable future*. Berkeley, CA: University of California Press.

REN21. (2017). *Renewables 2017 global status report*. Paris: REN21 Secretariat. Retrieved from http://www.ren21.net/gsr-2017/.

Rolston H., III. (1998). Challenges in environmental ethics. In M. J. Zimmerman, J. Callicott, G. Sessions, K. Warren, & J. Clarke (Eds.), *Environmental philosophy: From animal rights to radical ecology* (2nd ed., pp. 124–144). Upper Saddle River, NJ: Prentice Hall.

Rose, M. (1995). *Cities of light and heat: Domesticating gas and electricity in urban America.* University Park, PA: Pennsylvania State University Press.

Ross, M. (2001). Does oil hinder democracy? *World Politics, 53*(3), 325–361.

Ruotsalainen, J., Karjalainen, J., Child, M., & Heinonen, S. (2017). Culture, values, lifestyles, and power in energy futures: A critical peer-to-peer vision for renewable energy. *Energy Research & Social Science, 34*, 231–239. Retrieved from https://doi.org/10.1016/j.erss.2017.08.001

Sambo, J. (2016, January 31). East Africa: Ethiopia near to exporting power. *AllAfrica.com*. Retrieved from http://allafrica.com/stories/201602012148.html

Schumacher, E. F. (1973). *Small is beautiful: Economics as if people mattered.* New York, NY: HarperCollins.

Selin, C. (2008). The sociology of the future: Tracing stories of technology and time. *Sociology Compass, 2*(6), 1878–1895. Retrieved from https://doi.org/10.1111/j.1751-9020.2008.00147.x

Smil, V. (2016). Examining energy transitions: A dozen insights based on performance. *Energy Research & Social Science, 22*, 194–197. Retrieved from https://doi.org/10.1016/j.erss.2016.08.017

Smith, J., & High, M. M. (2017). Exploring the anthropology of energy: Ethnography, energy and ethics. *Energy Research & Social Science, 30*, 1–6. Retrieved from https://doi.org/10.1016/j.erss.2017.06.027

Sovacool, B., & Hess, D. (2017). Ordering theories: Typologies and conceptual frameworks for sociotechnical change. *Social Studies of Science, 47*(5), 703–750.

Sovacool, B. K. (2014). What are we doing here? Analyzing fifteen years of energy scholarship and proposing a social science research agenda. *Energy Research & Social Science, 1*, 1–29. Retrieved from https://doi.org/10.1016/j.erss.2014.02.003

Sovacool, B. K. (2016). How long will it take? Conceptualizing the temporal dynamics of energy transitions. *Energy Research & Social Science, 13*, 202–215. Retrieved from https://doi.org/10.1016/j.erss.2015.12.020

Sovacool, B. K., Burke, M., Baker, L., Kotikalapudi, C. K., & Wlokas, H. (2017). New frontiers and conceptual frameworks for energy justice. *Energy Policy, 105*, 677–691. Retrieved from https://doi.org/10.1016/j.enpol.2017.03.005

Sovacool, B. K., & Dworkin, M. H. (2015). Energy justice: Conceptual insights and practical applications. *Applied Energy, 142*, 435–444. Retrieved from https://doi.org/10.1016/j.apenergy.2015.01.002

Sovacool, B. K., & Geels, F. W. (2016). Further reflections on the temporality of energy transitions: A response to critics. *Energy Research & Social Science, 22*, 232–237. Retrieved from https://doi.org/10.1016/j.erss.2016.08.013

Spierenburg, D., & Poidevin, R. (1994). *The history of the High Authority of the European Coal and Steel Community: Supranationality in operation.* London: Weidenfeld and Nicolson.

Strezov, V., Evans, A., & Evans, T. J. (2017). Assessment of the economic, social and environmental dimensions of the indicators for sustainable development. *Sustainable Development, 25*(3), 242–253. Retrieved from https://doi.org/10.1002/sd.1649

Thompson, P. B. (2010). *The agrarian vision: Sustainability and environmental ethics.* Lexington, KY: University Press of Kentucky.

Tobey, R. C. (1996). *Technology as freedom: The New Deal and the electrical modernization of the American home*. Oakland, CA: University of California Press.

Verbong, G., & Geels, F. (2007). The ongoing energy transition: Lessons from a socio-technical, multi-level analysis of the Dutch electricity system (1960–2004). *Energy Policy, 35*(2), 1025–1037. Retrieved from https://doi.org/10.1016/j.enpol.2006.02.010

2 The history of concentrating solar power and large-scale engineering projects for the Mediterranean Region

Introduction: deferred dreams for solar power and the Sahara Desert

This chapter treats the vision for turning the Sahara Desert into an "energy powerhouse" as a deferred dream, tracing it back to visions for large-scale megaprojects to be built in the Sahara Desert in French colonial North Africa. Proposed large-scale Saharan engineering projects stemmed from the ambitious dreams of colonial governments that sought to reshape desert landscapes into large-scale technological systems, with profound consequences for both the environment and its human inhabitants, who were often written out of the landscape or at least imagined as not needing energy. While these visions remained unbuilt, they later re-emerged in different forms.

This chapter then overviews the history of concentrating solar power (CSP), which has been framed as an emerging but feasible technology for over a century and a half. People have envisioned Andalusia and the Sahara as attractive locales to build CSP since the 1800s. By examining CSP's history, this chapter illustrates how actors in different time periods have constructed CSP's emergence and feasibility and have situated CSP within particular contexts. Moreover, the term "emerging technology" is often used as a static, predetermined concept, but it should be analyzed as an actor category that is strategically constructed over time.

CSP's emergence relates not just to the technology itself but also to its relationship to electrical, social, and political systems. Mitchell (2011) and others have demonstrated that seemingly technical choices about energy technologies are always underpinned by sociopolitical controversies. However, there was a stronger entrepreneurial and private-sector focus on "black gold" (oil) discovered in Pennsylvania and the Middle East (Yergin, 2008) than there was for CSP. Actors' goals for CSP have been more ideological than financial—relating to colonization, nation-state pride, sustainability, climate change mitigation, and socioeconomic development. CSP required government support to be financially viable, which tied it to changing nation-state goals over time. Governments were first interested in CSP to provide energy in parts of colonies where coal was scarce and expensive. After the 1973 oil crisis, governments became interested in CSP for reasons of energy independence and national security. Recently, governments have become interested in CSP for its role in creating green jobs and mitigating climate change.

The history of CSP and large-scale systems for the Sahara relates to visions and ideas more so than to the material technology that was built. Historical and longitudinal analysis of energy alternatives can aid energy policy analysts in understanding why a technology has not emerged successfully in the past, how a technology has been interpreted by actors over time, and how a technology has related to inequity and social stereotypes. This can help us to avoid previous mistakes. Thus, this chapter is intended partly for an energy policy audience. This chapter also is intended for advocates of the scale-up of solar power and technological systems in deserts and regions with a history of inequity that today encompass large power disparities. Emerging technologies must be integrated within a complex socio-political terrain that has been shaped by histories that still matter to citizens and policymakers. Over time, proponents of CSP have envisioned building CSP in a landscape that was more socially and geographically complex than they realized. Their visions were often grounded in false assumptions and stereotypes.

Additionally, this chapter points to gaps in the history of CSP that historians could further explore. The secondary literature on CSP from 1860 to 1920 is contradictory and inconclusive. Furthermore, Africans' perceptions of and contributions to solar power have not been sufficiently studied (see Caille, 2017).

This chapter mainly focuses on North Africa and Western Europe but also includes a discussion of CSP development in California in the 1970s, since California serves as an energy policy test bed that is influential in global electricity system trends. First, this chapter describes the history of large-scale, imperialist engineering visions for the Sahara Desert, such as the trans-Saharan railway, the Sahara Sea, and the Atlantropa project. This chapter then describes the development of CSP in France and in the United States in the mid-1800s and describes how CSP related to colonialism. From World War I to the oil embargo of 1973, CSP received limited attention, although its advocates in the 1950s and 1960s situated solar power in a discourse relating to the socioeconomic development of "sunbelt" countries. Following the 1973 oil embargo, CSP was built in a number of countries, with the centers of activity in California and at the international Solar Research Platform in Spain. As the oil crisis waned, Luz Solar, the only commercial CSP company at the time, filed for bankruptcy in 1991. This chapter then describes the contemporary re-emergence of CSP in Spain and the United States due to interest in climate change mitigation and sustainable energy.

Large-scale engineering visions for a "coveted desert"

In this section I provide a historical overview of engineering visions for the Sahara Desert. While I will argue in Chapter 6 that North-South electricity integration is not intrinsically neocolonial, it is likely to be neocolonial if system imaginers do not understand this history and then work to chart a different course.

France occupied Algeria from 1848 to 1962 (with conquest beginning in 1830); Tunisia was under a French Protectorate from 1881 to 1955 and Morocco from 1912 to 1956. French colonialism in North Africa was not grounded only in "hard-headed economic logic" but also had ideological underpinnings (Swearingen,

1988, p. 3). Colonial powers used technology as a metric for judging civilization and social progress (Adas, 1990). The French "civilizing mission" reflected an ideology that Europeans had a normative duty to civilize Africans and that Africans were incapable of progress without European scientific and technological intervention (ibid.). The French government justified its occupation of North Africa partly by the ethnocentric assertion that French engineers could "properly" exploit North Africa's natural resources. (Comparisons could be made to contemporary assertions that Africa is well-endowed with energy resources, but Africans have failed to exploit them.) The French government believed that Morocco and other North African countries used to be the "Granary of Rome"—or major exporters of grain to the Roman Empire—but that Arabs had neglected the land and only French technology could return it to its former state and "make the desert bloom" (Davis, 2007; Heffernan, 2011; Swearingen, 1988). More recent evidence illustrates that although North Africa did provide grain to the Roman Empire, it was through traditional agricultural practices at nowhere near the scale assumed by the French government (Davis, 2007). Yet this myth, coupled with the assumption that Morocco had abundant water resources, did work in the world (Swearingen, 1988). The French Protectorate pursued an ambitious strategy to grow grain in Morocco for export to France during the 1920s (ibid.). French grain producers successfully lobbied for a ceiling on grain imports from Morocco so that the French market was not flooded with cheap grain (ibid.). Even with this ceiling, the French exported grain from Morocco to France in the 1920s even as Moroccans starved (Sater, 2010). The neocolonial concerns surrounding Desertec and similar visions relate to nation-states with less geopolitical power becoming electricity depots for more powerful countries, exporting electricity without meeting domestic demand. By 1930, it was apparent to the protectorate's leaders that Morocco's natural water resources depended on capricious rainfall and were subject to persistent drought (Swearingen, 1988). The granary vision reflected, at the least, an insufficient understanding of the landscape and local context. Leaders of the protectorate then developed a dam-building and irrigation vision that will be further discussed in Chapter 5 and Chapter 7.

The imaginary of deserts as empty frontiers also relates to colonialism. European colonial powers "coveted" the Sahara Desert and viewed it as *terra nullius*: virgin and unoccupied land belonging to no one (Bisson, 2003; Heffernan, 2011). Conquering the desert became the focus of imperial design and scientific curiosity (ibid.). Several early expeditions to the Sahara ended in shipwrecks along Morocco's Atlantic Coast, contributing to the imaginary of an inaccessible and remote Sahara shrouded in mystery (Bisson, 2003). As exploration of the Sahara began it was a "bitter surprise for the conquerors who believed the Sahara was empty of men because it was not only space that they had to conquer but also the Saharans themselves!" (Bisson, 2003, p. 61). The French government viewed the success of colonialism along the North African coasts as contingent on "pacifying" the Sahara Desert through military and technological control (Bisson, 2003). The so-called peaceful penetration of the Algerian Sahara took place from 1844 to 1852 (Brower, 2009). Nevertheless, some Europeans continued to view the Sahara as empty into the twentieth century. Muller (2005) states, "For most of the twentieth century, part of the [European] interest in Africa is the fact that it is deemed void, like a blank page provided to the brush of European engineering" (p. 360).[1]

European colonial powers envisioned a number of large-scale technological systems for the Sahara and pursued them over decades without success (Bisson, 2003). Examining these failed projects "allow[s] [for] an exploration of the outer edges of the French colonial imagination in Africa" (Heffernan, 2011 p. 619), and offers a lens for interpreting contemporary regional development projects. In *Engineers' Dreams*, published in 1954, Willy Ley described nine such macroscale engineering visions, arguing that "engineers' dreams are things that can be done as far as the engineer is concerned," but the projects remain "incomplete dreams" because of "political barriers." Four of the nine projects he described were planned for the deserts of the Middle East and North Africa (MENA) region. In 1997, Petroski argued that Ley's book has "staying power" and "continued currency" because of the grand scale of the projects and their demonstration of engineering's potential for major planetary change (p. 3). These "engineers' dreams" have in common immense scale and plans for the massive provision of water and electricity for irrigation and resource extraction in the region. The dreams' proponents often aim to improve connectivity between territories through technological integration. The goal of achieving connectivity and continuity in the Sahara through technological integration can also be seen in the Desertec vision. Each engineer's dream had behind it feasibility studies, publications, and advocacy work, which offers a partial explanation for why the projects endured over decades and later re-emerged in new forms.

One example is the Trans-Saharan Railway to connect French North Africa to West Africa, championed by French Minister of Public Works and engineer, Charles de Freycinet (Heffernan, 1989). While it was never constructed, the vision persisted for almost a century, from the 1870s to the 1960s (Heffernan, 2011). The railway would transport people and goods, such as cotton; improve access to the Sahara; and provide territorial continuity among the North African colonies (Bisson, 2003). Over 500 books and articles were written about the railway, a commission was formed to investigate its feasibility, and meetings were held to advertise its viability alongside economic and political benefits (Heffernan, 2011). The project received little attention during World War I and was stalled afterward because of its massive cost (Starostina, 2010). Interest in the railway was rekindled following the war as part of a broader plan for French national renewal (ibid.). Disturbingly, in 1941, construction started using Jewish slave labor but ceased in November 1942 because of Operation Torch, the Anglo-American invasion of French-held Algeria and Morocco (Heffernan, 2011).

The Sahara Sea is another example of Ley's persistent "engineers' dreams." In the 1870s, François Élie Roudaire, a French military geographer, hoped to find the mythical Lake Tritonis, described in Greek mythology as somewhere in present-day Tunisia (Bisson, 2003). French colonists hoped to use water from Lake Tritonis to irrigate cotton (ibid.). When Lake Tritonis was not discovered, Roudaire developed plans to engineer a Sahara Sea by piping Mediterranean water into a desert depression, but surveying missions deemed the plans infeasible because the depression was higher than sea level (Ghassemi & White, 2007; Ley, 1954). An architect employed by the German government, Herman Sörgel (1885–1952), later imagined that the Sahara Sea could be developed by filling dwindling Lake Chad with seawater (Ley, 1954). Sörgel argued that it would be cheaper and easier to access and navigate the Sahara by boat rather than by rail (ibid.). In 1957, a French company, called "Project ARTEMIS,"[2] hoped revenue from newly-discovered hydrocarbons in Algeria could be used to develop a Sahara Sea, but the results of the investigation remain unknown (Ghassemi & White, 2007). In 1983, the Tunisian and Algerian governments, with the participation of a Swedish engineering bureau, formed a company to study a Tunisian-Algerian Interior Sea, but it concluded in 1985 that the project would be unprofitable and would have negative social and environmental impacts (Ghassemi & White, 2007).

In the 1920s, Sörgel developed a vision called Panropa, later called Atlantropa, that framed North Africa as an unproductive wasteland waiting to be put to use as an agricultural and hydroelectric "hinterland" for Europe (Muller, 2005). Sörgel's racist vision included a massive 35-kilometer dam on the Strait of Gibraltar between the Atlantic and the Mediterranean, plus additional dams between Sicily and Tunisia and on the Dardanelles, totaling 365,000 MW (Edit Suisse Group, 2003; Ley, 1954). Sörgel imagined that Atlantropa would irrigate the Sahara to turn it into a "bread-basket for Europe," bearing resemblance to the Granary of Rome myth (ibid.). A picture used to advertise the project, archived at the Deutsches Museum, depicts a black African woman in a loincloth handing a breadbasket to a white European colonist, with a caption stating in German that "the over-industrialized, mechanized Europe and virgin, resource-rich Africa complement each other" (Sörgel, 1920). According to Ley (1954), Atlantropa would have reduced the water volume of the Mediterranean by 59,978 cubic miles, uncovering 220,000 square miles of land. It would have increased physical connectivity in the region, putting back in place the land bridges theorized to have previously existed between Spain and Morocco, Tunisia and Sicily, and Greece to the eastern end of the present sea (ibid.). Sörgel thought that creating more land would be a "peaceful" alternative to "living space" (*Lebensraum*, or the displacement of the supposed "inferior races" by the supposed "Aryan master race") (ibid.). Atlantropa was not the fantasy of a single engineer; a group of sponsors and supporters called the Atlantropa Institute promoted it until 1960 (Edit Suisse Group, 2003). Four books and roughly 1,000 articles were published on Atlantropa (ibid.).

Geopolitically, Atlantropa's proponents thought it would develop a "Eurafrica" region in which Europe controlled the energy, land, and water resources of Africa to compete against the rising forces of Asia and America (Trischler & Weinberger, 2005). However, Atlantropa was never politically supported (ibid.). More broadly, a Euro-African regional imaginary was politically supported and promoted in various forms from 1920–1960, ranging from a French-Maghreb union to uniting the entire continent of Africa with Europe (Muller, 2005). In addition to racism and Christian proselytizing, Muller explains that faith in technological progress played an important role in the Euro-African imaginary stating, "Pan-European imperialism often reveals a striking faith, pseudo-religious, in technology"[3] (p. 360). Proponents of Eurafrica saw Africa as providing space, materials, and energy to be exploited by Europeans using European technology. According to Muller (2005), the Euro-African community failed when its developers became sensitive to neocolonial critiques, Germany opposed privileging North Africa over other regions for investment, and former colonial powers expressed concern that they would fund unprofitable development projects in former colonies while other European countries reaped the benefits of developing the profitable projects (ibid.). Trischler and Weinberger (2005) argue that Atlantropa was shelved after the development of nuclear power. Nuclear power also offered promises for European integration, with Africa providing the uranium (Muller, 2005). The ideology that technology can bind Europe and North Africa together and provide Saharan connectivity can still be observed in discourses about regional infrastructural visions.

Ley's incomplete engineers' dreams focused on electricity for resource extraction and afforded no attention to who needed energy in the region and for what social good. His writing reflected the imaginary of a Sahara Desert barren of people, especially people deserving of electricity. About a proposed dam on the Red Sea Ley stated, "But nobody in the general area of the Strait of Bab el Mandeb needs such quantities of electrical current" (p. 156). About the proposed flooding of the Qattara Depression for hydroelectricity, Ley wrote, "There isn't anything in particular in the way of such a plan except an absence of customers for the current generated" (pp. 116–117). The myth that the Sahara is either uninhabited, or inhabited by people who do not require electricity, persists today and relates to a lack of recognition justice for North Africans, which will be discussed in the following chapter.

Overview of CSP technology

The remainder of this chapter discusses the history of CSP as situated within visions of the desert frontier, starting with a brief overview of solar technologies. Solar photovoltaic panels use the sun's light, generating electricity through the photoelectric effect, in which photons knock electrons loose from atoms. In contrast, CSP uses heat from the sun to generate electricity. Early versions of the technology used the sun's heat to cook food, pump water, and power a steam engine to achieve a variety of tasks (see Figure 2.1 for an example). Contemporary CSP plants are similar to conventional thermal power plants, but the heat comes from the sun rather than from fossil fuel combustion. The most widely deployed CSP design today is the parabolic trough design, a precursor of which can be seen in Figure 2.2. Simply put, parabolic mirrors reflect the sun's light onto tubes attached to the mirrors, and a heat-transfer fluid is used to transfer the sun's heat to water to produce steam for a steam turbine to drive a generator (see Figure 2.3). The second most deployed contemporary design is the CSP central tower (or power tower) plant, which consists of a field of heliostats, or garage-door-sized mirrors (or set of mirrors), that concentrate sunlight onto a central power tower receiver, using the heat to produce steam for a steam turbine. Another design is the concentrating linear Fresnel reflector, which uses long, thin mirrors to concentrate

Figure 2.1 Mouchot's solar engine, modified by Pifre to power a printing press, 1882
Originally from *Scientific American*, public domain image

Figure 2.2 Frank Shuman's Solar Engine I: Meadi, Egypt (Suburb of Cairo), 1909, public domain image

Figure 2.3 Kuraymat CSP trough field added to a natural gas plant (ISCCS), 2010, near Cairo, Egypt

Photo by author

sunlight and utilize its heat. The least deployed CSP design is the Dish Stirling engine, which looks similar to designs from the 1800s. The technology is a mirrored dish concentrator with a receiver attached to the dish's focal point. The Stirling engine uses the sun's heat to heat a gas that expands to drive a piston and crankshaft. It is highly efficient, but there are numerous barriers to the commercialization of Dish Stirling engines. Additionally, while these dishes could be deployed in a distributed configuration, such as in backyards, most CSP technologies today are designed to be integrated into a centralized grid system.

One of CSP's advantages is that the sun's heat can be stored using molten salts to generate electricity at night, which is currently cheaper than battery storage. However, CSP plants have a large land footprint, and some designs consume significant amounts of water for cooling and mirror cleaning. Air cooling can be used instead of water. This reduces the power plant's efficiency but cuts water usage by up to 90%, which is crucial in desert environments.

Solar power, colonialism, and desert frontiers

CSP was developed in relationship to concerns about coal running out and to colonial ambitions over a century before it became enrolled in environmental sustainability and climate change mitigation through the Desertec vision and similar initiatives.[4] In 1865, British economist William Stanley Jevons wrote *The Coal Question*, which predicted that British coal would soon run out (Jarrige, 2010). Additionally, France lacked coal deposits, which the French government viewed as impeding its industrial and international competitiveness, providing an impetus for French inventors and authors to imagine that solar heat could displace coal (Jarrige, 2010). The oil industry was still nascent, starting in 1859 in Pennsylvania (Yergin, 2008). Around this time, Augustin Mouchot (1825–1912), a young physics teacher at a high school (*lycée*) in France, was also concerned about coal running out. He began developing a solar device for distilling alcohol; a solar cooker; and a sun pump (*héliopompe*) that he patented in 1861, which utilized the heat of the sun to pump water (Butti & Perlin, 1980).[5] In 1865, Mouchot developed a parabolic mirror that he used to boil water (Jarrige, 2010; Butti & Perlin, 1980). In the summer of 1866, Mouchot presented his invention to Napoleon III (Butti & Perlin, 1980). The French government then provided him with workshop space for his experiments, and Mouchot built a solar receiver there. In 1869, Mouchot published a book entitled *Solar Heat and its Industrial Applications*, in which he argued that solar heat could provide free fuel and could aid in replacing imported coal from England, extending the lifespan of coal reserves (Mouchot, 1869). He summarized his book in an 1869 letter to the French Academy of Sciences (Paléo-Energétique, 2015). Mouchot's work was disrupted by the Franco-Prussian War from 1870 to 1871 (Jarrige, 2010). After the war, Mouchot promoted solar power as part of France's national recovery (Jarrige, 2010). In 1872, he received financial support from the French Association for the Advancement of Science. In 1873, he received support from the regional government—*conseil general d'Indre-et-Loire*—which was interested in the use of solar heat for distilling wine. These funds supported Mouchot's successful development of a solar steam engine, completed in 1874. Butti and Perlin (1980) assert that Mouchot then became concerned about the amount of space required for the 200 solar engines that would be needed to drive a 100-horsepower industrial motor.

Jarrige (2010) describes Mouchot as adopting "Orientalist rhetoric" after the Franco-Prussian war to promote the use of solar heat in French colonies. While Mouchot did not expect that coal reserves would be exhausted in the near term, he did see solar as a near-term solution for colonial governments that had to import expensive coal to remote and sunny colonies. For example, much of the coal consumed in Algeria had to be imported from England (ibid.). In 1875, the Director of Science and Letters in the French Ministry of Education provided 10,000 francs to Mouchot to undertake a scientific expedition to Algeria to test his solar oven, or *marmite solaire* (Royaumont, 1882; Jarrige, 2010). The oven was heated by the reflection of the sun on a silver plate. The Minister said that it was "indeed under the sky of our colony, a sky always ardent, much more than under the sky in Paris" that solar power should be developed (Royaumont, 1882, p. 233). This was among the earliest assertions that North Africa is the appropriate place for solar power because of its high solar insolation. In Algeria, Mouchot tested solar cookers and solar water pumps for irrigation (Perlin, 2013).

Mouchot's work in Algeria is not the only example of CSP's relationship to colonialism and empire. In 1870, William Adams, a British deputy registrar of the

High Court in Bombay, India, was among the first to imagine solar heat being put into service for the colonial mission (Johnson, 2015). Adams built a device in which mirrors concentrated the sun's heat onto a boiler connected to a 2.5-horsepower steam engine (Ragheb, 2011). He also patented a solar cooking device featured in *Scientific American* (Pope, 1903). In Adam's 1878 report, *Solar heat: A substitute for fuel in tropical countries*, he argued that the use of solar heat could save money by reducing the demand for coal imports to India (ibid.). Additionally, Charles Tellier, who was the inventor of refrigeration, also invented low-temperature solar collectors that vaporized ammonia to pump water. In 1890, he published *The Peaceful Conquest of West Africa by the Sun*, in which he forecast that Africa could become more industrialized and agriculturally productive if it used sun power (Tellier, 1890). In a vision similar to Desertec, Tellier imagined covering desert land in West Africa with solar collectors—or hot boxes—for water pumping and purification, irrigation, and making ice, and he even imagined the use of solar plows and boats (Caille, 2017).

Mouchot received 5,000 francs from the Consul General in Algeria, 5,200 francs from the French Association for the Advancement of Science, and 14,428 francs from the Minister of Agriculture to build a larger solar engine for the Algerian section of the 1878 World's Fair in Paris (Pope, 1903; Royaumont, 1882). Mouchot's device received a gold medal (Pope, 1903). At the World's Fair, Mouchot's assistant, an engineer named Abel Pifre, held a conference for the funders of Mouchot's work (Paléo-Énergétique, 2015). Pifre argued that the industrial applications of solar heat were not well-suited for cloudy Paris but for the colonies, with their abundance of sun and space. He discussed multiple applications for solar heat including cooking, heating liquids, making ice, desalinating seawater, producing alcohol and perfume, mining, as well as operating an engine and a generator using multiple solar dishes. He promised that high-quality steam could be produced without risk of fire and that storage would soon become possible. Mouchot developed a storage method by heating water in an insulated box during the day and then using this water to vaporize a liquid that would boil at a low temperature, but had limited success (Butti & Perlin, 1980). For unknown reasons, Mouchot decided to go back to teaching physics in 1880, and he sold his patents to Pifre, as well as Cordier who was an inventor in London (Royaumont, 1882; Jarrige, 2010).

Pifre's secretary Louis de Royaumont promoted Mouchot and Pifre's work in a colonial context in his 1882 book, *The conquest of the sun: Scientific and industrial applications of solar heat*. Royaumont's writing is ethnocentric and demonstrates a lack of understanding of the MENA region and its history. Royaumont's book begins with a derisive history of the "indigent natives" who "were desperate worshipers of the sun" (p. 6). He argued that "the inhabitants of the Arabian Peninsula" and some "Arab tribes" were backward because they did not relinquish this primitive sun worship when they converted to Islam (p. 6). Royaumont viewed solar engines as facilitating the scientific conquest of the sun and as emblematic of the human domination of nature. Royaumont stated, "Science and philosophy have joined forces; philosophy freed the slave, science enslaved the Sun" (p. 311). On the front page of the book (Figure 2.4), an anthropomorphized

Figure 2.4 The cover of Royaumont's *La Conquête du Soleil*, 1882

Source: Google Books (public domain image)

sun appears to be perturbed that it has been enslaved. Royaumont argued that CSP was feasible stating, "[CSP] is now a practical reality" (Royaumont, 1882, p. 227). Royaumont envisioned that French settlers and soldiers could carry solar *insolateurs* for pasteurization, water purification, cooking, and distilling wine far from the "civilized center." Royaumont thought solar power's primary application was to cultivate the desert through solar irrigation, particularly in Senegal. He also asserted that solar irrigation could enable Algeria to again become the Granary of Rome.

In the 1880s, French reporters expressed enthusiasm for the idea that solar heat could replace expensive coal in sunny and hot colonies (Jarrige, 2010). Royaumont quoted a French newspaper that called the invention of CSP "a glory for a civilizing and generous France" (p. 271). In 1881, Pifre formed the Central Society for the Use of Solar Heat (Société Centrale D'utilisation de la Chaleur Solaire) in France whose membership included Alphonse Couvreux, who helped plan the Suez Canal (ibid.). The society promoted solar as generous, free, and without fire danger (Ribeill, 2001). They declared CSP was feasible stating, "Our goal was to take the utilization of solar heat at its origin and follow it up to its complete development. This goal was reached" (Royaumont, 1882, p. 278). During the 1880s, Pifre publicly demonstrated that his solar engine could run a printing press, distill alcohol, make ice, and cook food (Jarrige, 2010).

Solar heat was also enrolled in other engineer's dreams during this time. In the late 1870s, Charles de Freycinet (French Minister of Public Works from 1877–1879) instituted a "solar heat" commission to investigate whether the sun could power the construction of the Trans-Saharan railroad (Royaumont, 1882). Mouchot's *insolateurs* were carried on a French expedition crossing the Sahara in 1880 and on a scoping expedition for the Sahara Sea (Jarrige, 2010). The commission approved the further use of solar *insolateurs* for the railway expedition. However, Pifre had insufficient financial resources to pursue this. Moreover, according to Royaumont, Freycinet was Pifre's greatest advocate, but he completed his term as Minister of Public Works in 1879.

A French government report in 1881 stated that the solar irradiance in France was insufficient to make solar devices practical, and that the feasibility of solar heat in the colonies depended on the future ability to easily transport the heavy devices and on the relative availability of coal (Butti & Perlin, 1980). Additionally, the mirrors were expensive to polish (Perlin, 2013). At the time, a chemist found that solar heat was too expensive but hoped the costs would come down by the time coal ran out (Jarrige, 2010). However, it remains unclear why the French colonial government altogether abandoned these efforts. Multiple advocates proclaimed that solar heat was feasible, but solar engines were not successfully commercialized or integrated into a major project.

Across the Atlantic, CSP inventors in the United States hoped the technology could irrigate dusty desert frontiers in the Southwest. In 1877, John Hittell and George Deitzler received the first US patent for a solar device (Pope, 1903). John Ericsson, a Swedish-American inventor known for designing a battleship called the Monitor, was also interested in developing an inexhaustible alternative to coal (Pope, 1903). Ericsson allegedly developed a solar engine resembling Mouchot's, discussed in *Scientific American* in 1870. The article described a vision similar to Desertec stating that over 22 million solar engines of 100-horsepower each could be deployed across a "rainless region extending from the northwest coast of Africa to Mongolia, 9,000 miles in length and nearly 1,000 miles wide" to capture "wasted" heat ("The Solar Engine," 1870, p. 323).[6] The exhaustion of coal in Europe will "cause great changes in regard to international relations, in favor of these countries which are in possession of continuous sun-power," such as "Upper Egypt" (ibid., p. 323). In 1884, Ericsson changed his design from one similar to that of Mouchot to parabolic trough mirrors with a coating that prevented them from tarnishing (Perlin, 2013; Pope, 1903). *Scientific American* published a drawing of Ericsson's new device in a colonial setting, depicting a man in a suit and wide-brimmed hat standing next to Ericsson's solar engine, with an African man and a hut in the background ("Ericsson's Sun Motor," 1884). Ericsson garnered interest among US farmers for using his design for irrigation, but he died in 1889 before he could carry out this work (ibid.).

In 1892, Aubrey Eneas, an English engineer living in Boston, received private funding to develop a solar steam engine (Butti & Perlin, 1980). Eneas used conical reflectors similar to Mouchot's design because he believed they would provide more power than the parabolic trough (ibid.). Eneas's patent stated that his solar generator "is especially intended for use in connection with irrigation of the arid plains of the West" (US670916 A, 1901). In 1901, Eneas built a solar irrigation system in California consisting of 1,788 mirrors that concentrated sunlight onto a boiler to run a water pump (Pope, 1903). Eneas stated, "The region ceases to be a desert, and can be made to blossom as the rose" (Pope, 1903, pp. 86–87). He displayed his solar engine at an ostrich farm in southern California, which, like the unusual birds, served as a tourist attraction (Pope, 1903). In 1903, Eneas built solar engines in Chandler and Tempe, Arizona for irrigation (Perlin, 2013; Pope, 1903), and he sold a solar pump to the British government in Egypt (Johnson, 2015). Johnson (2015) suspects that the Reclamation Act of 1902 removed any incentive for private inventors and entrepreneurs to develop their own means of irrigation. Reverend Charles Pope in his 1903 book, *Solar Heat: Its Practical Applications*, argued that solar heat was needed to replace coal because, unlike coal mining, solar energy could not be disrupted by labor strikes. He also advocated for the benefits of solar heat for irrigation stating,

> The grain-grower and the orchardist and horticulturist join [the farmer] in asking for [a] water-raising apparatus; and the Auteuil machine[7] and [Eneas's] Pasadena [solar] engine are his reply. *Pure country power* may be had as well as "pure country milk;" and *the costless engine* may replace the expensive ones now enjoyed by but a small part of the people who need them.
>
> (pp. 122–123)

Pope imagined that 320 acres covered with solar engines could power a railroad in the US West that would connect a manufacturing center rivaling the largest in the world. Pope stated that solar heat is

> not to be limited by our own republic's territory by any means. Many lands have large tracts which have heretofore passed as deserts because of the abundance of the sun's downpouring; this very fact is destined to make Algeria, the Soudan, Egypt, India, the table-lands of Ecuador, the arid coast of Peru . . . grow opulent when the utilization of solar heat is carried to the point of simplicity and economy.
>
> (p. 140)

Pope continued, "If the desert may be made to blossom, what has prevented the Sahara from being turned into one magnificent oasis?" (p. 142). Yet Eneas later concluded that the construction and installation costs of solar engines, two to five times other steam plants, were too high (Butti & Perlin, 1980).

In 1907, Frank Shuman, an inventor who had become wealthy designing reinforced glass, developed a small solar engine in his backyard outside Philadelphia. Shuman founded the Sun Power Company in 1908 to raise money for a larger solar plant (Perlin, 2013). He secured initial funding in the United States, but the funds he needed to build a large-scale power plant came from investors in London (ibid.). In 1910, Shuman tested a larger plant in his backyard, with 527 solar collectors that combined Tellier's hot boxes with mirrors, to make low-pressure steam to run a motor that Shuman designed (Perlin, 2013). (There were coal strikes from 1910 to 1914 in England, the world's largest exporter of coal (Mitchell, 2011), which may have spurred further interest in solar heat.) The Sun Power Company's investors asked a prominent physicist, Charles Vernon Boys, to review Shuman's design. Boys recommended a parabolic trough design similar to Ericsson's, and investor support was made contingent on Shuman's acceptance of Boys's parabolic trough design. (The parabolic trough remains the most utilized CSP design today.) Shuman built a 55-horsepower plant consisting of five parabolic troughs called "Solar Engine One" in British colonial Egypt using local material and labor in Al Meadi. The plant was competed in 1913 and was used to pump Nile River water at 6,000 gallons per minute to irrigate cotton, which was otherwise done manually ("American Inventor Uses Egypt's Sun for Power," 1916; Johnson, 2015).

After completing the plant, Shuman stated,

> Sun power is now a fact and no longer in the beautiful possibility stage. We have proved the commercial profit of sun power and . . . after our stores of oil and coal are exhausted the human race can receive unlimited power from the rays of the sun.
>
> ("American Inventor Uses Egypt's Sun for Power," 1916)

Again, advocates proclaimed that solar heat was feasible for use once fossil fuels were depleted. Whether solar was economically feasible in the early 1900s remains uncertain. I have not found a precise cost estimate of Solar Engine One in comparison to coal. In *Scientific American* in September 1911, Shuman explained that the interest rate on the upfront investment would determine whether his solar power plant would be profitable (Shuman, 1911a). In *Scientific American* in December 1911, Shuman argued that a 100-horsepower solar power plant would cost twice as much as a coal plant in upfront costs—$20,000 versus $10,000—and in operation and maintenance costs. Yet solar power could easily complete with coal in remote areas of colonies where coal ranged from $15 to $30 per ton, in comparison to $5 per ton at seaports (Shuman, 1911b). Shuman said that the soiling of the mirrors from sand would be problematic, but low-wage local laborers could be hired to continuously clean the mirrors (ibid.). In correspondence to *Scientific American* in 1914, Shuman asserted that solar power was already competitive with coal (Shuman, 1914). Butti and Perlin (1980) claim that the plant cost $8,200—twice the cost of a conventional plant—and that this could be paid off in several years, while other sources say Shuman spent around $40,000 but they are not clear what this figure included. Butti and Perlin say that the cost of coal ranged from $15 to $40, but a more specific figure and the interest rate on the investment would be required to calculate the payoff time.

Horatio Herbert Kitchener, British Consul General in Egypt from 1911 to 1914, offered Shuman a 30,000-acre plantation in Sudan to further test his system (Butti & Perlin, 1980). In 1913, Shuman gave a presentation on solar technology to the German Reichstag, which offered him 200,000 marks to build a solar power plant in German East Africa ("American Inventor Uses Egypt's Sun for Power," 1916).

Foreshadowing the Desertec vision, Shuman imagined a solar project of 20,250 square miles (a square with approximately 143-miles per side) in the Sahara Desert to generate 270 million horsepower (201 GW)—equivalent to the heat from all of the world's fuel mined in 1909 (Shuman, 1914). To illustrate his point, he used a box to represent the required land area, similar to the visuals Desertec proponents would use decades later (Figure 2.5). In a letter to *Scientific American* in 1914, he predicted this would cost $98.5 billion, but he also predicted that the engines would last 200 years, totaling "*almost nothing*" (Shuman, 1914). He concluded his letter by stating that "the human race must finally utilize sun power or revert to barbarism" (p. 179). Yet Shuman's Saharan solar vision did not come to fruition because World War I began in July 1914 and the British government reneged on its contract with Shuman and dismantled the system for scrap metal (Perlin, 2013). Shuman died of a heart attack in April 1918 before the war ended.

From 1860 to the beginning of World War I in 1914, the promoters of CSP declared that it was feasible. Proponents of CSP technology thought it would provide a substitute for coal once coal ran out. But Winston Churchill converted the Royal Navy's fuel source from coal to oil between 1912 and 1914, and a global economy based on free flows of oil was developed (Mitchell, 2011). An imperialist protectorate model congealed around oil (Mitchell, 2011) not solar. The increased exploitation of and reliance on oil, declining prices of coal and oil, World War I, the untimely death of prominent CSP advocates, cost, and an inability to produce electricity at night prevented CSP technology from taking hold. The post-war recession followed by the Great Depression likely prevented a post-war re-emergence of CSP (Butti & Perlin, 1980).

Figure 2.5 Shuman's map of the Sahara Desert

Source: Scientific American

Advocates of CSP technology from 1860 to 1914 also focused on its immediate feasibility in areas with abundant sunshine, a need for irrigation, and difficult and costly access to coal, which framed CSP as suitable for use in colonies where coal was expensive and sunlight abundant. Although US solar inventors initially developed CSP for domestic irrigation, Ericsson, Pope, and Shuman, as well as the editorial staff of *Scientific American*, connected solar heat to colonialism. Today's discourse on CSP in the Mediterranean region is similar to the historical discourse, in that both frame the desert as a barren and ideal place for Europeans to scale up solar power. While the failure of CSP technology could be interpreted as a form of "undone science" (Hess, 2016)—or a more sustainable path that could have been taken—narratives on solar heat from 1860 to 1914 were connected to the domination of nature and to imperialism, particularly visions that CSP plants built in colonies could fuel the West once coal ran out. The African people were largely excluded—treated as "backward" in Royaumont's book and as cheap mirror-washers in Shuman's vision. Historical visions for solar heat are glibly discussed in the literature as utopian, but utopian for whom?

CSP did not receive significant attention from governments from World War I until the 1973 oil embargo, nor did it receive notable private investment, resulting in its relative dormancy, although some research continued on CSP technologies.[8] More advancements were made in passive solar and solar photovoltaic panels during this time period, and the first silicon solar cell was invented at Bell Labs in 1954. A small group of advocates in the United States promoted solar power from post-World War II to the 1960s, led by "free market enthusiasts" rather than environmentalists (Laird, 2001). A discourse on solar energy in the "sunbelt" can still be observed in this postcolonial time period, but it shifted from an imperialist to a socioeconomic development framing (Johnson, 2015). For example, in the 1950s, the Association for Applied Solar Energy (AFASE) promoted solar for use in underdeveloped regions at international research conferences and at United Nations Educational, Scientific and Cultural Organization (UNESCO) conferences (Laird, 2001). At an AFASE meeting in 1954, Lewis Douglas, a former member of Congress, portrayed solar as a solution to Malthusian scarcity (ibid.). William Stead (1957), the former dean of the School of Business and Public Administration at Washington University, wrote an article for the *Bulletin of the Atomic Scientists* entitled "The Sun and Foreign Policy," arguing that solar power was appropriately scaled for "the solar belt" and its "technologically backward nations," while nuclear power was more appropriate for the West. Laird (2001) argues that solar energy was incompatible with US policy frames at the time, and more attention was afforded to nuclear power.

CSP's resurgence in the 1970s: La Plataforma Solar de Almería in Spain

Following the 1973 oil embargo, interest in renewable energy grew, and experimental CSP plants were developed in Italy (1 MW Eurelios), Japan (1 MW Sunshine), the United States (1 MW Central Receiver Test Facility at Sandia National Laboratories), France (2 MW THEMIS), and the USSR (5 MW SES-5) (Hernández, 2009). Policy goals for CSP related to national security, energy independence, and national technological prowess. Shortly after Francisco Franco's death, Spain arguably made the most concerted effort to develop a CSP R&D program. Juan Temboury, the general director of the Center for Energy Studies, advocated for the International Energy Agency (IEA) to choose Spain as the host country for the Small Solar Power Systems Project (SSPS) (ibid.). He was enthusiastic about hosting CSP research in Spain as the country transitioned to democracy. Additionally, Spain had high levels of external dependence on fossil fuels and lacked energy diversification. The IEA chose the German Aerospace Center (DLR) to be the operating agency and Spain to be the host country of La Plataforma Solar (PSA), or the Solar Research Platform, founded in 1977 (ibid.). My description of the PSA is based on interviews and accounts from PSA researchers and staff compiled in a 510-page book and a video by Valeriano Ruiz Hernández, a retired engineering professor and former president of the Spanish CSP advocacy organization Protermosolar, as well as in-person visits to the PSA I made in 2013 and 2014.

The PSA's founders interpreted CSP as part of the imaginary of the desert frontier. They hoped to site the PSA in the Tabernas Desert in Almería—a Mediterranean province in southern Spain and the only desert in Europe—because of its high solar insolation and flat land. Its main industry was spaghetti Western films. Running water, electricity, and telephone service lagged behind other areas of Spain (ibid.). After the researchers identified the Almería coast on a solar irradiance map, they met with the head of the provincial council who said, "You can have the whole coast" (Barrera, 2009). The PSA researchers viewed themselves as young, idealistic innovators who were "going to be the pioneers" who revived an economically deprived desert to make the solar "utopia become reality" (Hernández, 2009). To illustrate their commitment to rural uplift, the researchers used the Indalo, which was a good luck symbol in times of drought and economic downturn in Almería, as their logo (Andalucia.com, n.d.). SSPS—a 1 MW tower and trough demonstration—went into operation in 1982. The technologies researched and demonstrated at the PSA include CSP tower and trough (SSPS, CESA-1); concentrating linear Fresnel reflector; Dish Stirling engine; solar furnace; heat storage; and hybridization of CSP and natural gas. The involved companies included Martin Marietta, Interatom, Siemens, INITEC Energía, 3M, and Spanish engineering firm Abengoa.

There is a perception that Spain was left out of the Industrial Revolution, and advancing its progress in renewable energy was seen as a "Third Industrial Revolution" (Frolova, 2013). A researcher recounted, "Spain was aware from the very beginning that it had to be in the front row and not miss the train like it did in other processes of technological development" (Hernández, 2009, p. 146). José Luis Rodriguez Zapatero—President of the Government of Spain from 2004 to 2011 and member of the Spanish Socialist Workers' Party—wrote an introduction to Hernández's book that framed Spain as an innovative, technological, pioneering nation because of its work in CSP. He stated,

> For a long time, Spain's image has frequently been associated with sun and sand. A Mediterranean country, with a culture as diverse as it is rich, of friendly open people. Too often the enterprising character of its citizens and the talent of its researchers are forgotten. A trait which, however, thanks to

> the work and perseverance of persons such as those mentioned in [this book], has made it possible for Spain to also be seen today as a daring and innovative country.. . . Spain is now one of the pioneering nations in renewable energies.
>
> (Hernández, 2009, p. 19)

PSA researchers described being wounded by pejorative comments that the PSA was "that German thing in Almería" (Hernández, 2009, p. 48). One researcher reflected on how proud he was to be a Spaniard when they connected the SSPS project to the grid. Another said,

> The truth is I never felt dependent on the outside world. We considered this to be our own project and it used our own resources. It was our know-how and skills. Everything we needed was at hand and we threw ourselves into it.
>
> (Barrera, 2009)

Another researcher emphasized that CSP development at the PSA "was done with 100% Spanish technology, 100% Spanish workers and all of us really young. And it worked." (ibid.).

Yet the personal accounts in Hernández's book shift between framing the PSA as Spanish and framing it as a cosmopolitan international research center with nine participating countries, also recognized as a Large European Scientific Installation in 1990. The researchers recounted that work at the PSA was driven by fear that Spain was falling behind the United States in CSP development, even though SSPS was supported by US funding and corporations. The United States was the largest financial contributor to SSPS, followed by Germany and Spain. The project used US cylindrical parabolic collectors with German tracking technology. Hernández's book also described the international collaborative process at the PSA as being a source of pride. Additionally, a video I watched in the PSA's visitor center in 2013 stated that the researchers worked with different cultures and languages to learn "how to concentrate knowledge." One researcher asserted that the mutual trust developed between the Spanish and German researchers contributed to the project's success. Researchers called Germany "the motor of the SSPS project" (p. 197) and called the PSA a "Spanish-German plum cake" (p. 359). After a year of supporting the SSPS, the German government wanted to pull out to focus on solar PV and nuclear, but the PSA heads convinced Germany to continue supporting the project (ibid.). The project encompassed both the complexities of developing innovative technologies and of balancing international collaboration, and the researcher stated, "All this was like an enormous puzzle in which pieces were coming from different countries without any previous cross knowledge and with different companies supplying different components" (p. 198).

In 1986, the PSA faced a major setback when the sodium used for heat storage in the SSPS set fire. This occurred only five months after the Chernobyl nuclear reactor accident. Sodium was also used in fast nuclear reactors, and people living near the PSA were scared of being exposed to radiation. The PSA held a press conference to assure people in surrounding towns—who were already skeptical about the PSA's research and usefulness—that the sodium in the SSPS was not radioactive (ibid.). The PSA lost momentum as support "went up in smoke," and the Spanish Ministry of Industry, Trade, and Tourism questioned why this "outlandish" technology had been built "in the middle of nowhere" (ibid.). Moreover, the involved governments' interest in CSP was already waning because oil prices had fallen, and the United States and Germany wanted to tear down the experimental plants (ibid.). The PSA's leadership used a loophole in the PSA's founding documents to collect insurance money from the fire to continue their work. They conducted freelance work for the European Space Agency as well. International donors withdrew from the PSA, and it became a Spanish R&D platform led by the Spanish Center for Energy, Environment and Technology Research (or CIEMAT).[9] One of the PSA employees thought that this loss of international support solidified Spain as the world leader in CSP, stating that the PSA "later took off, leaving [the Germans and Americans] behind and making Spain the world leader in concentrating solar thermal energy technologies" (pp. 364–365).

Resurgence of CSP in Spain in the late 1990s

CSP research was dormant in Spain from the late 1980s to the late 1990s. The Kyoto Protocol on climate change was adopted in 1997, and it contributed to international interest in renewable energy, although support for renewables largely came from specific national government policies. In Spain, legislation and royal decrees spurred the development of a commercial solar sector, including, but not limited to, the Law of the Electricity Sector in 1997, Royal Decree 2818 in 1998, and the Plan for the Promotion of Renewable Energy in 1999, which set a target to achieve 29.4% renewable electricity consumption by 2010 (González, 2008). Several years later, the EU Directive 2001/77/EC on renewable energy sources promoted an increase in renewables in member states in order to mitigate climate change. In 2007, the EU set the "20-20-20 targets"—reducing greenhouse gas emissions by 20% relative to 1990 levels, implementing energy efficiency measures to reduce the 2020 projected level of energy consumption by 20%, and reaching 20% overall renewable energy consumption by 2020.

Spain's initial legislation focused on solar PV instead of CSP, which delayed the further development of CSP until 2004 (Hernández, 2009). In 2004, the Spanish government increased the feed-in tariff (FiT) subsidy, which was set in 2002 at 12 eurocents per kilowatt-hour (kWh) for 50 MW CSP plants, to 18 eurocents per kWh for 25 years, with even higher rates for the first 200 MW installed (Solar Thermal World, n.d.). In 2007, Royal Decree 661 again increased the FiT for CSP, to 26.9 eurocents per kWh (ibid.). The Spanish government hoped to increase green jobs by developing a domestic renewable energy manufacturing sector (González, 2008). Additionally, the Autonomous Community of Andalusia supported CSP development (Ridao, García, Escobar, & Toro, 2007). Falling agricultural prices at the time provided an economic incentive for farmers in Andalusia to rent or sell their land for renewable energy (Prados, 2010).

Abengoa Solar continued its work started at the PSA by launching the commercial CSP sector in Spain, completing the 11 MW PS10 facility in 2007 and the 20 MW PS20 facility in 2009. Michael Geyer, director for international business development at Abengoa Solar, who worked at the PSA for ten years, proclaimed, "We have brought from Andalusia solar thermal technology to the world" (Hernández, 2009, p. 244). Geyer calls himself "a German from Almería" (ibid., p. 246). At the commissioning of the PS20 plant, José Monzonís from ACCIONA, a Spanish infrastructural development company, stated, "The Mediterranean has significant potential not just for the countries in North Africa but in the medium term for all Arab countries and southern Europe" (Barrera, 2009). From 2008 to 2013, 50 commercial solar thermal power plants[10] were built in Spain with a total of 2,232 MW gross installed capacity (NREL, 2017). The levelized cost of electricity (LCOE)[11] for these plants is very high at roughly 30–35 US cents per kWh (IRENA, 2012, p. 32). In a preliminary analysis, Musi et al. (2017) found that the LCOE of CSP has gone up over time in Spain, rather than going down through learning effects as would be expected.

In 2011, Mariano Rajoy was elected President of the Government of Spain, replacing Zapatero, who was pro-solar. By 2013, the Eurozone crisis was well underway and the feed-in tariff was ridiculed as "the feed-in debt." With $37.4 billion of accumulated FiT debt, the Spanish government cut solar subsidies to control the deficit (Jordan, 2013, p. 52). In 2013, the government replaced the FiT for both new and *existing* CSP plants with a flat rate of return of roughly 7% over a plant's lifetime (Deign, 2014). In March 2014, the *CSP Today* industry magazine quipped that Spain can "say adios" to CSP investors as they look to the MENA region instead (ibid.).[12] Spain is facing lawsuits from renewable energy developers for reneging on the FiT. One benefit of a FiT is that it directly subsidizes electricity that is successfully produced, whereas in the 1970s investment subsidies for wind power in California funded some tax shelter experts who built wind turbines that never successfully generated electricity. However, in hindsight, Spain has been criticized for poorly planned subsidies that were too costly for the state budget.

The PSA is still operational. However, since 2016, it has been embroiled in political controversy because the Spanish government has blocked the PSA's ability to purchase equipment and curtailed temporary researcher positions as part of an effort to reduce Spain's deficit, in spite of the fact that much of the PSA's funding comes from the EU (Kraemer, 2017). In 2017, *El País* reported that the PSA's director, Sixto Malato Rodríguez, had resigned in protest (Domínguez, 2017).

Germany also implemented FiTs in the early 2000s. Under Germany's *Energiewende*, meaning energy turnaround or revolution, Germany has made significant strides toward increasing its renewable electricity capacity. It has sought to reduce its greenhouse gas emissions and to phase out nuclear and fossil fuel generation. However, Germany has focused on distributed solar PV and wind generation rather than on CSP, partly because of its lack of land area for CSP. The only CSP plant in Germany is a very small (1.5 MW) experimental power tower plant built in Jülich in 2008 to test air cooling and storage systems. The head of the Desertec Foundation stated in a 2013 interview with me,

> So then all the money that the companies put in research and development was put in wind and photovoltaic. Because the German [energy] market was so interesting. But CSP . . . is a kind of child without a father.

Different framings of solar: California in the 1970s

After the oil embargo, US-government interest in CSP focused on developing a centralized, scaled-up source of power as a substitute for foreign oil for reasons relating to national security and economics (Laird, 2001). System imaginers sought to turn part of Arizona and California into a National Solar Power Facility to generate enough electricity for the entire United States (Laird, 2001). Fractures grew between conventional advocates of solar focusing on achieving a techno-fix by plugging CSP into the existing grid to replace fossil fuels to scale, and ecological advocates focused on enacting broader societal change (ibid.). Ecological advocates wanted to decentralize energy production and put it in the hands of the people rather than large energy corporations and to promote individual responsibility, environmental awareness, and conservation. Perlin (2013) describes solar hot water heaters and solar PV panels built in the 1970s as "solar for hippies" (p. 354). British economist E.F. Schumacher's 1973 book, *Small is Beautiful*, advocated for small and locally accessible technologies. The book influenced the appropriate technology movement in the 1970s, which advocated for small-scale solar PV panels and passive solar (Laird, 2001). Amory Lovins's famous 1976 book, *Soft Energy Paths*, advocated for technology that was appropriately scaled for its use. Additionally, Langdon Winner argued that solar is fundamentally democratic, in contrast to nuclear, which necessitates a closed, authoritarian management structure (Winner, 1986).

Ecological advocates were unhappy with the federal government's centralized, large-scale approach to solar, which was focused on reducing cost through economies of scale. The Energy Research and Development Administration (ERDA) led the US government's CSP research program; the ERDA was the precursor agency to the Department of Energy (DOE). The ERDA was founded in 1974 when the Atomic Energy Commission was split into the ERDA and the Nuclear Regulatory Commission. The ERDA operated until the end of 1977. In an article in *Science Magazine*, critics of ERDA's R&D agenda complained that it focused on CSP power tower technology at the expense of solar PV panels (Hammond & Metz, 1977, p. 241). The ERDA failed to understand that solar was an inherently "democratic" technology (ibid.) The article argued that the ERDA's CSP program was "making solar after the nuclear model," as national laboratories specializing in nuclear power focused on building large solar power plants to plug into a centralized grid. They argued that CSP development resembled the breeder reactor program, scheduled to proceed from a small test to the first commercial plant in four stages (ibid.). A subsequent article in *Science* criticized the ERDA for focusing on CSP tower technology: 60 to 70% of the ERDA's solar R&D funding supported CSP power tower technology, and the contracts were awarded to large corporations, such as Martin Marietta and Boeing (Metz, 1977). Lovins called CSP developed by the ERDA "ingenious high technology," but he criticized CSP for supplying "energy in a form and scale inappropriate for most end-use needs" (Lovins, 1976 p. 42). This conflict over the appropriate scale of solar technology remains important to understanding solar power today.

California became the US testbed for CSP because of its openness to energy-policy experimentation and its high solar insolation. In 1982, the DOE, utility companies Southern California Edison and the Los Angeles Department of Water and Power, and the California Energy Commission completed Solar One, a CSP tower demonstration plant near Daggett, California in the Mojave Desert. Daggett also has been framed as an economically-depressed desert frontier town (see, Van Dyke & Wild, 1997). Solar One was a $108 million, 10 MW tower plant with a field of 1,818 heliostats (Boesman, 1994; Green, 1985). It provided electricity to the grid from 1984 to 1987. It was redesigned in 1995 with a molten salt storage system and operated though 1999 under the name Solar Two (Hernández, 2009). I saw it as it was being torn down in 2009 and replaced with a trough plant, as the CSP industry still alternates between trough and tower.

With the exception of Solar I, Luz International Limited was the only company to develop a commercial CSP facility during this time period. In the 1970s, Luz's founder Arnold Goldman, who was a computer and electrical engineer, published *A Working Paper on Project Luz* and *Moving Jewish Thought to the Center of Modern Science*, which illustrated his personal convictions for the technology that were influenced by what he called "Jewish thinking," or tying knowledge to concrete action ("Meet BrightSource's Arnold Goldman," 2010). Goldman imagined Luz as "a utopian eternal city" "[symbolizing] both light and consciousness" with CSP "turning rays of light into celestial power" (Berger, 1998, pp. 25–26). In 1979, Goldman incorporated Luz International and secured a contract with Southern California Edison to sell electricity from a CSP plant (Cumps & Cumps, 1991). In 1984, Luz completed the Solar Electric Generating Station I (SEGS I): a $62 million, 13.8 MW plant encompassing 82,960 square meters covered with parabolic mirrors in Daggett. Luz built nine facilities in this area collectively known as the 354 MW Solar Electric Generating System (SEGS) (Lotker, 1991). Goldman was quoted as saying, "If we can keep going strong past 1988, we will prove that we are not just an experimental company but the first new legitimate source of energy since nuclear" (Cumps & Cumps, 1991, p. 88).

The 1978 Public Utility Regulatory Policies Act (PURPA) supported the development of Luz's CSP plants to some degree because PURPA allowed small-scale independent power producers with less than one-half utility ownership to sell their electricity to the grid (Hirsh, 1999). At the time, most utility companies owned their own power generation. PURPA reflected a breakdown of consensus that utility companies are natural monopolies. Hirsh views the implementation of PURPA in California as the start of electricity deregulation, or the splitting apart of vertically-integrated utility monopolies by separating the generation from the sale of electricity (ibid.). Jerry Brown, California's Governor from 1975 to 1983, supported small-scale power before PURPA was enacted and strictly implemented PURPA (Hirsh & Serchuk, 1996). In 1983, the California Public Utilities Commission required utilities to sign contracts with "qualifying facilities" (QFs), or facilities of 30 MW or less, which provided them with rates of compensation for 20 to 30 years on the basis of the (faulty) assumption that fossil fuel prices would continue to increase (ibid.). Wind and solar producers, including Luz, sold their electricity to the grid through QF contracts, but much of the overall QF generation came from cogeneration (or combined heat and power) plants (Lotker, 1991).

Last-minute extensions in and reductions of federal and state tax credits contributed to Luz's failure. The federal Energy Tax Act of 1978 provided a 15% Energy Tax Credit for investment in renewables until the end of 1985. Congress extended the tax credit to the end of 1986, although the rate was reduced to 12%, and then Congress renewed the credit each year, reduced to 10%. The federal Investment Tax Credit was 10% from 1981 to 1985, reduced to 8.25% from 1986 to 1987, then to 6.5% from 1987 to 1988, and eliminated thereafter (Lotker, 1991). This annual Congressional renewal process left investors uncertain about the security of investment in CSP, and it compressed construction times for each SEGS plant (ibid.). California also provided a Solar Energy Tax Credit of 13.5% from 1981 to 1985, increased to 15% from 1985 to 1986 and reduced to 6.6% thereafter (ibid.). In 1989, Congress extended the 10% Energy Tax Credit but only for nine months. Additionally, California provided a property tax exemption for solar installations, but it became politically controversial since it only supported Luz (ibid.), which contravened policy goals in California to increase competition in the electricity sector (see Hirsh, 1999). The California State Legislature approved the tax exemption in 1990, but the Republican governor vetoed it; the legislature passed the bill again in early 1991 with a supermajority, but the governor delayed signing it into law until May (Lotker, 1991). Congress extended the Energy Tax Credit again from 1990 to 1991, but the shortened construction schedule for SEGS VIII and IX, on top of ongoing management difficulties, had already consumed Luz's net worth, and Luz was unable to secure financing for additional plants. Luz filed for bankruptcy in late 1991.[13]

According to Luz's former Vice President of Business Development, Michael Lotker, Luz built the SEGS plants in 30 MW phases because of PURPA's QF size limitations, which impeded the plants' economies of scale (Lotker, 1991). In 1987, Luz successfully lobbied the federal government to temporarily increase the limit from 30 to 80 MW, although Lotker claims that Luz would have preferred to build an even larger plant. In 1998 dollars, the LCOE fell from 24 cents per kWh for SEGS I to 9 cents per kWh for SEGS VIII and IX due to economies of scale (ibid.), although presumably also due to learning effects. Assuming that Luz would have been more successful without the QF size limitations also assumes it could have secured a more sizable upfront investment for a larger plant at an earlier date. Lotker believes this to be true, but it cannot be proven retroactively.

CSP's commercial failure related to uncertainty in the tax environment, cost compared to fossil fuels, and—in subtle ways—a conflict surrounding the appropriate scale of renewable energy technologies. While the federal government was interested in CSP for its ability to achieve economies of scale and to provide a large amount of electricity, a jumble of policy goals affected the government-incentive structure for CSP and resulted in an incoherent investment environment. More broadly, Laird (2001) found that support for solar power during this time period was not successfully institutionalized; the influence of solar advocates was limited and, below the surface, solar advocates espoused very different values.

California's "Solar Gold Rush" in the 2000s

During the mid-2000s in the United States, goals for CSP shifted somewhat because of an increased focus on sustainability and climate change. In the late 1800s/ early 1900s and in the 1970s, advocates of CSP argued that it would be needed as soon as coal or oil ran out. In contrast, in the 2000s some constituencies saw using fossil fuels until they run out as morally untenable because of the effects of fossil fuels on climate change. Climate change activists called for leaving fossil fuels in the ground in order to mitigate global climate change. In 2007, a subsidiary of the Spanish company ACCIONA restarted the US commercial CSP sector, completing a 64 MW parabolic trough plant in Nevada supported by a 30% federal Investment Tax Credit (Timilsina, Kurdgelashvili, & Narbel, 2012). In 2007, Arizona's Democratic Governor, Janet Napolitano, called Arizona the "Saudi Arabia of solar energy within the United States," and the Arizona Department of Commerce developed the Arizona Solar Electric Roadmap study (Shelton, 2007). US federal government support for renewables was furthered through the economic stimulus package in 2008 and 2009 (Timilsina et al., 2012). Republican Arizona Governor Jan Brewer (2009–2015) dubbed herself the "solar queen" for her support of large-scale solar. California's Republican Governor, Arnold Schwarzenegger (2003–2011), supported CSP for its potential to increase energy independence and produce green jobs (Schwarzenegger, 2011). The California state legislature instituted ambitious renewable energy and climate change mitigation targets. Headlines included "California's new gold rush," "solar gold," and "a twenty-first century El Dorado."

In 2004, Goldman had resurrected Luz as BrightSource Energy. But Goldman's utopian solar dream was again disrupted in 2007 when a multi-year controversy began over the site selection of BrightSource's Ivanpah Solar Electric Generating System (ISEGS) (Moore & Hackett, 2016). In this time period, it was clear that CSP was technologically feasible: It could be built. But activists questioned if it *should* be built and, if so, in what places and landscapes. They criticized Ivanpah and similar proposed solar projects for their industrialization of public lands; their effects on endangered and threatened species, such as the desert tortoise and rare plants; and their damage to sites of spiritual importance to Native Americans (ibid.). For these activists, Goldman's machine had entered the garden (or what some activists called the "old growth desert"), in a way that violated the activists' sacred values, rather than in a way that connected heaven and earth as Goldman had imagined in the 1970s (ibid.). Conflicts over the scale of solar power re-emerged. Local environmental activists advocated for distributed solar PV instead of CSP, while some national environmental organizations supported CSP because of the scope and urgency of climate change (ibid.). Shortly after his 2011 election as governor, Jerry Brown—who supported small-scale technologies when he was governor in the 1970s—is said to have vowed to "crush" the opposition to large-scale solar on public lands in California (Weirub, 2011).

After a lengthy environmental review and siting process, BrightSource completed ISEGS in 2014. It has a gross capacity of 392 MW including natural gas for startup and intermittent cloud cover. It is sited on 3,400 acres of public land, it cost $2.2 billion, and it was partly financed by a $1.6 billion federal loan guarantee. During the construction of ISEGS, it became clear that it would affect the tortoise population more significantly than expected: 660–880 tortoises would be harmed on the project site, rather than the original estimate of 30 (Moore & Hackett, 2016). As ISEGS started operation, observers gawked as "streamers"—birds, insects, and debris—lit on fire in the air over the mirror field (ibid.) In Spain, similar siting controversies arose over protecting rural communities and ensuring benefit for them, as well as spatial planning, environmental and cultural impacts, and the visual effects of the "sprawl" of renewable energy (Prados, 2010). Currently, skeptics in the United States and abroad are questioning whether CSP meets sustainability criteria or whether it is a technological juggernaut. The ISEGS case, which cuts across environmental, social, and technological domains, illustrates that the challenges associated with the construction of sociotechnical electricity systems are integrated challenges.

During the 2000s, most CSP companies focused on a centralized, plug-in, scalable design. Modular Dish Stirling engines, which could be sited in backyards, received little attention. Tessera Solar, one of the only companies promoting Dish Stirling, focused on covering large swaths of land with 25 kW Dish Stirling engines, rather than selling the expensive technology to households. One challenge with distributed siting is that Dish Stirling could legally qualify as an "attractive nuisance" in backyards because children could use it to start a fire. Additionally, the dishes are difficult to make because the process requires brute-force welding next to fragile mirrors (Andraka, Rawlinson, & Siegel, 2012). Unlike BrightSource, Tessera did not receive a federal loan guarantee, and it filed for bankruptcy in 2011.

Recently, the nexus of CSP's development has shifted to the Global South—including the MENA region, South Africa, Chile, and China—as well as Australia. CSP costs remain high in the United States, at approximately 14 cents per kWh in 2015, although the DOE's SunShot Initiative set a goal to achieve an LCOE of 6 cents per kWh by 2020 (Feldman, Margolis, Denholm, & Stekli, 2016). It is difficult for the price of CSP to compete with the reduction in natural gas prices caused by the US shale gas revolution, the development of cheap and efficient natural gas combined-cycle power plants, as well as the reduction in the price of solar PV panels and onshore wind plants. However, the cost of CSP plants in Global South has fallen, as will be discussed later in this book.

Conclusion

Although it is often assumed that CSP was invented during a period of rising environmental awareness in the 1970s, it was, in fact, developed during the mid-1800s and became entangled within a search for cheap and abundant sources of fuel to power resource extraction in North Africa. When CSP re-emerged during the 1970s, system imaginers sought to replace fossil fuels in electricity systems to increase countries' energy independence and sovereignty, but CSP did not gain traction. In the 2000s, CSP became wrapped up in a complicated debate in the United States about global climate change, ethics, sustainability, and preservation.

A longitudinal approach to analyzing an emerging technology allows for comparing and contrasting actors' goals and priorities for the technology in different time periods. Actors have connected CSP to varied goals and ideologies, including finding energy sources to fuel colonial enterprises, solving perceived resource scarcity, improving national security and competitiveness, fostering energy independence, and mitigating climate change. Government financial support for CSP has been necessary but also has been tied to problems that have received episodic attention. Advocates of CSP consistently have associated it with irrigating and developing the desert frontier, discursively framing CSP as being constructed by pioneers in depressed desert zones in order to make the desert bloom using irrigation powered by solar farms that harvest abundant solar rays in virgin landscapes.

In the seventeenth century, Francis Bacon argued that scientific knowledge would allow humans to control nature and would act as an instrument of self-control by means of rational intelligence (Leiss, 1972). Bacon attempted to remove ethical strictures that impeded human control of nature (Merchant, 1989). In the United States today, democratic culture has been paired with a liberal attitude toward technological progress (Ezrahi, 1990). This attitude tends to ideologically oppose ethical limits placed on technological progress. Additionally, Winner observes that people think artificial kinds are easier to control than natural kinds, partially as a result of the dualism between nature and society. Yet technologies are often seen as having their own agency as they become integrated into systems that are beyond any one individual's control (Winner, 1977). For the activists opposed to the Ivanpah plant, CSP was not sustainable because it resulted in a perverse trade-off between preserving an "old growth desert" teeming with life and reducing greenhouse gas emissions. As activists look at green technologies such as the Ivanpah power plant with skepticism about whether they meet the criteria of energy justice and sustainability, they essentially seek to impose ethical restraints on technological change.

These debates are far from closed or clear-cut. Energy system builders are faced with a more complex challenge than displacing fossil fuels—that of shifting energy systems toward the often ill-defined concept of sustainability, which requires ethical principles. Today's debate about CSP's value is intertwined with complicated questions about what counts as a sustainable technology across the triple bottom line of environment, economy, and society as well as the role of renewable energy in sustainable development. Solar is often assumed to be intrinsically sustainable. More recently, activists are not asking "Can we?" (e.g., is it feasible?) but "Should we"? How do the ideological promises made for technologies change as they re-emerge, and are the past politics of the technology cemented in it? Is imperialism intrinsic to the politics of CSP, or merely part of an artifact left behind in Shuman's time? For example, the solar farming metaphor used today implicitly frames the technology as benign but still evokes a problematic frontier metaphor. Yet Winner (1986) argues that *artifacts* have politics, rather than technologies, suggesting that redesigns of technologies could result in different sets of politics and societal outcomes than in the past.

The following chapters continue to explore these themes. Chapters 3 and 4 examine the re-emergence of CSP in Europe in the mid-2000s as a "mature technology" subsumed within a vision for a Mediterranean-wide electricity system and valued for its crucial system property of dispatchability. The Desertec system imaginers framed CSP as a technology that could bridge continents in the postcolonial world, even while reflecting in their discourse certain aspects of colonial-era visions for megaprojects in the Sahara. Chapters 5 through 7 explore the success of CSP in Morocco under a government-supported program. In Morocco, the question has become whether CSP can deliver multiple ancillary benefits to society, such as the development of supply chain industries, job opportunities for both educated and undereducated populations, and a green electron export commodity, as well as an overall improvement in energy equity. Chapter 8 discusses ethical principles for energy systems design.

Notes

1 Translated from the original French.
2 *Association de Researches Technique pour L'Etude de la Mer Intériure Saharienne.*
3 Translated from the original French.
4 In 1767, Horace-Bénédict de Saussure invented a hot box, essentially a shoebox-sized glass greenhouse (Perlin, 2013).
5 Several Italian inventors also worked on CSP during this time, but less is known about them: see Silvi (2010).
6 A similar claim was made in *Scientific American* decades later: see Kaempffert (1913).
7 The Auteil machine was a device designed for using solar heat to pump water. It was made in Auteuil, France in 1885 ("The utilization of solar heat for the elevation of water", 1885).
8 Some research was funded by US government agencies, including the: Department of the Interior, Air Force Cambridge Research Center, Wright Air Development Center, Naval Research Laboratory, National Bureau of Standards, and Weather Bureau (Stead, 1957), as well as the National Science Foundation. In 1930s, Dr. C.G. Abbot, Secretary of the Smithsonian Institution from 1928 to 1944, studied solar radiation; patented solar cookers, distillers, and hot water heaters; and built a small solar trough (Devorkin, 1998; Ley, 1954). Abbot thought the practical application of solar energy would increase funding for the Smithsonian's astrophysical research (Johnson, 2015). Additionally, Daniele Gasperini, an Italian working in Libya, designed the Elio Dinamic solar motor, a prototype of which was presented in Tripoli in July 1936 (Silvi, 2010). The USSR experimented with concentrating and passive solar in Uzbekistan in the 1930s and 1940s (Avezov, Avezova, Matchanov, Suleimanov, & Abdukadirova, 2012). In 1949, Félix Trombe built a solar furnace in the French Pyrenees Mountains. A solar furnace called Héliodyne was built in Algeria at Bouzaréah in 1954, and the Institute for Solar Energy was created at the University of Algiers in 1960 (Caille, 2017).

9 CIEMAT stands for Centro de Investigaciones Energética, Medioambientales y Tecnológicas.
10 This count includes multiple plants built on a contiguous site.
11 The LCOE typically includes the installed cost, fuel costs, interest rate on capital, the amount of electricity generated each year, and operation and maintenance costs. Solar radiation levels, accumulated dust on mirrors, the choice of wet-cooling versus dry cooling, and storage affect the economics of CSP.
12 Abengoa filed for bankruptcy in 2016 and is undergoing debt restructuring, but the reasons it went bankrupt are complex and remain unclear at this date.
13 In 1992, many Luz engineers took the intellectual property rights to form a company called Solel to build parabolic troughs. Siemens bought Solel in 2009, and it went bankrupt when Siemens left the CSP sector in 2012.

Works cited

Adas, M. (1990). *Machines as the measure of men: Science, technology, and ideologies of Western dominance*. Ithaca, NY: Cornell University Press.

American inventor uses Egypt's sun for power. (1916, July 2). *New York Times*. Retrieved from www.debarel.com/BSB_Library/1916_shuman_ny_times.jpg

Andalucia.com. (n.d.). *Almeria Province: Indalo man*. Retrieved April 5, 2014, from www.andalucia.com/province/almeria/indalo/home.htm

Andraka, C., Rawlinson, K. S., & Siegel, N. (2012). *Technical feasibility of storage on large dish Stirling systems (No. SAND2012-8352)*. Sandia National Laboratories. Retrieved from http://citeseerx.ist.psu.edu/viewdoc/download?doi=10.1.1.452.6449&rep=rep1&type=pdf

Avezov, R., Avezova, N., Matchanov, N., Suleimanov, S., & Abdukadirova, R. (2012). History and state of solar engineering in Uzbekistan. *Applied Solar Energy*, *48*(1), 14–19.

Barrera, F. R. (2009). *Solar thermal power: History of a research success (La electricidad termosolar: Historia de éxito de la investigación)*. Sevilla, Spain: Protermosolar.

Berger, J. J. (1998). *Charging ahead: The business of renewable energy and what it means for America*. Oakland, CA: University of California Press.

Bisson, J. (2003). *Mythes et réalités d'un désert convoité: le Sahara*. Paris, France: L'Harmattan.

Boesman, W. (1994). *Big science and technology projects: Analysis of 30 selected US government projects*. Washington, DC: Congressional Research Service.

Brower, B. C. (2009). *A desert named peace: The violence of France's empire in the Algerian Sahara, 1844–1902*. New York, NY: Columbia University Press. Retrieved from www.jstor.org/stable/10.7312/brow15492

Butti, K., & Perlin, J. (1980). *A golden thread: 2500 years of solar architecture and technology* (1st ed.). Palo Alto, CA and New York, NY: Cheshire Books.

Caille, F. (2017). *L'Afrique solaire ou le récit oublié. Représentations sociales et expérimentations en matière d'énergie solaire en Afrique 19ème–20ème siècles*. Retrieved from https://hal.archives-ouvertes.fr/hal-01502785

Cumps, J. L., & Cumps, K. L. (1991). *Bargain for profit: A practical guide to bid for power*. Belgium: Garant.

Davis, D. (2007). *Resurrecting the Granary of Rome: Environmental history and French colonial expansion in North Africa*. Athens, OH: Ohio University Press.

Deign, J. (2014, March 21). Masdar versus Spain: Say adios to CSP investors? *CSP Today*. Retrieved from http://analysis.newenergyupdate.com/csp-today/markets/masdar-versus-spain-say-adios-csp-investors

Devorkin, D. (1998). *Charles Greeley Abbot 1872–1973: A biographical memoir*. Washington, DC: National Academies Press.

Domínguez, N. (2017, November 22). Dimiten los jefes del principal centro de energías renovables por las trabas del Gobierno. *El País*. Retrieved from https://elpais.com/elpais/2017/11/21/ciencia/1511291694_919819.html

Edit Suisse Group. (2003). Atlantropa. *Cabinet*, Spring, (10). Retrieved from www.cabinetmagazine.org/issues/10/atlantropa.php

Eneas, A. G. (1901, March 26). *US670916 A*. Washington, DC: United States Patent Office.

Ericsson's sun motor. (1884, May). *Scientific American, 50*(20), 310.

Ezrahi, Y. (1990). *The Descent of Icarus: Science and the transformation of contemporary democracy*. Cambridge, MA: Harvard University Press.

Feldman, D., Margolis, R., Denholm, P., & Stekli, J. (2016). *Exploring the potential competitiveness of utility-scale photovoltaics plus batteries with concentrating solar power, 2015–2030* (No. NREL/ TP-6A20–66592). Golden, CO: National Renewable Energy Laboratory. Retrieved from www.nrel.gov/docs/fy16osti/66592.pdf

Frolova, M. (2013, June). *Energy and landscape*. Presentation for Arizona State University study-abroad program. Granada, Spain.

Ghassemi, F., & White, I. (2007). *Inter-basin water transfer: Case studies from Australia, United States, Canada, China and India*. Cambridge, UK: Cambridge University Press.

González, del río P. (2008). Ten years of renewable electricity prices in Spain. *Energy Policy, 36*(8), 2917–2929. Retrieved from https://doi.org/10.1016/j.enpol.2008.03.025

Green, T. (1985, March 10). Three "Sun Giants" loom in the desert. *Los Angeles Times*, p. I19.

Hammond, A. L., & Metz, W. D. (1977). Solar energy research: Making solar after the nuclear model? *Science, 197*(4300), 241–244. Retrieved from https://doi.org/10.1126/science.197.4300.241

Heffernan, M. (1989). The limits of utopia: Henri Duveyrier and the exploration of the Sahara in the nineteenth century. *The Geographical Journal. 155*(3), 342–352.

Heffernan, M. (2011). Shifting sands: The trans-Saharan railway. In S. Brunn (Ed.), *Engineering earth: The impacts of megaengineering projects*. Berlin, Germany: Springer.

Hernández, V. R. (2009). *Solar thermal power: History of a research success (La electricidad termosolar: Historia de exito de la investigacion)*. Seville: Protermosolar.

Hess, D. J. (2016). *Undone science: Social movements, mobilized publics, and industrial transitions*. Cambridge, MA: MIT Press.

Hirsh, R. F. (1999). *Power loss: The origins of deregulation and restructuring in the American electric utility system*. Cambridge, MA: MIT Press.

Hirsh, R. F., & Serchuk, A. H. (1996). Momentum shifts in the American electric utility system: Catastrophic change-or no change at all? *Technology and Culture, 37*(2), 280. Retrieved from https://doi.org/10.2307/3106817

IRENA. (2012). *Renewable energy technologies: Cost analysis series: Concentrating solar power*. International Renewable Energy Agency. Retrieved from www.irena.org/documentdownloads/publications/re_technologies_cost_analysis-csp.pdf

Jarrige, F. (2010). "Mettre le soleil en bouteille": Les appareils de Mouchot et l'imaginaire solaire au début de la Troisième République. *Romantisme, 150*, 85–96.

Johnson, C. (2015). *"Turn on the Sunshine": A history of the solar future* (Doctoral dissertation). University of Washington. Retrieved from https://digital.lib.washington.edu/researchworks/bitstream/handle/1773/33187/Johnson_washington_0250E_14240.pdf?sequence=1&isAllowed=y

Jordan, P. G. (2013). *Solar energy markets: An analysis of the global solar industry* (1st ed.). London: Elsevier.

Kaempffert, W. (1913, April 5). Harnessing nature can the free energy of space be utilized? *Scientific American*, 308–309.

Kraemer, S. (2017, November 20). *Spain's government kneecaps research at Plataforma Solar de Almería*. Retrieved January 11, 2018, from https://cleantechnica.com/2017/11/20/spains-government-kneecaps-research-plataforma-solar-de-almeria/

Laird, F. (2001). *Solar energy, technology policy, and institutional values* (1st ed.). Cambridge, UK: Cambridge University Press.

Leiss, W. (1972). *The domination of nature*. Montreal, CA: McGill Queen's University Press.

Ley, W. (1954). *Engineers' dreams*. New York, NY: Viking Press.

Lotker, M. (1991). *Barriers to commercialization of large-scale solar electricity: Lessons learned from the Luz experience* (No. SAND91–7014). Albuquerque, NM. Retrieved from www.nrel.gov/csp/troughnet/pdfs/sand91_7014.pdf

Lovins, A. B. (1976). *Soft energy paths: Toward a durable peace*. San Francisco, CA: Friends of the Earth International.

Meet BrightSource's Arnold Goldman. (2010, May 1). *Green prophet*. Retrieved from www.greenprophet.com/2010/05/arnold-goldman-brightsource/

Merchant, C. (1989). *The death of nature: Women, ecology, and the scientific revolution*. New York, NY: Harper & Row.

Metz, W. D. (1977). Solar thermal electricity: Power tower dominates research. *Science*, *197*(4301), 353–356. Retrieved from https://doi.org/10.1126/science.197.4301.353

Mitchell, T. (2011). *Carbon democracy: Political power in the age of oil*. London: Verso.

Moore, S., & Hackett, E. J. (2016). The construction of technology and place: Concentrating solar power conflicts in the United States. *Energy Research & Social Science*, *11*, 67–78. Retrieved from https://doi.org/10.1016/j.erss.2015.08.003

Mouchot, A. B. (1869). *La chaleur solaire et ses applications industrielles*. Paris, France: Gauthier-Villars.

Muller, K. (2005). Reconfigurer l'Eurafrique. *Matériaux pour l'histoire de notre temps*, *77*(1), 52–60. Retrieved from https://doi.org/10.3406/mat.2005.1017

Musi, R., Grange, B. Sgourdis, S., Guedez, R. Armstrong, P. Slocum, A. & Clavet, N. (2017). Techno-economic analysis of concentrated solar power plants in terms of levelized cost of electricity. *AIP Conference Proceedings*, *1850*(160018). Retrieved from https://doi: 10.1063/1.4984552

NREL. (2017, March 20). *Concentrating solar power projects in Spain*. Retrieved January 11, 2018, from www.nrel.gov/csp/solarpaces/by_country_detail.cfm/country=ES

Paléo-Energétique. (2015). *Abel Pifre L'utilisation de la chaleur solaire*. Livroscope, Paléo-Énergétique et L'œil d'Or. Retrieved from www.paleo-energetique.org/wp-content/uploads/2016/01/abel-pifre-lutilisation-de-la-chaleur-solaire-tablettes.pdf

Perlin, J. (2013). *Let it shine: The 6,000-year story of solar energy* (Revised ed.). San Francisco, CA: New World Library.

Petroski, H. (1997). Engineering: Engineers' dreams. *American Scientist*, *85*(4), 310–313.

Pope, C. (1903). *Solar heat its practical applications*. Boston, MA: Colonial Press.

Prados, M-J. (2010). Renewable energy policy and landscape management in Andalusia, Spain: The facts. *Energy Policy*, *38*(11), 6900–6909. Retrieved from https://doi.org/10.1016/j.enpol.2010.07.005

Ragheb, M. (2011). *Solar thermal power and energy storage historical perspective*. Retrieved on April 1, 2015 from http://mragheb.com/NPRE%20498ES%20Energy%20Storage%20Systems/Solar%20Therm%20Power%20and%20Energy%20Storage%20Historical%20Perspective.pdf

Ribeill, G. (2001). De l'objet technique à l'utopie sociale: Les ressorts de l'imaginaire technologique des ingénieurs au XIXe siècle. *Réseaux*, *109*(5), 114–144.

Ridao, Á. R., García, E. H., Escobar, B. M., & Toro, M. Z. (2007). Solar energy in Andalusia (Spain): Present state and prospects for the future. *Renewable and Sustainable Energy Reviews*, *11*(1), 148–161. Retrieved from https://doi.org/10.1016/j.rser.2004.12.007

Royaumont, L. de. (1882). *La conquête du soleil, applications scientifiques et industrielles de la chaleur solaire (héliodynamique)*. Paris, France: C. Marpon et E. Flammarion.

Sater, J. N. (2010). *Morocco: Challenges to tradition and modernity* (1st ed.). London: Routledge.

Schwarzenegger, A. (2011, March 30). Energy ideas Democrats and Republicans agree on. *The Atlantic*. Retrieved from www.theatlantic.com/technology/archive/2011/03/energy-ideas-democrats-and-republicans-agree-on/73228/

Shelton, S. (2007, July 15). Suddenly, solar's hot: Energy prices have the sun looking good. *Arizona Daily Star*. Retrieved from http://tucson.com/business/technology/suddenly-solar-s-hot/article_f1fe4ce4-f64f-5510-afc6-b50bd0027ed9.html

Shuman, F. (1911a, September 30). Power from sunshine a pioneer solar power plant. *Scientific American*, *105*(14), 291–292.

Shuman, F. (1911b, December 2). The cost of power from the sun and from coal a comparative estimate. *Scientific American Supplement*, *72*(1874), 359.

Shuman, F. (1914, February 28). Feasibility of utilizing power from the sun. *Scientific American*, *110*(9), 179.

Silvi, C. (2010). The use of solar energy in human activities throughout the centuries, *Annals of Arid Zone*, *49*(3&4), 157–174.

The solar engine. (1870, November 19). *Scientific American*, *23*(21), 323–324.

The utilization of solar heat for the elevation of water. (1885, October 3). *Scientific American*, *53*(14), 214.

Solar Thermal World. (n.d.). Success Story: Feed-in Tariffs Support Solar Thermal Power in Spain. Retrieved January 12, 2018, from http://www.solarthermalworld.org/sites/gstec/files/CSP%20FIt.pdf

Sörgel, H. (Illustrator). 1920. *Atlantropa*. Munich, Germany: Deutsches Museum.

Starostina, N. (2010). Ambiguous modernity: Representations of French colonial railways in the Third Republic. *Journal of the Western Society for French History*, *38*. Retrieved from http://hdl.handle.net/2027/spo.0642292.0038.012

Stead, W. (1957). The sun and foreign policy. *Bulletin of the Atomic Scientists*, *13*(3), 86–90.

Swearingen, W. D. (1988). *Moroccan mirages: Agrarian dreams and deceptions, 1912–1986*. London: I. B.Tauris.

Tellier, C. (1890). *La conquête pacifique de l'Afrique occidentale par le soleil*. Paris, France: Librairie centrale des sciences, mathématiques, électricité, arts militaires et industriels, agriculture, etc.

Timilsina, G. R., Kurdgelashvili, L., & Narbel, P. A. (2012). Solar energy: Markets, economics and policies. *Renewable and Sustainable Energy Reviews*, *16*(1), 449–465. Retrieved from https://doi.org/10.1016/j.rser.2011.08.009

Trischler, H., & Weinberger, H. (2005). Engineering Europe: Big technologies and military systems in the making of 20th century Europe. *History and Technology*, *21*(1), 49–83. Retrieved from https://doi.org/10.1080/07341510500037503

Van Dyke, D., & Wild, P. (1997). *Daggett: Life in a Mojave frontier town*. Baltimore, MD: Johns Hopkins University Press.

Weirub, A. (2011, August 10). *Challenges ahead: Brown's 12,000 MW local renewables target*. Retrieved January 15, 2018, from https://ilsr.org/challenges-ahead-browns-12000-mw-local-renewables-target/

Winner, L. (1977). *Autonomous technology: Technics-out-of-control as a theme in political thought*. Cambridge, MA: MIT Press.

Winner, L. (1986). *The whale and the reactor: A search for limits in an age of high technology*. Chicago, IL: University of Chicago Press.

Yergin, D. (2008). *The prize: The epic quest for oil, money & power*. New York, NY: Simon and Schuster.

3 The critical geopolitics of renewable energy and spatial energy justice

Envisioning the Mediterranean, perceiving Desertec

> [Desertec is] rather like what Mark Twain said of Wagner's music; it's not as bad as it sounds. This idea is catching on. This idea is the idea of the future and it's probably going to be the biggest single project which will bring together North Africa, the countries of the Arab League, and the European Union.
>
> — Sir Graham Watson, Member of the European Parliament, third Dii conference, 2012

Introduction: critical geopolitics and energy justice

In 2012, Sir Graham Watson suggested that the deferred Desertec dream was ahead of its time but would be the political integration project of the future. Watson was a British Liberal Democrat in the European Parliament from 1999 to 2014 and a proponent of deepening European integration. In 2013, just one year after Watson's optimistic prediction about Desertec, I watched from the audience of the fourth Desertec Industrial Initiative (Dii) conference in Rabat, Morocco as Bertrand Piccard stood on stage in a black flight suit telling skeptical investors that Desertec could still take off despite political setbacks. Piccard and his team had developed a solar airplane that can fly without fossil fuels even at night. His maiden voyage from Switzerland to Spain to Morocco in June 2012 traced the path of the transmission lines under the Strait of Gibraltar that connect Spain and Morocco: a possible future pathway for solar electricity exports from North Africa to the EU. A team of Moroccan dignitaries met Piccard when he successfully landed in Rabat, illustrating the political significance of his trip. Piccard viewed himself as an energy "refugee" from Europe who fled across the Strait of Gibraltar to escape a European energy policy that was foundering, thereby reversing the typical direction of the immigrant journey and portraying himself as immigrating to a more progressive energy policy in Morocco.

Until refugees came streaming across the Mediterranean through Turkey to escape Syria in 2015, the Strait of Gibraltar was perceived as the main passageway for illegal immigration within the Mediterranean region and a focal point for regional human insecurity. It has been called the Strait of Death because immigrants have died trying to cross it in ramshackle boats. Although the Desertec vision is seemingly a technical project, Desertec should be understood within the context of actors' changing perceptions of the Mediterranean region and Mediterranean integration, integration that has recently been impeded by the refugee crisis, the potential closing of borders within the Schengen Area, and the Brexit vote.

This chapter explores the system imaginers' geographical portrayal of the Mediterranean region and the Sahara Desert in order to better understand Mediterranean electricity integration from the perspectives of energy justice and human security. This chapter discusses the Desertec discourse as a whole, whereas the following chapter will provide more explanation about the actors involved and the variations on the broader vision. Although Mediterranean electricity integration has stalled, the export of renewable electricity across nation-state lines is under consideration in many areas of the world, so the possible implications of electricity exports ought to be further examined.

In the field of international relations, the intersection between energy and geography is often studied through the lens of geopolitics. A realist conceptualization of geopolitics defines geopolitics as the literal geography of where energy reserves are located across the globe and the undue power and influence that these physical reserves bring to countries (see, for example, Correljé & van der Linde, 2006; Klare, 2009; Rodrigue, 2004; Wesley, 2007). This conceptualization of geopolitics assumes that nation-state energy politics often trumps market-based actions because of the important role that energy resources play in international political order. Moreover, realists conceive of conflict over energy resources as zero-sum, meaning they view one nation-state's energy gain as another's loss.

In contrast, Goldthau and Witte (2010) have argued, reflecting the liberal international relations school of thought, that the concept of energy geopolitics is outdated because the global market structure for oil, and increasingly for natural gas, renders the actions of any one nation-state largely irrelevant on a global scale. They view complex global markets and institutions as being able to compensate for the uneven geographical distribution of natural resources and, to a certain degree, for nation-states' manipulation of energy markets through the practices of resource nationalism and energy diplomacy. Moreover, they think that scholars of energy geopolitics make a "myopic and erroneous presumption that global energy politics is necessarily a zero-sum game, in which one country's energy security is another's lack thereof" (ibid., p. 2). They argue that more focus is needed on improving the institutions and markets that craft the "rules of the game" for governing energy, such as information sharing, laws, and rules and standards. While Goldthau and Witte do not think this process of changing rules is apolitical, they do have faith in improvements in market-based rules to render win-win outcomes.

The constructivist school of thought offers a third overarching perspective on energy geopolitics. Constructivists argue that threats to energy security are not "objective entities waiting to be discovered and counteracted" but are instead socially constructed (Trombetta & Dyer, 2013 p. 7). Their approach calls for an examination of how the messy norms and perspectives of the actors that govern electricity systems shape geopolitical threats and opportunities. From their perspective, geopolitics is still a relevant concept because it continues to influence the thinking of political elites who govern energy (Peters & Westphal, 2013). This field of critical geopolitics frames geopolitics as constructed and shaped over time by the ways in which space is imagined, "represented and contested" (Agnew & Corbridge, 1995, p. 7). As Agnew and Corbridge (1995) state, "To understand the world, therefore, requires that we understand its changing geography," seeing it as socially constructed and dynamic rather than static (p. 14). Space is not an "unchanging 'essence' that constantly produces the same effects" (ibid., p. xi).

My analysis in this chapter of the Desertec vision's geography supports the assertion that the concept of geopolitics is often used in simplistic and literal ways that do not employ a wide enough lens and do not sufficiently acknowledge how energy is governed today, particularly in the context of a transition toward low-carbon energy sources. It is often assumed that the concept of geopolitics will become irrelevant if countries transition away from oil and natural gas and toward solar and wind energy. Therefore, limited exploration has been undertaken of the critical geopolitics of renewable electricity systems. Although renewable energy technologies are often thought to be technologies that promote independence, a large-scale transition to renewable energy could reorder the geopolitics of energy systems rather than doing away with geopolitics. The Desertec vision provides an avenue for anticipating the geopolitical dynamics of energy transformations. My analysis of the geographical shaping of Desertec treats the geography of energy as imagined, manipulated, perceived, and rewritten by the actors in power rather than as static. It suggests that the ways in which powerful actors imagine geography and landscapes will continue to be relevant even as more electricity is generated from sun and wind. Moreover, actors' perceptions of geopolitics will likely result in resistance to transnational renewable power systems.

In this chapter, I describe the processes through which actors framed the Mediterranean region in the context of the Desertec vision, as well as the implications of this process for energy-justice outcomes. Through these framings, the actors negotiated a tension between constructing a logical and feasible argument for increasing techno-economic connectivity within the Mediterranean region and assuaging fears that further integration would lead to conflict and also that solar electricity would be a geopolitical resource within a fractious region. They did so through four key processes. First, system imaginers portrayed North Africa as possessing a geographical advantage over Europe for generating solar electricity on specific parcels of land—which I refer to as renewable energy terroir—while downplaying the geopolitical risk of European energy dependence on North Africa by visualizing the land as abundant, barren, and devoid of people. This framing also played on a historical archetype of the Mediterranean region related to trans-Saharan trade routes in order to lend historical legitimacy to a vision for Mediterranean electricity trade. Second, they framed a modern energy system as one that is highly interconnected, and they emphasized that the Sahara could become an energy bridge and the Mediterranean could become an energy hub, although they left ambiguous the degree to which Mediterranean peoples and cultures themselves would be connected through this process. Third, they utilized the metaphor of an island essentially to frame connectivity to future "premium networked spaces" as sustainable and efficient and to frame detachment from a future centralized, transnational grid as marginalizing and inefficient. Fourth, through cartographic methods they erased nation-state boundaries and replaced them with transmission interconnections, while problematically arguing that technological integration facilitates political integration.

The second part of this chapter incorporates data from fieldwork I conducted in North Africa. It illustrates how critical geopolitics and security relate to energy justice, exploring recognition justice for people living in desert landscapes slated for solar development, energy access for people living in sub-Saharan Africa, and human security relating to the politics of immigration. By bringing together the spatial politics of renewable energy and energy justice, I emphasize that the challenges of energy governance today go beyond addressing the insecurities stemming from unevenly distributed resources. Goldthau and Sovacool (2012) characterize the energy governance challenge not only in terms of achieving energy security but also in terms of redressing energy injustice and transitioning to low-carbon energy systems.

Goldthau and Witte (2010) argue that the zero-sum framing—that one country's energy security gain is another's loss—ought to be replaced by a framing that considers mutual benefit from improvements in energy security. Similarly, the Desertec Foundation argued that win-win solutions that improve "global energy security" are possible. It is indeed clear that increased energy security in one Mediterranean country does not necessarily cause decreased energy security in another Mediterranean country. Yet the tenets and principles that constitute global, or even regional, energy security have not yet been clearly defined, especially in relationship to human security. Karlsson-Vinkhuyzen & Jollands (2013) developed the concept of "deep energy security," which they define as "energy security that contributes to human security over space (from the local to global) and time (that is, now and for future generations)" (pp. 513–514). This perspective calls attention to the possibility that a win for security at the scale of the nation-state could be a loss for security at the scale of the individual. Win-wins can be achieved only if renewable energy is viewed as a public good for citizens, rather than as a prize to be gained for nation-state security (Karlsson-Vinkhuyzen, Jollands, & Staudt, 2012).

The concept of "deep energy security," as well as the concept of energy justice, could aid in developing criteria for qualitatively assessing global energy security. This requires understanding who wins and loses across scales from new energy systems development. The evaluation of energy justice should focus on energy equity rather than the lower standard of energy equality. To achieve global energy security on the basis of energy equality, everyone should have the same access to and quantity of electricity. In contrast, energy *equity* and "deep energy security" require a much higher standard—including, for example, the fair negotiation of complex issues related to land and water, a just process for recognizing local populations and engaging them in energy siting processes, and the negotiation of sociopolitical differences. While Goldthau and Witte are optimistic that improved market rules and transparency will ensure win-win outcomes, the analysis in this book suggests that win-win outcomes cannot be achieved unless those who govern energy transitions consider potential security and equity outcomes at both the nation-state scale and the human scale.

Renewable energy terroir

Partly because sun and wind are available across the globe, less attention has been given to exporting renewable electricity than to exporting unevenly distributed fossil fuel resources. To substantiate the logic and benefits of exporting renewable electricity, Desertec system imaginers argued that renewable electricity would be best generated in desert landscapes—where solar resources are strongest and the climate is most advantageous—and then exported. At the Dii's 2012 conference, Cornelia Pieper, then minister of state at the German Federal Foreign Office, explained that it is, in fact, "obvious" that Europe will import electricity from North Africa because North Africa has a more favorable climate for generating renewable electricity. In the Desertec feasibility studies, North Africa is often framed as the best place to site concentrating solar power plants, as CSP in Egypt would yield more electricity than in Madrid, Spain and even more than in Freiburg, Germany. This reflects a Ricardian Model of trade, or trade that is ostensibly based on a single factor of production, in this case a geographical advantage for CSP generation in North Africa over Europe. This discourse frames energy as a commodity and establishes a logic for exporting electricity in a globalized world, in contrast to the nation-state-based model of energy security that frames energy as a strategic resource.

Wine growers in France use the term *terroir* to describe the precise geographical conditions in which specific grapes are grown in order to yield a high-quality wine appealing to an international audience. In conversations at energy industry conferences and in technical arenas, a set of siting specifications emerged that were said to yield high-quality solar electricity. This constitutes a techno-geographic and techno-economic framing of renewable energy terroir. Renewable energy terroir is, first and foremost, the sunniest available site because high solar insolation makes the energy cheaper; other aspects of the terroir include flat and stable land, thousands of acres of contiguous land without conflicting uses, high altitude, good drainage, access to complementary infrastructure such as transmission lines and water sources, and access to a nearby workforce. According to the actors' discourse, all of these elements together could enable the generation of higher-quality electricity in North Africa than could be produced in crowded and cloudy Germany. Here, high-quality means clean, dispatchable, and relatively cheap. I observed this discourse while studying the CSP industry over 10 years, from the siting process for the US Ivanpah Solar Electric Generating System to the siting process for the Moroccan Noor CSP plants and the Egyptian Kuraymat ISCCS plant (a natural gas plant integrated with CSP). Earlier versions of this discourse can also be observed in historical documents (see Chapter 2).

To identify renewable energy terroir, the solar industry relies on technologies of measurement, such as abstract geographic information system (GIS) methods, meteo stations, and shadowband pyranometers. A technician stated,

> Before we build a solar thermal [CSP] plant we need to know the exact conditions at the respective locations. This includes direct solar radiation above all. . . . [Meteo stations with pyranometers] are how we establish certainty with regard to planning and investments, because we know exactly where the sun shines most.
>
> (Solar Millennium AG and Tamia Film &TV, 2010)

The resulting solar radiation maps identify the sites that developers see as best suited for solar power development.

Yet the geographical advantage argument implies that renewable energy could, in fact, be a geopolitical resource because high-quality renewable resources are not evenly distributed and the sites seen as ideal for harvesting renewable energy—the best terroir—constitute discrete plots of land, mostly outside Europe. The justification for exporting renewable energy from North Africa to Europe conflicts with the need both to alleviate the perceived insecurities associated with reliance on other countries for electricity and to demonstrate that a surfeit of land is available for renewable energy scale-up. This conflict was more pronounced after renewable energy siting controversies unfolded in Europe and the United States (see, for example, Devine-Wright, 2011; Moore & Hackett, 2016). CSP developers utilized techniques of visual abstraction to emphasize to the public that a surfeit of land is available for CSP plants. While the GIS maps that identify specific renewable energy terroir are typically proprietary and costly, system imaginers widely distributed maps that identified large regions of the globe as appropriate for CSP development. Advocates of the Desertec vision framed the world's deserts as appropriate for large-scale CSP plants by depicting the world's "sunbelt," or the area with the most solar radiation, from an abstract viewpoint. The sunbelt framing constructs a contiguous, alternative source of energy wealth—ripe for export to sun-poor areas—on an abstract, empty landscape (see Figure 3.1 and Frank Shuman's similar depiction in Figure 2.5).

Thus, while CSP's proponents emphasize in technocratic forums that CSP requires large swaths of contiguous land that meets specific requirements, they publicly trivialize the total amount of land required. This serves tacitly to reconcile the contradiction that the renewable energy terroir, which would by definition constitute a set of unique places, is not a scarce, geopolitical resource. For example, TSK Flagsol, a German CSP company that developed the Kuraymat ISCCS plant, stated in its advertising materials that there are a "*nearly unlimited number* of local options" for siting solar power. "On a worldwide scale there is an *abundance of vast deserts* and semi-arid regions. Less than three percent of the Sahara desert would suffice to meet the world's energy demand with parabolic trough power plants" (emphasis added) (Flagsol, n.d.). A 2011 Dii brochure described Desertec as "a long-term vision: using the *abundance of space*, wind and sun to generate renewable energy in North Africa and the Middle East" (emphasis added). The small boxes on the bottom left of the map in Figure 3.1 reinforce this framing by emphasizing how little land would be needed to meet the world's electricity demand, the demand of 25 EU countries, and the demand of the MENA region in 2005 as well as the generation

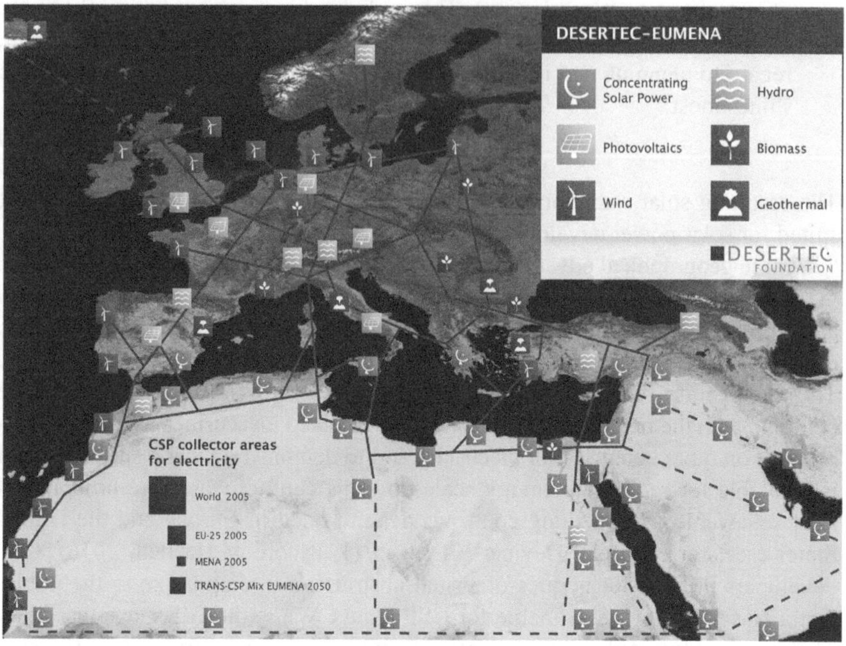

Figure 3.1 Desertec Foundation's representation of the Desertec vision

Source: Desertec Foundation, www.desertec.org, Creative Commons license

from CSP required in a scenario from a report that will be discussed in the next chapter. The European continent is shaded in green, while the MENA region is in yellow, indicating barren land appropriate for CSP. The concept of renewable energy terroir is invisible at this scale; the map suggests that nearly all North Africa and the Middle East would be appropriate for solar power.

De Souza et al. (2018) argue that European perceptions of Desertec's feasibility were influenced by "institutional orientalism," which they define as a "complex set of distorted perceptions" and stereotypes informed by European othering of North Africa, including the fear that a solar cartel in the MENA region would cut off electricity to Europe (p. 96). In fact, actors also used Orientalist stereotypes to counter claims that solar power would be geopolitical.[1] They downplayed geopolitical risk by framing the Sahara as an empty, barren landscape where no one lives who could disrupt energy supplies. The promotional materials for a renewable energy conference in Cairo depicted massive solar panels in empty, sandy deserts, with diminutive camel caravans below them. This image reinforced Orientalist stereotypes rather framing the desert as a place where people live. Such visuals not only portray the renewable energy terroir as devoid of humans or governments who could disrupt the supply of energy, they also portray it as devoid of people who would benefit from energy access.

Connectivity for a modern energy system

Pace (2006) explored the discourse related to imaginaries of the Mediterranean region and found that there is no one shared understanding of the region, which may not even be a single region given its diversity. The Mediterranean is not an objective and coherent region. Rather, various archetypes of the Mediterranean region and the Sahara exist, and these are reflected in the Desertec discourse. In one archetype, which emerged after World War II, the Mediterranean region is depicted as a shatterbelt, characterized by political fragmentation; linguistic, ethnic, and epistemic chaos; and contested boundaries ("Shatterbelt," 2010). It would be necessary for this archetype to be overcome in order to achieve political buy-in for greater regional energy integration. In the second archetype, the Sahara is depicted as a bridge and an empty space to be traversed to reach the North African coasts and eventually Europe, representing "a *trans*-Saharan rather than *intra*-Saharan trade" (McDougall & Scheele, 2012, p. 5). In this archetype, the Sahara Desert is seen as a path of connectivity to the north of the Mediterranean Sea, but a barrier to the south of the Sahara,

> dividing the Mediterranean world from "real" (i.e., *sub*-Saharan) Africa, isolating the countries of North Africa from their southern and eastern neighbors, and demarcating entirely distinct areas of study for scholars and students.
> (McDougall & Scheele, 2012, p. 4)

A third archetype, stemming from more recent scholarship on Mediterranean history, portrays the Mediterranean region as groupings of interconnected microecologies where life depends on connectivity, exchange, and widespread networks of support (Horden & Purcell, 2000). Building on this newer framing, McDougall and Scheele (2012) view the Sahara both "as part of densely interdependent networks" (p. 6) and as "a global borderland" (p. 7). There is more intra-Saharan trade than trans-Saharan trade (Scheele, 2012). There are still constraints on motion, such as national boundaries, but people defy them by smuggling goods across them (McDougall & Scheele, 2012). Each framing can be seen at one point or another in visions for Mediterranean electricity systems. However, the predominant archetype has been the Old-World, Orientalist view characterized by trans-Saharan trade routes, in which the empty Sahara Desert acts as an energy bridge to the north of the Mediterranean and a boundary to sub-Saharan Africa.

The Mediterranean and the adjacent Sahara have long been imagined as "barriers, bridges, or borderlands" (McDougall & Scheele, 2012, p. 4). System imaginers used maps and metaphors to frame deserts as connective links and the Mediterranean as a hub for a future modern energy system. The Desertec vision's promoters have not been alone in their efforts to engineer the desert as an infrastructural link. For example, the Planck Foundation's DesertCorp project seeks to boost economies in Africa and Asia by turning deserts into economic productivity zones through seawater irrigation for growing halophyte crops, creating "man-made salt Amazons." They state, "deserts now often are large barriers. Deserts could become connections. Infrastructure is what makes the difference" (Planck

Foundation, n.d, p. 19). Similarly, the Sahara Solar Breeder project proposed by the Science Council of Japan aims to create a "global clean energy super highway *connecting* GW [gigawatt] scale PV power stations located at various desert areas with high-temperature superconductor (HTSC) transmission lines" (www.ssb-foundation.com/). Finally, the Sahara Forest project, advanced by a Norwegian limited-liability company, seeks to "green the desert" using CSP plants and greenhouses to address food-energy-water nexus challenges (http://saharaforestproject.com/).

To establish this connectivity, system imaginers reframe desert frontiers metaphorically from boundaries and remote spaces to bridges that can be technologically "put into service" (Zickfeld, Wieland, Blohmke, Sohm, & Yousef, 2012). The original illustration depicting the Desertec Foundation's vision used lines to illustrate transmission corridors connecting continents, rather than nation-state borders (see Figure 3.1). The Desertec feasibility studies then described the connection of the region through a "backbone HVDC grid" linking the Middle East, North Africa, and Europe. HVDC lines connecting the continents would act as long-distance interstate highways, while local AC grids would serve as country roads and city streets (DLRa, 2006). Dii's *Desert Power 2050* study, published in 2012, presented a different vision of desert and Mediterranean connectivity (see Figures 3.2 and 3.3) (Zickfeld, Wieland, Blohmke, Sohm, & Yousef, 2012). This study envisioned markets and transmission systems that are so deeply integrated that electricity would be instantly available where it is needed, when it is needed. According to the study, this

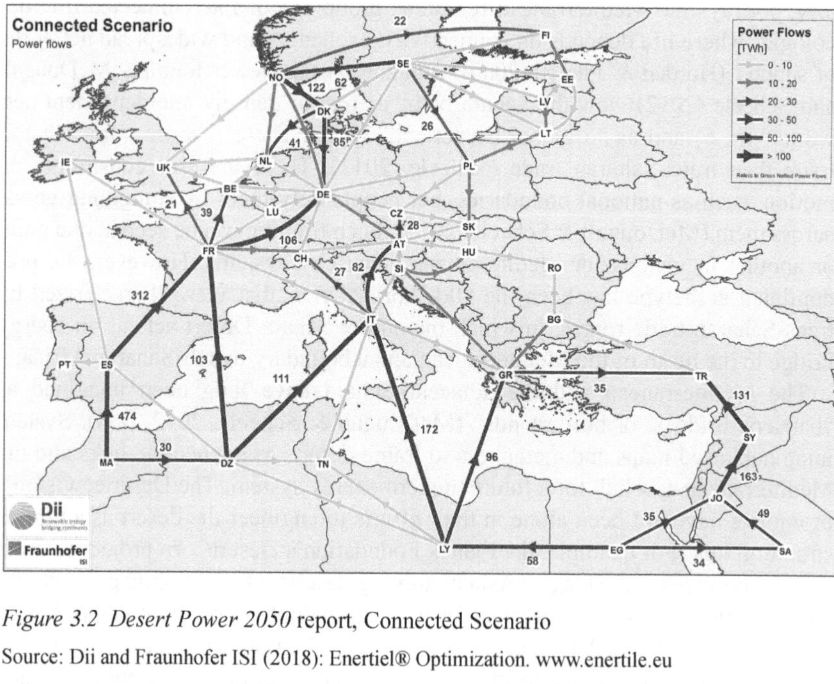

Figure 3.2 Desert Power 2050 report, Connected Scenario

Source: Dii and Fraunhofer ISI (2018): Enertiel® Optimization. www.enertile.eu

system would efficiently allocate fluctuating renewable energy supply to reliably meet demand across the region. Figure 3.2 illustrates deep integration but also significant abstraction, constructing the region as a whole, which visually negates the political challenges entailed in transnational transmission systems development.

The front cover of the *Desert Power 2050* report is even more abstract than Figure 3.2, illustrating energy use and energy flows (see Figure 3.3). It does not show nation-state borders, and it emphasizes the massive scale of the vision by depicting the edges of the globe. Through its nighttime view, it also emphasizes areas of demand, but it depicts today's energy demand rather than the demand projected for 2050—if it were showing 2050 projections it would represent MENA's electricity demand as roughly equal to that of Western Europe.

In Dii's scenario, once electricity from North Africa "bridges" the Sahara, the Mediterranean Sea then serves as a transportation hub through which electricity

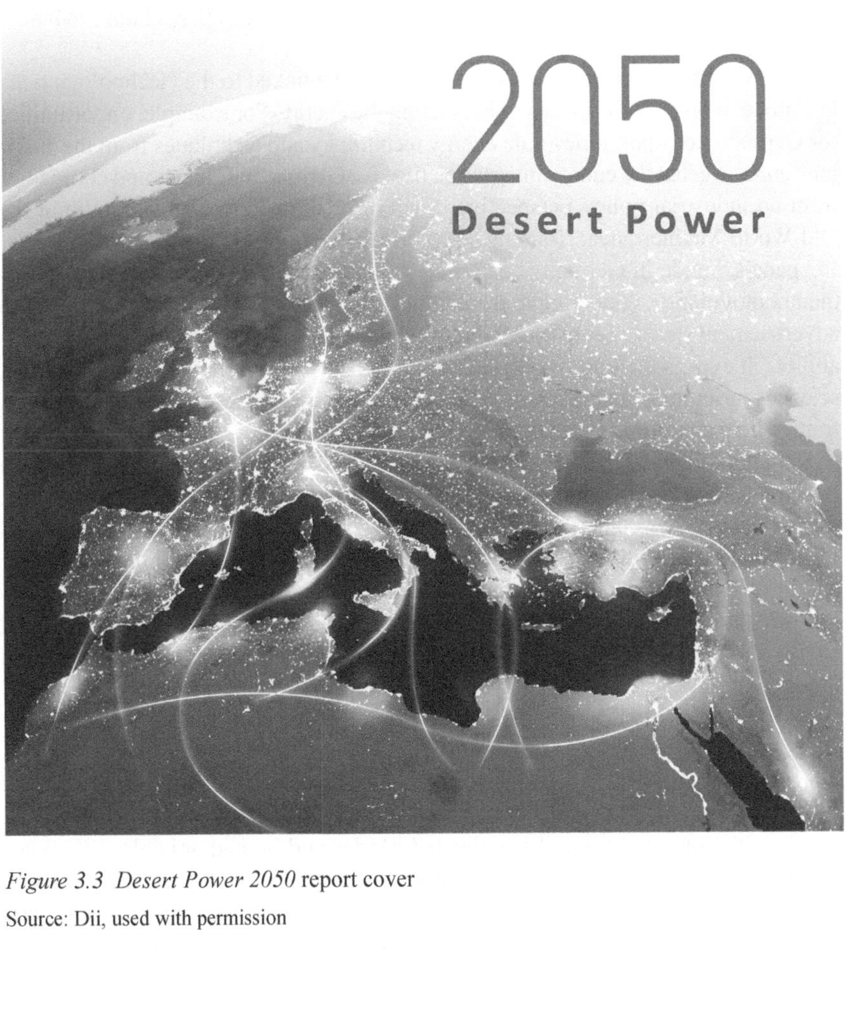

Figure 3.3 Desert Power 2050 report cover

Source: Dii, used with permission

flows. To aid in justifying a vision for energy cooperation, the actors discursively reframed the Mediterranean as a hub rather than a boundary. Describing the beginning of the Desertec visioning process, an interviewee stated, "And they painted the first picture of an EU-MENA region—Europe, Middle East, North Africa—as a whole. Understanding the Mediterranean Sea not as a boundary but as a transport means" (see Figure 3.1). Desertec vision proponents referred to the Mediterranean as "a bridge between two continents," Dii's slogan for its first three years was "renewable energy: bridging continents," and the *Desert Power: Getting Started* report asserted there is "no choice but to understand the Mediterranean as a hub rather than a border." At Dii's 2012 conference, Cornelia Pieper stated, "[The Mediterranean] is not only a water body that separates us, but it can also be a bridge between our two continents." The first transmission line built underneath the Mediterranean Sea also reflected this framing. In *Transmission & Distribution World* magazine, the Moroccan utility company ONEE and the Spanish Transmission System Operator (TSO) Red Eléctrica de España referred to the transmission line as "a bridge between two continents" (Granadino & Mansouri, 2007) and "the first link in the [MEDRING] chain" (Manuel Rodriguez Garcia, 2010).

According to the vision, the sunbelt must also be linked to the "technology belt" in Europe to provide the means of harvesting the energy. For example, opportunities for Germany to export renewable energy technology and techniques from the *Energiewende* sweeten the energy integration deal for German politicians and companies from an industrial policy perspective. Such efforts are further situated within the Old World Mediterranean framing. In a print advertisement by Centrosolar, solar PV panels "made in Germany" cap the Pyramids of Giza, tacitly comparing German photovoltaic engineering to the engineering feat of building the pyramids. The advertisement states, "Innovative solutions, award-winning quality, and proven reliability." It evokes the framing of the Mediterranean as the birthplace of civilization, science, agriculture, and technological progress, although technological progress is framed here as European rather than North African.

System imaginers frame the Mediterranean as an ideal region in which to develop a modern electricity system, and they frame CSP plants as energy farms (rather than power plants) developed in the Fertile Crescent, which is sometimes regarded as the birthplace of modern agriculture. This solar farming metaphor has been used for at least a century and a half (see Chapter 2). Despite evoking this Old World archetype, one of the Desertec feasibility studies discussed energy farms connected to a Mediterranean transmission grid as the epitome of a modern energy system. The study explained that thousands of years ago humans evolved from nomadic hunter-gatherers to settlers and farmers—except in the energy sector:

> Today's civilization is still based on gathering different forms of fossil energy, just like our ancestors, that collected berries and hunted animals . . . Today, fossil energy resources are still sought and gathered until the last drop is spent. It becomes more and more evident that this is not a civilized behaviour, and certainly not a sustainable one, because there is no other planet in view to

> move to after resources are depleted and the atmosphere is spoiled. However, our hunting and gathering ancestors found a solution to that dilemma: they became farmers, sowing seeds in springtime and harvesting corn and fruits in autumn, making use of technical know-how and the abundance of solar energy for their survival. That's exactly what is overdue in the energy sector: we must become farmers for energy. . .
>
> (DLR, 2006b, p. vii)

Though it uses the slippery terms *civilized* and *sustainability*, this quotation turns on its head the typical conception of fossil-fuel-based electricity systems as society's most modern infrastructure, implying that farming solar energy is the epitome of modernity and civilization. Moreover, it indirectly disputes a competing vision for a sustainable energy system that is small-scale and distributed, arguing instead that large-scale farming connected to regional markets would be more compatible with nature and the needs of a growing population, as well as less geopolitical than gathering fossil fuels, provided that the terroir of renewable energy in North Africa is deeply integrated into a regional grid and electricity market. Through its advocacy of extending the Desertec vision worldwide, the Desertec Foundation framed energy farming not merely as a solution to the energy challenges of the Mediterranean region, but also as part of global technological progress toward modern and low-carbon energy systems.

The harvest from these remote desert farms must be brought to market through increased Mediterranean infrastructural and market connectivity. The solar sites themselves would be distributed, but the connection to markets would be facilitated through a regional and centralized system, bringing an unevenly distributed solar energy resource to market. These visions use metaphors related to connectivity because system imaginers view the most sustainable, efficient electricity system as one that would expand the centralized system benefits to adjacent regions through deep interconnection. Following criticism, promoters of the Desertec vision emphasized that it would consist not of a single, monolithic project in a single box, but of numerous renewable energy farms distributed across the region and connected to a regional grid. Dii emphasized that this geographic diversity of supply coupled with deep market interconnections would improve energy security. The Desertec Foundation showed the CSP collector area boxes in Figure 3.1 splintering into many smaller boxes, with a disclaimer that the Desertec Foundation never presumed all the projects would be sited in one place.

The techno-politics of islanding

Desertec proponents used the metaphor of an island to emphasize further that a highly interconnected electricity system would be more efficient and sustainable. An "islanded" power system is one that is isolated from neighboring systems. The Iberian Peninsula, for example, is an electricity island because it is not fully synchronized to the same frequency and interconnected with the grids of the rest of continental Europe. From a technical standpoint, the islanded grid must match the frequency and phase of adjacent grids in order to interconnect with them.

According to electricity industry experts, "islanding has been seen by utilities as an undesirable condition due to concerns about safety, equipment protection, and system control" (Greacen, Engel, & Quetchenbach, 2013 p. 13). According to promoters of grid interconnectivity, islanded grids are inefficient because they are more likely to shed the load since electricity cannot be sourced from a neighboring grid, or to curtail (waste) excess generation that cannot be transmitted elsewhere because of lack of interconnectivity. System imaginers thought that regionally interconnecting the centralized electricity system—with its perceived benefits of efficiency, reliability, control, carbon dioxide mitigation, and safety—would be more advantageous than building a distributed electricity system.

At the 2013 Dii conference, Dii's CEO Paul van Son emphasized that countries cannot generate electricity in an isolated way unless they are on an island in the ocean. Furthermore, he said there is nothing worse than living on an island because of its lack of economic diversification. Islands also lack the "vast" area of deserts and are not well suited for hosting large-scale power generating facilities. In other words, they lack renewable energy terroir. System synergies could be captured through cooperation and connection that cannot be gained on electricity islands. In sum, islands are the most marginalized components of a system that prioritizes interconnectivity, cut off from "premium networked spaces."

In fact, policy officials from the island of Malta saw an opportunity to gain power through interconnection. At the 2012 Dii conference, the Maltese minister of foreign affairs said, "Due to its geographical location, Malta has historically been a bridge between the North and Southern shores of the Mediterranean on a political, cultural, and social level." Malta attempted to increase its geopolitical weight by reframing itself from a marginalized island to a bridge.[2] It would both gain access to electricity generated on North African land and serve as a hub in the Mediterranean region for electricity flows.

This islanding discourse is not specific to Desertec, but has spread to other projects. For example, the website for the "EuroAsia Interconnector," an HVDC submarine cable under development to connect Greece, Cyprus, and Israel, states, "We believe no one and no nation should be considered an island." The European Energy Union, announced in February 2015, sought to avoid islanding within Europe, and discussions about electricity integration in the US Southwest in 2016 also referred to eliminating electricity islands (Sommer & Joyce, 2016). Even the Caribbean islands have an electricity integration vision, and a study has been conducted on the feasibility of connecting the grids of 19 islands using HVDC cables.

Some other countries in the Mediterranean region have been wary of ending islanding, because deep integration reduces individual nation-state control and would make it difficult to later disentangle grids that cross nation-state lines. The political challenges of avoiding islanding through connectivity are significant at best and insurmountable at worst. Extensive political collaboration would be required to harmonize policies, standards, and codes to enable islanded grids to become interconnected. Achieving deep integration may require a history of political collaboration, an increase in mutual trust, and a gradual scale up of integration. Grids that operate in islanded mode may mirror a lack of political integration with neighbors and trust. For example, from 2007 to 2012, Turkey exported electricity to Syria through a single 124-km, 400 kV transmission line operating in islanded mode because the countries' grids were not synchronized (Hafner, Tagliapietra, & Andaloussi, 2012). In 2011, Turkey threatened to withhold water and electricity exports in retribution for attacks on Turkish diplomatic missions in Syria (Ozerkan, 2011). In 2012, Turkey cut electricity exports to Syria, but Turkish officials blamed a technical malfunction (Khawaja, 2012). Doubting the veracity of Turkey's claim, Syria began importing electricity from Iran instead (Haydar, 2012).[3] Since integration between Syria and Turkey was limited and in islanded mode, it was relatively easy for the countries to extricate their grids. Deeply integrated electricity markets and grids would be difficult to disentangle, and good political relations between and among countries would likely be a prerequisite in order for government officials to perceive integration as secure. (This point is further explained in Chapter 6).

Strategic geographical reframing and the technological integration–political integration binary

The original Desertec studies envisioned the creation of a new political region called "EU-MENA," which would require the regional negotiation of new laws, subsidies, and regulations to govern this regional, liberalized electricity infrastructure and market. Actors sometimes redrew political boundaries to accommodate the technological system. A map from Dii's *Desert Power 2050* report (see Figure 3.4) divided up the Mediterranean on the basis of existing electricity

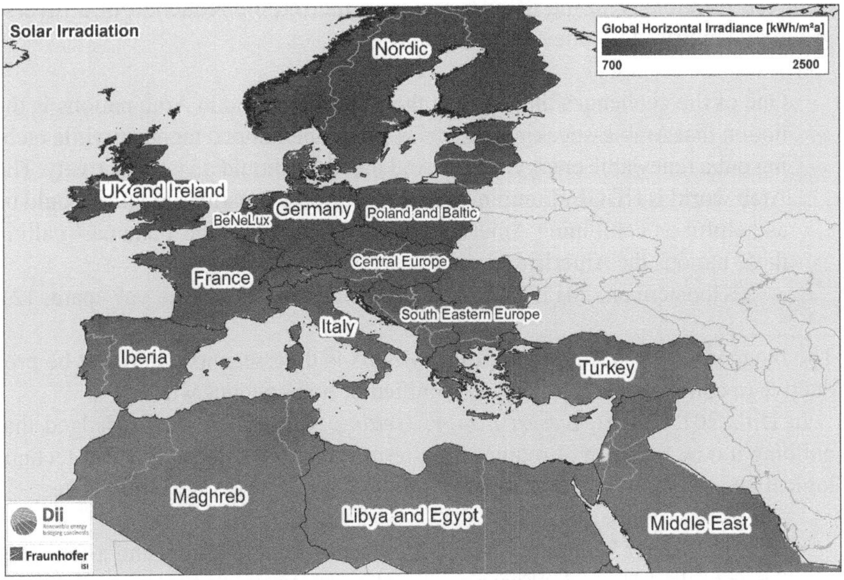

Figure 3.4 Desert Power 2050 report, portrayal of EU-MENA solar resources—country labels have been enlarged

Source: Dii & Fraunhofer ISI Enertiel® Optimization. www.enertile.eu, used with permission

integration, to emphasize the need to interconnect existing grids and avoid islanding. For example, islanded "Iberia" appeared as one country on the map. Morocco, Algeria, and Tunisia were drawn as the Maghreb country, even though relations between Morocco and Algeria have long been acrimonious. Actors also strategically renamed various parts of EU-MENA to give them a Mediterranean identity. Within the Mediterranean electricity systems integration discourse, Southern Europe is often framed as the northern shore of the Mediterranean, while North Africa is framed as the southern shore. In an EU-funded study on Mediterranean electricity integration, southern Europe, including Portugal, Spain, France, and Italy, was framed as the "Northwestern Mediterranean." Part of the Middle East was called the Eastern Mediterranean, encompassing Syria, Lebanon, Israel, Palestine, and Jordan. North Africa was called the Southern Mediterranean, including Egypt, Libya, Tunisia, Algeria, and Morocco (Med-EMIP, 2010).[4] These framings chart a more unified vision of the Mediterranean by framing the northern coast of Africa as Mediterranean rather than as African, Middle Eastern, or Arab. While maps can be easily redrawn in the visioning phase of a sociotechnical system, these reconfigurations of nation-state boundaries may not gain traction in reality. For example, the Spanish government was uninterested in the reframing of Spain as a bridge for Moroccan energy to Europe (see Chapter 4).

Within the attempt to frame EU-MENA as one region, MENA's regional coherency often goes unquestioned despite a lack of South-South political integration. In 2013, the Regional Center for Renewable Energy and Energy Efficiency (RCREEE) in Cairo released the first Arab Future Energy Index, which assessed the progress of MENA countries in renewable energy. The *Green Prophet*, a blog on green technology in the MENA region, criticized the reduction of a diverse group of countries into one region:

> One of the challenges that I see in the Arab world and in Arab nations is the notion that Arabic-speaking countries should be lumped together while etching out a renewable energy future. And that they should do it collectively. The Arab world is HUGE. Lumping all Arabic-speaking nations together could be as helpful as combining America with Mexico, Cuba and Panama—calling these nations the Americas.
>
> (Kloosterman, 2013, "Arabs need to think more German, ya?" para. 12)[5]

The blogger raises an important point: MENA is diverse, and it may not be productive to conceive of the region as a unified or homogenous whole.

In Dii's 2013 report, *Desert Power: Getting Started*, Dii acknowledged that building a new region will require long-term effort but emphasized that technological integration will help to achieve political integration. The report states,

> In 1950, Robert Schuman [French minister of foreign affairs and a founding father of the EU] said: "Europe will not be made all at once, or according to a single plan. It will be built through concrete achievements which first create a de facto solidarity." This report was researched and written in the spirit

> of identifying the **concrete achievements** capable of creating the solidarity needed to build a sustainable power system for EUMENA. And in the same spirit, Dii calls on the stakeholders and decision makers in the ongoing processes **to act now in favor of more renewables and electricity system integration** around the Mediterranean and in the Middle East. . . . A sustainable power system for EUMENA can only be built in partnership, with mutual respect, curiosity, and open-mindedness towards new solutions.
>
> (Dii 2013, p. 27, emphasis in original)

In this quotation, Dii does acknowledge that constructing EU-MENA would require more than metaphorical and cartographic work, contravening its visualizations.

Regional energy systems proponents frame the construction of transmission lines as the first concrete step toward stitching together the vast EU-MENA region. In so doing, they have established a binary between technological integration and political integration. The Barcelona Process that launched the Union for the Mediterranean sought to develop a "ring of friends" around the Mediterranean (Del Sarto & Schumacher, 2005). Several years later, the EU funded the MEDRING studies, which envisioned transmission integration gluing the region together to create this "ring of friends." Some system imaginers assert that technological integration will eventually result in political integration. In other cases, the technological system itself is framed as a proxy for political cooperation across this diverse region. The *Desert Power 2050* report envisioned a system, not of Mutually Assured Destruction, but of mutual reliance—"in which no single country is dependent on another but instead each country is reliant on the system as a whole" (p. 83). Technological dependence is viewed here as a substitute for political dependence.

In international relations, the idealism paradigm asserts that economic integration will facilitate peaceful relations among countries (Guzzini, 1998). Likewise, regional electricity integration discourses often assume that technological integration will lead to cooperation, dual-sided dependence, and peace. While Woodrow Wilson thought the League of Nations would bring peace through the development of international policy, in the Desertec vision, EU-MENA countries would be economically intertwined through a liberalized electricity grid that proponents assert would lead to peaceful political cooperation. Creating a binary between political integration and technological integration is problematic, as the two are themselves difficult to extricate (van der Vleuten & Kaijser, 2005). In fact, common regulatory frameworks came to be acknowledged as an adhesive necessary for energy integration. According to industry experts, "A key unmet need is for the regulatory 'glue' that connects technical standards and the players involved" (Greacen et al., 2013, pp. 15–16). In 2009, the EU developed such a regulation: a "cooperation mechanism" called Article 9. Article 9 allowed for renewable electricity imports from "third countries" (countries not part of the EU) to meet the EU's 20-20-20 targets (described in Chapter 2). As of today, no Article 9 projects have been implemented, and this regulatory "glue" has not resulted in transmission interconnection owing, in part, to a lack of political integration and trust (Lilliestam, Ellenbeck, Karakosta, & Caldés, 2016). The focus on technological integration assumes that electricity is a technoeconomic rather than a sociopolitical challenge, and this assumption has resulted in the skirting of the policy and justice questions at stake in electrical power systems development. Expanding the reaches of the EU would involve significant political challenges because it would require the creation of a "regional society" (Hettne & Söderbaum, 2000) subject to transnational rules. Assumptions that increasing technological integration would lead to peace, rather than to unprecedented challenges for cooperation and security, are likely unfounded. Therefore, drawing the EU-MENA (or EUMENA) region on paper does insufficient work to overcome the political challenges entailed in the construction of a politically integrated region.

Spatial energy justice and "deep energy security"

Regional geographic depictions omit particular populations and problem-framings through cartographic methods, the use of metaphors, and visual abstraction. These omissions draw attention to the need for system imaginers and policy makers to recognize local-scale perspectives. Next, I discuss three energy justice issues implicated in the vision, partly by supplementing the previous visual analysis with field observations made at the local scale. Through the lens of spatial energy justice, this section begins to explore the justice concerns that would be associated with the construction of transnational electricity systems.

Recognition justice and sociocultural terroir

In order for affected populations to be included in a just procedure for designing a Mediterranean energy system, they must first be recognized. The visual depiction of the Sahara as a vast, empty wasteland ignores complex land and water rights, does not recognize the people living there, and could result in uneven development that excludes local populations. As Edward Said stated,

> The earth is in effect one world, in which empty, uninhabited spaces virtually do not exist. Just as none of us is outside or beyond geography, none of us is completely free from the struggle over geography. That struggle is complex and interesting because it is not only about soldiers and cannons but also about ideas, about forms, about images and imaginings.
>
> (1994, p. 7)

Said (1994) illustrates that the colonial past continues to shape "perceptions of the present and the future." This tendency can be observed in the case of Desertec, in which system imaginers were influenced by the Old World Mediterranean archetype. They viewed North Africa as a barren frontier ripe for energy development, a perspective stemming from the colonial-era energy visions discussed in Chapter 2 (p. 7). Yet whether one conceives of the Sahara as empty depends on one's perception of what counts as empty space and what populations matter.

In 2011, I observed differences in perspectives of empty landscapes when I visited the Kuraymat ISCCS plant outside Cairo near Beni Suef. On a warm October day, I boarded a tour bus with a group of Desertec-vision enthusiasts and investors. From the comfort of our bus, the only sign of Egypt's political revolution was the military transport that followed us. As we drove out of Cairo, the landscape became increasingly sparse and sandy. I thought back to more colorful desert landscapes slated for solar development that I had visited in California, and I wondered if the siting situation could be different here. I sat next to an Egyptian engineer who supported CSP. She cautioned me not to assume that a sandy landscape would be free of conflict, stating, "We will not give away our land or sell our sand." I asked others in the group about who lived in the vicinity of the power plant and whether there were similar solar-siting controversies to those underway in the deserts of the US Southwest. They informed me that this plant was built on uninhabited lands free of conflicting uses. However, once we reached the power plant site I scaled the viewing platform and saw neither huts nor nomadic tents but actual apartment buildings. The perception of the landscape as empty contradicted what I saw. What is perceived as "uninhabited land" is constructed by the use of cartographic methods and metaphors as discussed previously, but also through a literal lack of seeing and valuing certain populations, admittedly including my own presumption that a sandy desert might have lower conflict potential than the seemingly greener Mojave Desert. Chapter 7 further explores issues of unseen and unrecognized populations living in the vicinity of CSP facilities and addresses how the trade-off calculus for CSP siting decisions is partly couched in conflicting perspectives of the land.

The Desertec vision reflects the practice discussed in Chapter 2 of defining the Sahara by its technological potential and its physical, rather than human, characteristics. Bisson (2003) argued that the boundaries of the Sahara are often inappropriately delineated on the basis of geographic determinism rather than human populations and the management of the Sahara by humans, through irrigation for instance. Scheele (2012) stated, "If we believe in statistics, the Sahara is perhaps the fastest changing, most dynamic, and wealthiest region of the African continent" with rapid urbanization and population growth and significant energy resources (p. 7). Rapid urbanization is shaping the Sahara as Saharan people respond to the demand for products, such as dates, in nearby cities as well as remote European markets (Bisson, 2003). The Sahara should be defined not only by its physical characteristics but also by its human and social characteristics, including the effects of this rapid urbanization in coastal regions on cities in the outskirts of the Sahara—the so-called "doors" to the Sahara—from which products are shipped through complex transportation and energy networks (ibid.). Bisson viewed Ouarzazate—a Moroccan city that is the focus of Chapter 7—as a key example of such a Saharan city, outside the geographic boundaries of the Sahara, but indelibly shaped by it.

As the visioning process proceeded, it became apparent that building CSP plants in the imagined middle of nowhere—in unpopulated, sandy Saharan dunes—would be infeasible. The first CSP plants are being built instead in the

Sahara's "doors," including Ouarzazate, which have at least some of the necessary infrastructure—including existing transmission lines and access to water, worker populations, supply chain industries, and stable ground. An interviewee from the oil industry explained that his company operates natural gas facilities in remote areas of the Sahara Desert only because these facilities are exceptionally profitable, even considering the significant risks and the costs associated with lodging workers and trucking in needed goods. Since electricity from CSP currently exceeds the grid price, it clearly would not offer sufficient payoff in remote areas. Moreover, building CSP plants in the Sahara's doors would have different societal implications than would building solar power plants in an empty desert. This suggests that CSP's feasibility will depend not only on geographic and technical factors but also on community acceptance and local socioeconomic development and human security.

The Desertec dream of exploiting vast energy resources in an empty desert ran aground as some people questioned who owns the sun, wind, and land in a desert that is not, in fact, empty. Sunlight may be a free resource, but it is not evenly distributed across the globe and its use is intertwined with other natural resource uses. An interviewee from a Dii shareholder company stated,

> In my own words, Algeria, for example, didn't want Desertec—they were in the beginning critical—had a critical view on Desertec, to give their sun power away to other people. [They said] it's our sun, our land, so we use it for ourselves. Morocco was the easiest country and most stable.

Algeria was not alone in its resource sovereignty concerns. For example, a Moroccan energy expert I interviewed perceived that Moroccans saw the sun and wind as free and belonging to everyone, but not their land and water.

Partha Dasgupta's work on natural capital aids in conceptualizing these conflicts. Dasgupta (2010) argued that environmental resources should be accounted for in economic growth, wealth, and development as natural capital. Natural capital overall is not fixed since it can be degraded. Solar heat and light are renewable resources that cannot be degraded, but solar irradiance is only accessible for "harvest" via specific plots of land under private or nation-state property rights. In some cases, harvesting sunlight uses significant water resources, and, in all cases, the plants require materials across their life cycle, such as steel. These resources can be degraded. Dasgupta (2010) refers to hidden resource uses as "implicit subsidies" on exports that people affected by the production of the export must pay. Who pays is always a difficult issue (ibid.), but it is particularly challenging in the case of renewable power plants, which sometimes have difficulty economically competing with fossil fuels. This is a dilemma because putting a price on the implicit subsidies would likely jeopardize solar power development under current economic conditions but failing to compensate people for their loss of natural capital, especially water usage, could reduce the human security of marginalized populations, thus impeding "deep energy security." Would siting Europe's energy infrastructure on another continent make those suffering from the externalities (implicit subsidies) increasingly invisible? Would North African countries choose to utilize these resources for domestic use—to trend toward what international relations scholars derisively call "resource nationalism"—or for export and the globalization of electricity? In order for electricity to be globalized, or even regionalized, numerous countries must decide that electrons are commodities appropriate for export. Along with the electrons, they will virtually export the land, water, and other natural resources that are a part of renewable energy terroir.

When I visited the tourism office in Ouarzazate, Morocco in 2016, the analytical category of renewable energy terroir became an actor category, and a more sociocultural, historic, and subjective concept of terroir emerged that aids in thinking through virtual exports. A political appointee responsible for promoting tourism in the Ouarzazate region spoke of its "terroir" as a broader compilation of the region's agricultural products—including saffron, roses, and dates—its carpets and handicrafts, its history as a commerce route, and most importantly its culture. This terroir is celebrated and sustained through festivals called *moussem*. He said that the terroir was also influenced by the confluence of two images of Morocco—that of a Mediterranean country with its blue and white kasbahs and hand-painted tile, and that of the gateway to the Sahara. In sharp contrast to the shatterbelt framing of the Mediterranean, he perceived Moroccans as embodying a pluralistic and tolerant Mediterranean identity. He said that historically the trade routes brought people together and created a Moroccan mosaic of diversity, also reflecting the Old World archetype of the Mediterranean. At first the exchange was limited to economic trade, but later it deepened to cultural and social change through the cosmopolitan mixing of Arab, African, and Amazigh cultures. "We lived together in harmony," he stated. He explained that the Moroccan mosaic is part of the strength of the country. Its power relates to its continuity in time and space, solidifying Morocco as a strong, unified country with a long history as a nation-state.

He argued that it is important for human security for CSP plants built in the Ouarzazate region to be integrated into this cultural terroir, rather than being a separate project. A solar electron has the power to uplift communities, to sustain their culture, and to strengthen the Moroccan mosaic. But it also has the power to damage the fragile mosaic and the cultural terroir if the infrastructure is implemented in a way that does not respect the region's inhabitants. He used the analogy of bringing tourists to the region. Rather than resorting to protectionism by trying to keep tourists out, the tourism bureau sought to advertise Ouarzazate to the type of cosmopolitan tourist who would appreciate the culture and respect the resources of this fragile region. Interviewees in Midelt, Morocco, the next area slated for large-scale solar development, also spoke of a sociocultural version of terroir that incorporated *both* geographic specifications *and* sociocultural ones.

While the system imaginers also focused on Mediterranean connectivity, it is not clear whether only electricity infrastructure would be connected or whether people living in the terroir would be connected with people on the other side of the Mediterranean who would enhance and respect the mosaic. The program for Dii's 2011 conference in Egypt stated, "It's about connecting people, cultures, and continents." However, it pictured an empty, sandy desert and pyramids. In contrast to the slogan, the imagery visualizes deserts as interstitial zones through which energy will travel to cities around the Mediterranean. Many of the abstract images and maps used to promote the vision serve to obscure the sociopolitical complexity of this renewable energy terroir, or to reduce it to Orientalist depictions of people on camel caravans. In Margaret Heckel's book *Desertec—Energy for Everybody*, which is a journalistic account of the Desertec vision's history, a fictional character named Anna travels to North Africa to see the power plant she helps fund.

> [Anna] owns enough mirrors of a power plant in Tunisia to supply her with all of the energy she uses. So she has an idea: Why not use the money to make a trip with her friends to see the solar power plant? As with all plants with a Desertec label, the people that built it were very keen to ensure that the local population benefited as much as possible from it. Anna knows that nearby her plant there are schools, which train technicians for installing the mirrors and even a small university with different programs. So why not go and look for herself how the power is generated in the desert? It's a cool idea—a trip to see where your power supply is generated.
>
> (2011)

Heckel imagined that energy integration would spur interest in cultural integration and exchange. But would Anna's voyage represent green tourism, or cultural voyeurism that results in the "othering" of North Africans? The local population ought to be seen as true participants and collaborators in the energy system rather than as an exotic other. More broadly, what would the role of citizens be in a system that seeks to achieve global energy security? A conception of terroir that is based on social rather than merely techno-geographic specifications could shape a cosmopolitan vision of Mediterranean connectivity that does indeed relate to connecting people and cultures and to improving the human security of those living in the terroir.

Electricity for Africa

Chapter 2 illustrated that CSP's developers have historically assumed that people in the vast, empty sunbelt regions would require little electricity in the future. Electricity from solar power facilities built in MENA would need to be transmitted to and consumed in load centers in Europe. But the facts oppose this assumption: Electricity demand in the MENA region is projected to reach European levels by 2050 as demand stagnates in Europe and skyrockets in MENA. When it became apparent that MENA would not have surplus electricity for decades, the Desertec visioning process was disrupted as people questioned the idea of importing electricity from energy-scarce countries.

Perspectives of energy justice and energy poverty shift again when the scale of the African continent is considered. In one archetype of the Mediterranean region, the Sahara is perceived as a boundary between the Mediterranean region and sub-Saharan Africa. While the Desertec vision frames deserts as connections and the Mediterranean as an electricity hub, it treats the Sahara as a boundary to sub-Saharan Africa, and this could impede electricity integration between North Africa and sub-Saharan Africa. Compared to their southern neighbors, North Africans enjoy high rates of electricity access: 75% of sub-Saharan Africans lack access to electricity compared to only 1% of North Africans (World Bank, 2011). Desertec's proponents consider the vision a win-win for all parties, including the energy impoverished, while Simeland and Juma (2011) criticize the vision for investing in relatively wealthy North Africa while sub-Saharan Africa lacks electricity. According to the BBC, Daniel Ayuk Mbi Egbe, a proponent of solar energy development in sub-Saharan Africa, stated, "When you go to many African countries there are constant electricity cuts—if you [Desertec] want to help then you need to think not just about exporting to Europe but about supplying Africa as well" (McGrath, 2012, para. 23).

A multi-scalar perspective leads to several questions not often considered in the Mediterranean-region context. Who should have the right to be linked to "premium networked spaces"? What technologies are appropriate in the perceptions of people who lack energy access? Proximity to regional hegemons, such as South Africa in the case of the Southern African Power Pool or the EU in the case of Desertec, could expand "premium networked spaces," while access to solar lanterns is framed as sufficient for people living in the periphery. This is, in part, a distributive justice issue that considers who wins and loses from such systems, but it is also a "spatial justice" issue (Soja, 2010) because the way the Sahara is imagined serves to erase certain nationalities and social groups.

Integrated and equitable or separate but equal?

The geographical construction of the vision for a Mediterranean electrical power system also reflects sociopolitical controversies related to immigration, connectivity, and barriers between Europe and MENA. An early paper on the Desertec vision stated that "tensions from poverty and immigration" in the Mediterranean region could be reduced through North African development (Wolff, n.d.). In 2013, an interviewee from the Desertec Foundation explained that the Desertec vision could address immigration and refugee challenges because solar power could improve people's lives domestically so that they are not forced to immigrate. This scenario would deeply integrate technological infrastructure and provide equal access to it but would keep Mediterranean politics—and Mediterranean peoples—largely separate. Later on, Desertec promoters promised to deeply integrate Mediterranean infrastructure and policy to bind the region together to address shared problems. This intention is reflected in Sir Graham Watson's quotation (from the beginning of this chapter) asserting that Desertec would be the biggest single project bringing together the EU, North Africa, and the Arab League. In the best-case scenario, the first vision would offer energy equality, ensuring that everyone has equal access to energy through reliable, interconnected infrastructure while the region remains culturally and politically fragmented. The second vision could offer broader energy equity and justice outcomes that strike at the core of cultural and political integration issues in a region encompassing deep sociopolitical differences and inequities. Both depictions of the Desertec vision require actors to grapple with issues of political and social feasibility. The first assumes that technological integration could be achieved without political will. The second assumes that unprecedented political and technological integration would yield overwhelmingly positive, win-win outcomes. Yet recent events, such as the immigration crisis and the Brexit vote, expose the significant political challenges that would need to be negotiated in order to achieve and maintain political integration.

Immigration debates often reflect the framing of the Mediterranean as a shatterbelt, with little political connectivity and deep cultural divides. The Desertec vision was under development as Islamophobia increased in France in the 1990s and after 2000. Samuel Huntington's inflammatory idea that future world conflicts will involve world civilizations—or the highest-order social groupings on the basis of cultural identity, language, history, and religion—was taken up in France after 9/11 (Scott, 2007; Huntington, 2011). The French government sought to halt immigration, concerned that an influx of Muslims would impinge on French culture (Scott, 2007). The French government banned the hijab in public schools, ID photos, and public buildings (ibid.). The French concept of *laïcité*—which refers to the protection of the state from religion, or the opposite of the US conception of freedom of religion—was seemingly at stake (ibid.).

The situation worsened following the terrorist attacks in Paris in 2015. In January, twelve journalists and cartoonists were killed in a terrorist attack on the office of *Charlie Hebdo* magazine, an event that heightened political tensions between North and South Mediterranean. Expressions of solidarity in France—voiced online through the *Je Suis Charlie* (I am Charlie) hashtag—were rebutted by some online activists in the MENA region with the hashtag *Je Ne Suis Pas Charlie* (I am not Charlie). These activists argued that the media was affording disproportionate attention to the lives of French citizens over the lives of Syrian refugees of war and victims of terrorist attacks in the Global South. Later in 2015, terrorists associated with Daesh (ISIL/ISIS) coordinated a series of distributed attacks in Paris against bars, a concert hall, and a stadium where young people had been enjoying a carefree weekend, leaving 130 people dead and hundreds wounded. The Paris *banlieue* (literally meaning suburbs, but pejoratively used to refer to "slums" where immigrants from North Africa and other countries live) were the subject of anger and fear, as police conducted early-morning raids searching for perpetrators. Many North Africans expressed on social media that they were being unfairly stereotyped as terrorists, and they pointed out that the Paris attacks had overshadowed a suicide bombing in Lebanon in which more than 40 people were killed, as well as the ongoing civil war in Syria in which at least 470,000 people have died (Specia, 2018). These activists argued that European lives were being valued above those of people from the MENA region.

Simultaneously, conflicts in the MENA region, particularly in Syria, Libya, and Iraq, caused more than a million immigrants and refugees to stream into Europe in 2015 alone. Germany arguably has been the country most open to refugees. However, in early 2015, 18,000 people took part in a protest in Germany against immigration and what protestors termed the "Islamization of Europe"—despite pleas from Chancellor Merkel to desist so that Germany would not be seen as xenophobic (Smale, 2015). Xenophobic attitudes were further inflamed when various media reported that more than 1,000 women in various locations in Germany had been sexually assaulted in crowds of revelers on New Year's Eve 2015: a story that police later said was fabricated (Christensen, 2017). The attacks were blamed on Arab immigrants or refugees or people of North African descent, stirring up further xenophobia in Germany. Some Europeans also perceive that immigrants will strain the social welfare system (Lehne, 2016). Because of pressure to close borders to stem the flow of refugees and guard against terrorism, the European press predicted the end of the Schengen Area, or the zone of borderless travel essential to European integration. Britain's 2016 Brexit referendum vote to leave the European Union further impeded the European integration project. The future of continued energy integration within Europe itself would be impaired by the breakdown of the Schengen Area and the rise of populism.

I observed the human security dimension of the politics of integration during a trip to Morocco with a group of students from Arizona State University in the summer of 2015. Our tour bus parked in the large parking lot of Marjane, a sprawling big-box store and a center of mass consumption in a developing country. A sub-Saharan African immigrant initiated a conversation with some of the students in the parking lot near our bus. Sub-Saharan immigrants come to Morocco hoping to reach Europe but are often stranded in Morocco. This man was merely searching for spare change. Everyone got back onto the bus, but as it pulled away we heard a sickening clunk. Our Moroccan colleagues disembarked, and we watched them gesture underneath the bus, as if shooing away a stray animal, visibly upset. After what felt like an eternity, a young Moroccan man crawled out from underneath the bus. He had mistakenly thought our bus filled with privileged Americans was bound for the Tanger-Med port, to cross the Strait of Gibraltar. He hoped we were his one-way ticket to Spain. We drove away, with students, faculty, and Moroccan colleagues equally stunned. I wondered what he had to gain from going to Spain. How would he make a living? How deep the anomie of young people must be for him to take such a risk with his life, as he surely could have been run over by our bus, beaten and arrested at the Moroccan port, or deported once he reached Spain. It was a jarring awakening for many of the students on our trip, who became aware for the first time of the privilege of mobility their American passports provide, a mere circumstance of place of birth.

Once we reached Spain the following day, our colleagues (who regularly take American students to Morocco) informed us that this habitually occurs. Once they had *twelve* people hiding under their bus at the port. The Tanger-Med Port Authority invariably discovers these stealthy passengers through infrared detection.

Success is more likely on a makeshift boat, but so is death. Across the strait, I saw sub-Saharan Africans who had successfully made the journey to Spain selling wares off blankets that were designed to be quickly scooped up at the first sign of a police officer who might charge them for illegally selling goods on the street. The impermanence of such a life—even for those who successfully make the journey—reflects the burden of placelessness in a globalized world. North Africans living in the *banlieue* of Paris are stereotyped as riffraff, rapists, Islamic fundamentalists, and vagrants (Swedenburg, 2015). Immigrants in the *banlieue* lament the melancholy of living a hybrid life, caught between two cultures and not feeling at home in either one (ibid.).

Considering this context, it is important to question whether the goal of regional energy integration is to bind people together equitably through a cosmopolitan identity, or to achieve energy equality for each separate region (MENA and the EU). Is the envisioned Desertec system an "integration miracle," as Dii labeled it, or a segregation solution, to arrange a regional power system so as to replace flows of migrants across borders with flows of electrons? On one hand, by providing equal access to technology on both sides of the Mediterranean, even in an integrated system the region could be technologically equal and improve human security but separate in terms of immigration and better sociocultural understanding. On the other hand, an integrated Mediterranean electricity system could, in theory, connect groups with different interests across the Mediterranean region to jointly negotiate their conjoined energy futures in an imagined Mediterranean community, resulting in greater understanding across diverse groups.

Procedural justice will be important to ensuring that equitable energy systems are built in the Mediterranean region. In fact, the quotation discussed earlier from *Desert Power: Getting Started* reflects a framing of political integration loosely related to procedural justice: "A sustainable power system for EU-MENA can only be built in partnership, with mutual respect, curiosity, and open-mindedness towards new solutions" (Dii, 2013, p. 27). Scott (2008) argues that instead of focusing on political oneness and unanimity, Mediterranean policymakers should be focusing on how diverse groups of people negotiate their differences on the basis of common ground rather than trying to eliminate these differences. Ramadan (2008) opposes the "outdated integration discourse" in the Mediterranean region because it highlights the differences between North and South. South-South conflict is also prevalent, including ongoing tensions between Morocco and Algeria (discussed in Chapter 6). Additionally, in the next chapter, I will discuss how North-North politics within the EU disrupted the Desertec vision. Awareness of the human security issues related to immigration and the implicated cultural divides will be an important aspect of energy justice to consider as electricity integration projects are pursued across the globe.

Conclusion

In this chapter, I have illustrated that large-scale solar energy is not entirely immune to geopolitical concerns. However, the intersection between geography and politics for a Mediterranean solar power future is not as straightforward as is the literally uneven geographical distribution of energy resources among Mediterranean countries. Instead, multiple actors crafted visual and metaphorical representations of the geography of a future Mediterranean grid. They did so using a variety of techniques to frame the region and its resources in a way that attempted to metaphorically and visually unite the region to establish the vision's logic. These techniques were influenced by long-standing archetypes of the Mediterranean region. The first technique related to the terroir of renewable energy, which constitutes the solar insolation levels, land grade, altitude, and geographical landscape in which solar energy is best produced, as measured through GIS methodologies. The assertion that North Africa has better solar terroir than Europe was intrinsic to the vision's logic, yet it also posed potential problems for system imaginers, as it suggests that solar energy terroir is unevenly distributed and is a potential geopolitical resource. Therefore, system imaginers also worked to portray the land available for CSP plants as abundant and barren through visual techniques such as abstraction and mapping. The second technique framed deserts as connections, bridges, or borders, rather than as boundaries—areas that could support a modern energy system on the basis of farming solar resources rather than hunting and gathering fossil fuels. The third technique used the metaphor of an electricity island to frame connectivity as more sustainable and efficient and isolation as the cause of marginalization and economic decline. Using the fourth technique, Desertec's promoters worked to depict the Mediterranean as a future, unified EU-MENA region that could be constructed through concrete steps over time. They visualized a region that would be knit together through power infrastructure rather than broken up along nation-state lines.

While some of these spatial representations of the Mediterranean projected a positive view, others served to erase or obscure the human security aspects, and therefore the "deep energy security," of the vision. One issue was the lack of recognition justice for stakeholders in North Africa, related to the framing of the Sahara Desert as barren. In reality, sunlight is abundant, but its successful exploitation is linked to other resources over which local populations seek sovereignty. Renewable energy terroir delineates certain parcels of land as appropriate for solar power generation, and one characteristic of these parcels is access to population centers that provide both workers and necessary infrastructure. The considerations for renewable energy terroir in North Africa are not merely techno-geographic but are also sociocultural. Writing these populations out of the landscape perpetrates both recognition and procedural injustices, as unrecognized populations are unlikely to be consulted in the process of designing and building the sociotechnical system.

While many of the geographical techniques utilized by the system imaginers conveyed Mediterranean connectivity, the cultural and societal aspects of this connectivity were underappreciated. The Desertec vision could be enacted as a vision to bind the region together technologically as a hub for electricity while keeping it socially and culturally fragmented. Or it could be deployed as a cosmopolitan vision to integrate the Mediterranean mosaic, promoting cross-cultural awareness and understanding. In the current time period, the politics of immigration and cultural difference are centrifugal forces pulling the region apart. The processes through which the actors constructed the geopolitics of the Mediterranean region could be influential in the rules of the game that are crafted for energy transitions. Such an understanding ought to be added to a governance process that too often views energy as merely a technical or techno-economic domain.

Although geopolitical framings of the Mediterranean region are significant in the discussion of Mediterranean electricity integration, energy security and energy justice are not necessarily zero-sum games. However, the Desertec case study illustrates that the outcomes for energy security and energy justice look different from different perspectives and scales. A win-win framing risks simply leaving the losers out of the frame. What is needed instead is what Agnew and Corbridge (1995) call "representational spaces" that empower communities and consider space in the context of human security, turning "space into places for securing human life against the depredations of coercive hierarchy" (p. 227).

Notes

1 I use Orientalism and Orientalist in the way that Edward Said does in his famous book *Orientalism*. I have followed Said's convention of capitalizing Orientalism and Orientalist.
2 Pace (2006) found that Malta generally presents itself as a bridge between North Africa and Europe.
3 Iran had been coping with international sanctions on its oil by using the oil to generate electricity and then selling this electricity regionally (Mirsaeedi-Glossner, 2013).
4 In the MEDRING study, Southeastern Europe included Slovenia, Croatia, BiH (Bosnia and Herzegovina), Serbia, Montenegro, FYROM (Former Yugoslav Republic of Macedonia), Bulgaria, Albania, and Greece. Turkey was listed separately.
5 This is, in fact, exactly how those regions are being framed within regional electricity integration discourses. See, for example, the US State Department's "Connecting the Americas 2022" program.

Works cited

Agnew, J., & Corbridge, S. (1995). *Mastering space: Hegemony, territory and international political economy*. London: Routledge.

Bisson, J. (2003). *Mythes et réalités d'un desert convoité: le Sahara*. Paris: L'Harmattan.

Christensen, C. (2017, February 17). Europe's biggest paper ran a bogus refugee "sex mob" story. What now? *The Guardian*. Retrieved February 20, 2017, from www.theguardian.com/commentisfree/2017/feb/17/europes-biggest-paper-bild-bogus-refugee-sex-mob-story?CMP=share_btn_fb

Correljé, A., & van der Linde, C. (2006). Energy supply security and geopolitics: A European perspective. *Energy Policy*, *34*(5), 532–543. Retrieved from https://doi.org/10.1016/j.enpol.2005.11.008

Dasgupta, P. (2010). Nature's role in sustaining economic development. *Philosophical Transactions of the Royal Society B: Biological Sciences*, *365*(1537), 5–11. Retrieved from https://doi.org/10.1098/rstb.2009.0231

de Souza, L. E. V., Bosco, E. M. G. R. L., Cavalcante, A. G., & da Costa Ferreira, L. (2018). Postcolonial theories meet energy studies: "Institutional orientalism" as a barrier for renewable electricity trade in the Mediterranean region. *Energy Research & Social Science*, *40*, 91–100.

Del Sarto, R., & Schumacher, T. (2005). From EMP to ENP: What's at stake with the European Neighbourhood Policy towards the Southern Mediterranean? *European Foreign Affairs Review*, *10*(1), 17–38.

Devine-Wright, P. (Ed.). (2011). *Renewable energy and the public: From NIMBY to participation*. Oxford: Earthscan.

Dii Desertec Industrial Initiative. (2013.) Desert power: Getting started. Munich, Germany: Dii GmbH.

DLR. (2006a). *TRANS-CSP: Trans-Mediterranean interconnection for concentrating solar power*. Stuttgart, Germany: German Aerospace Center Systems Analysis and Technology Assessment Section. Retrieved from www.dlr.de/tt/trans-csp

DLR. (2006b). *AQUA-CSP: Concentrating solar power for seawater desalination*. Stuttgart, Germany: German Aerospace Center Systems Analysis and Technology Assessment Section. Retrieved from http://www.dlr.de/tt/ AQUA-CSP-Full-Report-Final.pdf

Flagsol. (n.d.) Solar thermal power plants from vision to realisation. Retrieved from http://www.flagsol-gmbh.com/flagsol/cms/upload/download/Flagsol_Brochure.pdf

Goldthau, A., & Witte, J. M. (Eds.). (2010). *Global energy governance: The new rules of the game*. Washington, DC: Brookings Institution Press.

Goldthau, A., & Sovacool, B. K. (2012). The uniqueness of the energy security, justice, and governance problem. *Energy Policy*, *41*, 232–240. Retrieved from https://doi.org/10.1016/j.enpol.2011.10.042

Granadino, R., & Mansouri, F. (2007, May 1). A bridge between two continents. *Transmission & Distribution World*. Retrieved on July 16, 2012, from http://tdworld.com/underground-tampd/bridge-between-two-continents

Greacen, C., Engel, R., & Quetchenbach, T. (2013). *A guidebook on grid interconnection and islanded operation of mini-grid power systems up to 200 kW*. Berkeley, CA: Lawrence Berkeley National Laboratory.

Guzzini, S. (1998). *Realism in international relations and international political economy: The continuing story of a death foretold* (1st Ed.). London: Routledge.

Hafner, M., Tagliapietra, S., & Andaloussi, E. H. (2012). *Outlook for electricity and renewable energy in Southern and Eastern Mediterranean countries* (SSRN Scholarly Paper No. ID 2156872, MEDPRO Technical Paper). Rochester, NY: Social Science Research Network. Retrieved from http://papers.ssrn.com/abstract=2156872

Haydar, Z. (2012, October 19). Syria turns to Iran for electricity, leaves regional power grid. *Al-Monitor*. Retrieved November 17, 2014, from www.al-monitor.com/pulse/business/2012/10/syria-signs-electricity-deal-with-iran-to-deal-with-power-shortage.html

Heckel, M. (2011). Desertec- energy for everybody. Potsdam: published by author.

Hettne, B., & Söderbaum, F. (2000). Theorising the rise of regionness. *New Political Economy*, *5*(3), 457–472. Retrieved from https://doi.org/10.1080/713687778

Horden, P., & Purcell, N. (2000). *The corrupting sea: A study of Mediterranean history*. Oxford: Wiley-Blackwell.

Huntington, S. P. (2011). *The clash of civilizations and the remaking of world order*. New York, NY: Simon & Schuster.

Karlsson-Vinkhuyzen, S. I., & Jollands, N. (2013). Human security and energy security: A sustainable energy system as a public good. In M. J. Trombetta & H. Dyer (Eds.), *International handbook of energy security* (pp. 507–525). Cheltenham, UK: Edward Elgar Publishing.

Karlsson-Vinkhuyzen, S. I., Jollands, N., & Staudt, L. (2012). Global governance for sustainable energy: The contribution of a global public goods approach. *Ecological Economics*, *83*, 11–18. Retrieved from https://doi.org/10.1016/j.ecolecon.2012.08.009

Khawaja, M. (2012, October 11). Turkey cuts off power supply to Syria: Denies political motive. *Arabian Gazette*. Retrieved November 17, 2014, from www.arabiangazette.com/turkey-syria-power-supply/

Klare, M. T. (2009). *Rising powers, shrinking planet: The new geopolitics of energy*. Basingstoke: Palgrave Macmillan.

Kloosterman, K. (2013, October 3). 13 Arab nations get energy future mapped in helpful new report. *Green Prophet*. Retrieved October 4, 2013, from www.greenprophet.com/2013/10/13-arab-nations-get-energy-future-mapped-in-helpful-new-report/

Lehne, S. (2016). *How the refugee crisis will reshape the EU*. Retrieved November 3, 2016, from http://carnegieeurope.eu/2016/02/04/how-refugee-crisis-will-reshape-eu-pub-62650

Lilliestam, J., Ellenbeck, S., Karakosta, C., & Caldés, N. (2016). Understanding the absence of renewable electricity imports to the European Union. *International Journal of Energy Sector Management, 10*(3), 291–311. Retrieved from https://doi.org/10.1108/IJESM-10-2014-0002

Manuel Rodriguez Garcia, J. (2010, February). *The Medring: Reality or fiction?* Presented at the 1st Solar Thermal Electricity Industry Forum, Seville. Retrieved from www.estelasolar.eu/fileadmin/ESTELAdocs/documents/STEI_Forum/Presentations/Session_2.2.C_-_Juan_Manuel_Rodriguez_Garcia_-_The_Medring_Reality_or_Fiction.pdf

McDougall, J., & Scheele, J. (2012). *Saharan frontiers: Space and mobility in Northwest Africa*. Bloomington: Indiana University Press.

McGrath, M. (2012, November 19). Desert solar power plan in doubt. *BBC*. Retrieved November 19, 2012 from www.bbc.co.uk/news/science-environment-20357167

Med-EMIP (Euro-Mediterranean Energy Market Integration Project). (2010, April). Medring update: Volume-I overview of the power systems of the Mediterranean Basin. Retrieved from http://www.iemed.org/observatori-en/arees-danalisi/documents/arxius-externs/2011/2010-update-of-the-medring-study-regarding-mediterranean-electricity-interconnections-overview-of-the-power-systems-of-the-mediterranean-basin

Mirsaeedi-Glossner, S. (2013). Iran's flourishing regional influence. *Science & Diplomacy*. Retrieved from www.sciencediplomacy.org/article/2013/irans-flourishing-regional-influence

Moore, S., & Hackett, E. J. (2016). The construction of technology and place: Concentrating solar power conflicts in the United States. *Energy Research & Social Science, 11*, 67–78. Retrieved from https://doi.org/10.1016/j.erss.2015.08.003

Ozerkan, F. (2011). AFP: Turkey hits Syria with energy sanctions. *Agence France-Presse*. Retrieved November 19, 2011 from www.google.com/hostednews/afp/article/ALeqM5jJRI1JuHWljLOzXL97ADprKCorAg?docId=CNG.e4e461402118db348f30ab6275a98e25.d91

Pace, M. (2006). *The politics of regional identity: Meddling with the Mediterranean*. Hove, UK: Psychology Press.

Peters, S., & Westphal, K. (2013). Chapter 5: Global energy supply: Scale, perception and the return to geopolitics. In M. J. Trombetta & H. Dyer (Eds.), *International handbook of energy security* (pp. 92–113). Cheltenham, UK: Edward Elgar Publishing.

Planck Foundation. (n.d.). *Deserts become economic areas*. Retrieved from www.planck.org/downloads/Emerging-Nations-Desert-Development-PPP.pdf

Ramadan, T. (2008). *Face à nos peurs: Le choix de la confiance*. Lyon: Tawhid.

Rodrigue, J-P. (2004). Straits, passages and chokepoints: A maritime geostrategy of petroleum distribution. *Cahiers de Géographie Du Québec, 48*(135), 357–374.

Said, E. W. (1994). *Culture and imperialism* (Reprint ed.). New York, NY: Vintage Books.

Scheele, J. (2012). *Smugglers and saints of the Sahara*. Cambridge: Cambridge University Press.

Scott, J. W. (2007). *The politics of the veil*. Princeton, NJ: Princeton University Press.

Shatterbelt. (2010). *Encyclopedia of geography terms, themes, and concepts*. Santa Barbara, CA: ABC-CLIO.

Simeland, T., & Juma, M. (2011). Desertec: A testimony for deliberate divide of the African continent. In *Regional integration in Africa: Bridging the North- sub-Saharan divide* (pp. 34–44). Pretoria: Africa Institute of South Africa.

Smale, A. (2015, January 5). Anti-immigration rallies in Germany defy calls to desist. *The New York Times*. Retrieved March 19, 2015, from www.nytimes.com/2015/01/06/world/europe/pegida-rally-dresden-germany.html

Soja, E. W. (2010). *Seeking spatial justice*. Minneapolis, MN: University of Minnesota Press.

Solar Millennium AG and Tamia Film & TV. (2010). *We are developing the future!* Promotional Film. Erlangen Germany: Solar Millennium AG.

Sommer, L., & Joyce, S. (2016, April 19). *With too much solar, California looks to West for markets*. Retrieved November 3, 2016, from http://insideenergy.org/2016/04/19/with-too-much-solar-california-looks-to-west-for-markets/

Specia, M. (2018, April 13). How Syria's death toll is lost in the fog of war. *New York Times*. Retrieved May 1, 2018 from https://www.nytimes.com/2018/04/13/world/middleeast/syria-death-toll.html.

Swedenburg, T. (2015). Beur/Maghribi musical interventions in France: Rai and rap. *The Journal of North African Studies*, *20*(1), 109–126. Retrieved from https://doi.org/10.1080/13629387.2014.983737

Trombetta, M. J., & Dyer, H. (2013). *International handbook of energy security*. Cheltenham, UK: Edward Elgar Publishing.

van der Vleuten, E., & Kaijser, A. (2005). Networking Europe. *History and Technology*, *21*(1), 21–48. Retrieved from https://doi.org/10.1080/07341510500037495

Wesley, M. (2007). Chapter 1: The geopolitics of energy security in Asia. In M. Wesley (Ed.), *Energy security in Asia* (pp. 1–12). New York, NY: Routledge.

Wolff, G. (n.d.). *The Desertec concept*. Retrieved March 19, 2015, from www.trec-uk.org.uk/resources/desertec_talk.pdf

World Bank. (2011). *Energy-electricity access*. Retrieved September 9, 2012, from http://web.worldbank.org/WBSITE/EXTERNAL/TOPICS/EXTENERGY2/0,contentMDK:22505004~pagePK:148956~piPK:216618~theSitePK:4114200,00.html

Zickfeld, F., Wieland, A., Blohmke, J., Sohm, M., & Yousef, A. (2012). *Desert power 2050: Perspectives on a sustainable power system for EUMENA*. Munich, Germany: Dii GmbH & Fraunhofer.

4 The life cycle of a vision

Desertec system designs

> The idea to harvest the energy from the deserts has been around for decades in one form or another—a fantastic idea. In fact, so fantastic that the chances of this idea being realized must surely be zero. Implementing a system that supplies whole continents with renewable energy from deserts would require an extreme amount of commitment and intelligence by an unusually large number of stakeholders, by governments, by industries, by NGOs, by scientists, and by private citizens.
>
> —Aglaia Wieland, Managing Director, third Dii conference, 2012

The Desertec system imaginers have sought to extend the centralized electricity system configuration across a complex region in order to build an integrated system based almost entirely on renewable electricity. In contrast, some advocates of renewable energy transformations, such as the German NGO Eurosolar, envision a paradigm shift to a distributed system in which electricity is generated primarily on rooftops and in backyards. This chapter examines visions for a future renewable electricity system that would largely leave the centralized system configuration in place but would still result in dramatic reconfiguration. Since electrical power infrastructure already exists in many parts of the world, electrical power systems are not being built from scratch. Rather, the old systems are being refurbished and rearranged and new infrastructure is being grafted onto the old. The construction of new power systems for the mid-twenty-first century will respond to the existing sociotechnical and sociopolitical landscape. Developing complex, regional renewable-energy-based systems will require cross-sector management, as referenced by Dii Managing Director Aglaia Wieland in 2012, as well as regional management. Regional and cross-sector management pose new risks even as the new system that is under construction attempts to address existing risks. The innovation involved is not only technological but also social and political.

The export of electricity across nation-state borders is not itself a new phenomenon, but it is now being considered at much greater scales. According to nation-state-reported data compiled by the International Energy Agency, 725.28 TWh (terawatt-hours) of electricity were exported across nation-state lines worldwide in 2016, representing only 3.6% of total electricity generation (IEA, 2017). However, this figure does not sufficiently reflect the activity and trends in transnational grid integration. On the basis of extensive online research, I have found that there are at least 668 cross-border transmission lines worldwide, with additional lines planned for all regions of the world and all but 11 countries worldwide have cross-border lines or are part of a vision for cross-border lines. Examples of cross-border transmission projects under development include the Eastern Electricity Highway project between Ethiopia and Kenya, the highly complex CASA-1000 project to connect Central Asia to Afghanistan and Pakistan, a subsea 3 GW HVDC transmission link between Saudi Arabia and Egypt, and the North Sea link, which will transport electrons from Norway to the UK. These projects are examples of "shallow integration," in which transmission lines connect two or more countries without the presence of a single electricity market (ESMAP, 2010; Moore, 2017). In contrast, "deep integration" is characterized by a single grid and market for electricity in a region (ibid.). There are 17 examples of emerging or existing regional power pools worldwide, from each continent of the globe, with nearly all countries involved. There are visions for deep electricity integration that involve scaling up intermittent renewable energy (Cottier & Espa, 2017), although in reality most existing projects focus on hydroelectricity. In the long term, high levels of renewable energy consumption could be achieved through large-scale cross-border export or deeply interconnected supergrid models (Grave, Paulus, & Lindenberger, 2012).

Oseni and Pollitt (2014) assert that "international power pools are simply an example of the opening up to trade of a previously non-traded commodity" (p. 23). I argue that this assertion oversimplifies the issue. The political, regulatory, and technological integration needed for electricity exports is much higher than what is needed for exports of primary fuels and other commodities. Moreover, electricity systems do not operate like other commodity supply chains because of the temporally specific nature of electricity supply and demand and because of power systems' relationship to nation-state pride and politics. Therefore, it is important to take the social and political aspects of sociotechnical system design as seriously as the technological and economic aspects.

Visions for new energy systems

Hughes (1983) argued that electrical power systems are better characterized as sociotechnical systems than as technological systems because they are shaped by social factors—including powerful actors, politics and political institutions, and economics—as well as by geography. These systems develop through exercises of power to reflect the interests, values, and politics of powerful social groups. Once systems have achieved momentum, they are difficult to transform. It is important to update Hughes's work to better understand how contemporary large-scale sociotechnical systems evolve, especially as they undergo dramatic change.

Geels added to Hughes's conception of sociotechnical systems the multi-level perspective theory of sociotechnical regime change. He defines *niches* as incubators in which radical innovations form, innovations that could carve out a place in a path-dependent system. These technologies can break through the sociotechnical regime up to the landscape level (Geels, 2004), but breaking through is difficult (Verbong & Geels, 2007). Niches could also be viewed as spaces for the incubation of future system designs that system imaginers then occupy as they work to shape the future.

It is important to understand the role that visions play in changing sociotechnical energy systems (Brown, Rappert, & Webster, 2000).Visions play an important role in innovation by setting expectations for the technology and its future benefit (Selin, 2008). A sociotechnical vision is more sub-level than a collectively held sociotechnical imaginary because a vision is imagined but not yet collectively held by government and society (Bergman, Schwanen, & Sovacool, 2017). Selin (2006) stated that visions are "speculative framings and expectations that accompany and fortify a new technological field" so as to "colonize the future" (p. 3). These expectations are "performative" because they attract and coordinate actors who influence or align themselves with the vision (Borup, Brown, Konrad, & Van Lente, 2006). Embedded in these imaginations are powerful ideographs—popular but empty terms, such as *technological progress*—that legitimate expectations for future social benefits from the envisioned technology or technological system (Van Lente, 2000). These expectations for social benefits become requirements after they are used to recruit supporters to a shared vision, and the promises of benefits can later lead to disappointed expectations, in a promise-requirement cycle (Van Lente & Rip, 1998).

Expectations also shape actors' depictions of and assumptions about future societal outcomes, although often in ways that are unrealistic or lack systematic methods of foresight. Narratives of futures reflect ideologies, just as interpretations of the past do (Brown et al., 2000). Geels and Smit (2000) argue that imagined futures for an emerging technology often reflect static conceptions of society and social practices and are biased by the cultural framings of the period in which the technology is developed. Means of representing the future are often utopian, promising win-win outcomes greater than what can be achieved in reality (Brown et al., 2000). Visions are rarely, if ever, achieved as drawn up on paper. Yet energy visions, forecasts, and models themselves do work in the world, not as unbiased predictions of what the future holds, but rather as social constructs that contribute to self-fulfilling prophesies and that shape, expand, or even constrain actors' ability to imagine and construct future energy systems. While it can be methodologically difficult to examine "the [project] that was never built" (Brooks, 2006, p. ix), it is also crucial to understand the often-unseen processes that shape the energy futures that current and future generations will inhabit.

This chapter explores sociotechnical visions for energy transformations by analyzing how a vision for a Mediterranean sociotechnical electricity system was developed and negotiated over time and how regional politics and other intervening factors shaped and disrupted this vision. It analyzes system imaginers' framings of technologies, and their benefits and drawbacks, within a systems context. Given the Desertec vision's ambition and scale, actors must depict it as feasible. Means of proving feasibility include framing the end goal as an inevitable and proximal future, framing the technologies to be deployed in the vision as mature, and framing the system imaginers as credible and objective. Sociotechnical systems construed as elegant and sustainable in the design phase are reimagined and problematized as they become situated in existing sociopolitical and geographic contexts. What happens to this network of supporters when a vision must conform to the realities of disruptive events, media framings, and financial challenges, as a result of which certain expectations are not met? And what would facilitate or impede the vision from becoming a sociotechnical imaginary that is collectively held at the regional scale?

Table 4.1 presents a qualitative model of a vision for a sociotechnical energy system and the factors that shape it, drawing from the literature and this empirical study. Understanding the evolution of a vision for a power system almost entirely based on renewable energy contributes to a more nuanced understanding of why it has been so challenging to scale up renewable energy in existing power

Table 4.1 Model of a vision for a socio-energy system

Vision components	*Examples*
Technology & its framing	• Technological mix (e.g., solar PV, CSP, wind, HVDC) • Valued system properties the technology fulfills, e.g., dispatchability • Framing of technological maturity vs. innovation • Technological neutrality vs. technological advocacy
Geopolitical construction (see Chapter 3)	• Constructing a shared region • Visualization methods • Imagined boundaries/barriers/connectivity
System properties	• Dispatchability • Associated system benefits, (e.g., water desalination) • Mutual reliance • Baseload and peak power
Depiction of modernity	• Energy farming • Temporal control
Managing organizations	• Transmission System Operator • A UN Global Energy Interconnection Cooperative Alliance • Regional regulatory body
Characterization of societal benefits and risks	• Risk to investors • Meeting rising energy demand and mitigating energy poverty • Climate change mitigation
The construction of feasibility	• Methods of proof-of-concept (transmission line, reference project) • Feasibility studies • Technological maturity or neutrality
Cost & economics	• Lack of price for the service of dispatchability • Low price of natural gas • Falling prices of solar PV • Investor risk perceptions • Cost of capital and equity share

(*Continued*)

Table 4.1 (Continued)

Vision components	*Examples*
SHAPERS	**EXAMPLES**
System imaginers	• Companies and consortiums involved • Participating governments • Funders
Sociopolitical context	• Regional political process on energy (e.g., Union for the Mediterranean) • EU electricity policy developments (Article 9) • National renewable energy targets • Foreign policy, region building • Energy security definition/framing, national versus global
Timescale	• Perspective of the future: proximal or distant • Grappling with vast timescales
Wildcards	• Arab Spring, Eurozone crisis • Falling price of solar PV and changing energy economics • Events that disrupt feasibility (e.g., BP Algeria crisis)
Ideology underpinning the vision	• Limits to growth • Neoliberalism • Culturally situated framings of sustainability (see Chapter 5)
Stakeholder and media perceptions and response	• Neocolonial critique • Misunderstandings, different framings • Depictions of Dii

systems, as high cost is not the sole reason. Changes in energy systems always relate to sociotechnical controversies rather than representing apolitical technological progress (Mitchell, 2011). Moreover, visions for energy transformations unmask complexity in existing energy pricing and economics. The contestation underway in visions for energy systems change illustrates that several previously closed concepts, such as the idea of baseload power and the assumption that AC power is preferable to DC, are being reopened to scrutiny. I find that while Hughes (1983) says that sociotechnical power systems are centrally planned, the process of developing visions for new energy systems illustrates that central planning is now difficult because of scale, political complexity, and path dependence.

A number of studies have explored the feasibility of the Desertec vision in its various instantiations (DLR, 2006; Ummel & Wheeler, 2008; Williges, Lilliestam, & Patt, 2010; Zickfeld, Wieland, Blohmke, Sohm, & Yousef, 2012). I discuss this existing research throughout this chapter. It is clear that the Desertec vision both offers benefits and poses risks (Patt, Komendantova, Battaglini, & Lilliestam, 2011). Lilliestam, Ellenbeck, Karakosta, and Caldés (2016) examine the macrodimensions of renewable electricity imports to the EU (i.e., political support and macroeconomics), the microdimensions of electricity imports to the EU (i.e., the business case for investing in renewable electricity imports), and the public discourse relating to the lack of electricity imports to the EU. Beneking, Ellenbeck, and Battaglini (2016) find what is defined as a strength and what is defined as a weakness in regard to renewable energy cooperation between the EU and North Africa depends on the perspective of the actor who is considering renewable energy cooperation. Backhaus, Gausling, and Hildebrand (2015) examine how the assumptions made and the parameters set in various Desertec models influence the outcomes. The different inputs include whether wind and solar PV are included or only CSP, different values used for solar irradiation levels, different financing structures, different operating time estimates for the CSP plants (25 versus 30 years), different assumptions about the number of hours of molten salt storage available at night from CSP plants, and different costs of capital and subsidies. These assumptions and inputs determine the required public investment and the system's overall cost (e.g., 4.5 trillion euros over 40 years, 200 billion euros over 30 years, 43 billion euros over 21 years, 20 billion euros over 10 years). The assumptions made and the parameters set were so different that comparing the models required "compar[ing] the incomparable." The authors conclude that none of the feasibility studies is sufficient to assure investors that Desertec is a feasible and low-risk project.

As I demonstrate in this chapter, these studies actually were proposing very different system designs. I discuss the social construction of the system designs, rather than just the models behind them, and I explain the social process surrounding their evolution, addressing the different sets of actors and different subvisions and the internal conflict among them. I take this approach rather than treating Desertec as a single discourse coalition as Rothe (2016a) does. This chapter provides a theoretically inflected narrative on Desertec, contributing to the science and technology studies (STS) literature on visions and to social science research on energy transformations. Instead of viewing Desertec as having been killed by government instability in North Africa (Robinson, 2017) or by the uncertainty of cost estimates (Backhaus et al., 2015), I illustrate the multiple, complex aspects that shaped, influenced, and impeded the vision. Rather than assessing the Desertec vision's feasibility, I consider how feasibility is socially constructed. I argue that the vision ended in a feasibility trap, in which the social construction of feasibility across a variety of measures was insufficient to gain enough traction for the monumental sociopolitical task of achieving deep regional electricity integration, and therefore the vision did not become a collectively held sociotechnical imaginary.

The methodology for this analysis, extending from 2009 to 2017, is based on 32 in-person or video call interviews with government officials and European energy company personnel, including employees of the Desertec Industrial Initiative (Dii), Dii shareholder companies, and Desertec Foundation employees; content analysis of 400 news articles; extensive document analysis of planning and visioning documents related to Desertec and to Mediterranean energy systems (e.g., feasibility studies from the German Aerospace Center and from Dii); in-person event ethnography at two Dii conferences and online observations of a third; and three trips to southern Spain as part of a study-abroad program to visit energy infrastructure and meet with people working in the energy industry. (See Chapter 1 for more specifics.) I focused on Dii and Desertec rather than the related

French-run Medgrid or the Mediterranean Solar Plan because of the Desertec proponents' interesting attempt to form an industrial network of power to advocate for this massive energy transition.

In this chapter, I describe three system designs under the umbrella of the Desertec vision and explore the factors that shaped them. I call Design I "integrated solutions for a crowded planet." It is a vision for the export of electricity from CSP in North Africa through "solar pipelines" to Europe. Design II I call "the integration miracle." It is a vision for a deeply integrated supergrid and a single market for electricity between Europe and MENA. Design III I call "virtual exports" because this vision would develop a virtual carbon market instead of constructing material cross-continental transmission infrastructure. For each design, I discuss the changing elements of the vision as laid out in Table 4.1. I discuss how Design I was shaped both by public criticism and the changing constellation of actors advocating for it. This narrative ends not within the MENA region but in China, as actors there envision developing a global grid. If the Desertec vision failed, then rising from its ashes is an even broader and more ambitious plan for a global renewable energy grid—a plan that comes from a nation-state and corporation with the capital and power to execute it in certain parts of the world. I conclude this chapter by discussing issues that should be addressed up front in the development of transnational grids, before visions progress further.

Design I: integrated solutions for a crowded planet, 2003 to 2007

After the Chernobyl nuclear reactor accident in 1986, physicist Gerhard Knies made a back-of-the-envelope calculation that the world could be powered by renewable energy from deserts. Thanks to Knies's extensive social networking, the idea took off. According to an interviewee, Knies started the MENA renewable energy conferences that convened engineers and politicians in Ghana, Amman, Cairo, and Damascus to discuss renewable energy development in deserts.

In 2003, the Club of Rome and the Hamburg Climate Protection Fund founded the Trans-Mediterranean Renewable Energy Cooperation (TREC) to promote what I refer to as Design I of the Desertec vision. The Club of Rome's *Limits to Growth* report from 1972 shaped Design I. It defined the "world problematique" facing all nations as "poverty in the midst of plenty; degradation of the environment; loss of faith in institutions; uncontrolled urban spread; insecurity of employment; alienation of youth; rejection of traditional values; and inflation and other monetary and economic disruptions" (Meadows, Randers, & Meadows, 2004, p. 10). It argued that the world's population would overshoot the "carrying capacity" of the globe. *Limits to Growth*, and other works that followed it, such as Paul Ehrlich's *The Population Bomb*, have been critiqued for neo-Malthusian assumptions that treat the world's low-income populations as "surplus population" (Hughes, 2000). *Limits to Growth* was also criticized for failing to consider the role of technology in increasing the efficiency of resource use and in mitigating environmental harm (ibid.). Design I reflects an unusual commingling of what Mitchell (2011) dichotomizes as the Malthusian worldview that emphasizes scarce resources and the technologist worldview that presumes technological progress will yield limitless societal and environmental benefit. Thus, Design I could be seen as a version of "weak sustainability," in which technological development allows welfare to be maintained across generations. In Design I, economic growth is tacitly treated as allowable for the South provided that the South hosts technology that facilitates the sustainable consumption of the North.

Before an international renewable energy development conference in Germany in 2006, the German Federal Ministry for the Environment, Nature Conservation, and Nuclear Safety was searching for an ambitious project to showcase Germany's commitment to renewable energy. The ministry funded the German Aerospace Center (DLR) to conduct three feasibility studies on TREC's vision. In the MED-CSP, TRANS-CSP, and AQUA-CSP reports, DLR assessed the feasibility of an 80% renewable energy future for the EU-MENA region, with 15% of the EU's electricity sourced from MENA by 2050. In an interview with me, the scientist who headed the project stated,

> We got an order from the German Ministry of the Environment to check this idea for plausibility and feasibility. . . . Then we sat down with our partners, and we tried to kill this concept, but we did not succeed in killing it.

In 2007, TREC released the first Desertec *White Book*, which promoted Design I, with the support of El Hassan bin Talal, prince of Jordan and the former president of the Club of Rome, who in the introduction presented Desertec as a solution to power and water provision for a growing population. The *White Book* framed global climate change and energy poverty as key challenges that could be solved by North-South cooperation between the "sunbelt and the technology belt" (TREC, 2007, p. 65). The book was presented in Berlin to an over-capacity audience of 400 people in an evening presentation that ran four hours longer than expected, illustrating the increasing interest in the vision.

Design I of the Desertec vision maintained individual-country-level electricity markets, while CSP provided the essential service of firm power on demand (i.e., dispatchable power) for Europe. Overall, the actors constructed Design I as feasible by framing the involved technologies as mature, presenting particular definitions of sustainability and security, and presenting win-win societal benefits.

DLR envisioned a system of 20 HVDC transmission corridors, each of 5 GW capacity, to provide a "backbone" or "highway" for renewable energy export in the EU-MENA region. These corridors are somewhat analogous to natural gas pipelines, in which fuel is funneled from one point to another (Trieb et al., 2016). These lines would provide *connectivity* by linking the best sites for CSP in North Africa to European load centers. The *Trans-CSP* report framed HVDC as a *"mature"* technology, as opposed to an innovative technology, emphasizing previous experience with HVDC (e.g., 90 projects collectively with 75 GW of capacity worldwide at the time). Advocacy for HVDC transmission through Desertec and other transnational grid projects may indicate a reopening of the

1887—1892 "battle of currents" between Edison and Tesla that Hughes (1983) discussed (p. 106). The Design I system imaginers argue that HVDC is more *efficient* than HVAC for long distances, with half the transmission losses and a lower cost for lines exceeding 500 km. They argue HVDC can improve *stability* because it can help to prevent cascading AC failures, and if part of a DC circuit is disrupted the other part continues carrying half of the current. Unlike HVAC, HVDC can be buried, maintaining the relative *invisibility* of power systems in "premium networked spaces" in Europe. DLR suggested that because of these benefits an HVDC backbone grid "will probably be implemented in Europe within this century to enhance security and stability" (DLR 2006, p. 112). By discussing HVDC as a mature, efficient, stable, and invisible technology, the system imaginers framed it as feasible and ready for deployment.

The system imaginers emphasized the importance of *dispatchable* CSP generation, with electricity exported through these HVDC transmission corridors. A dispatchable power source can be turned up or down on demand (Greacen, Engel, & Quetchenbach, 2013). They also argued that CSP was a mature technology on the basis of previous experience with the technology (this experience was described in Chapter 2). DLR analyzed 50 countries in the Mediterranean region and found that each country could reach an 80% renewable electricity future in terms of renewable resources, but not in terms of *reliable* system functioning. While the countries could theoretically produce their own fluctuating (intermittent) renewable energy from wind and solar PV, they would require a replacement for the baseload power provided by coal and nuclear power plants that generate a constant flow of electricity close to 24-7, 365 days a year. This became more important after the Fukushima nuclear accident in 2011, after which Germany opted to retire its nuclear capacity. However, baseload power is inflexible, whereas CSP plants, with cheap molten salt storage, can generate electricity 20 or more hours per day and can be easily ramped up or down. A DLR scientist explained in an interview with me, "Some countries needed this import of solar electricity from the South to the North because they had a lack of flexible renewable energy . . . and this high quality of energy is only available in North Africa." The value of dispatchability is not priced in the power sector, as demand has long been deliberately increased to match supply (see Nye, 1990) and power plants have been added to cover peaks in demand, but DLR argued that the service of dispatchability would have intrinsic value in a future with high dependence on fluctuating renewable energy sources.

Additionally, fossil fuels played some role in Design I as mature technologies that improved system *reliability* by providing electricity during peak hours of consumption and ensuring 25% spinning reserve capacity (i.e., turbines that can be brought online quickly to maintain the grid's frequency without generating electricity). System imaginers framed energy farming from CSP as modern, whereas they compared fossil fuel consumption to a hunter-gather mentality in which humans are "like squirrel[s] eating in the summer the nuts gathered for wintertime" (DLR, 2006, p. 32).

Design I reflected the *Limits to Growth* report's focus on scarcity. For example, the *Med-CSP* report included a life cycle assessment to explore the supply sufficiency of the necessary materials (DLR, 2005). The *Trans-CSP* report defined sustainability on the basis of the triple bottom line of economy (i.e., internalizing the externalities of fossil fuels and keeping electricity affordable), socioenvironmentalism (i.e., "compatibility with society and the environment"), and security (i.e., security of supply through international cooperation and interdependence rather than autarky) (DLR, 2006, p. 29). The security pillar prioritized network stability from HVDC and diversification of Europe's energy portfolio by increasing the number of energy sources in Europe from five in 2003 to ten in 2060. The report viewed security as *inter*dependence, stating,

> The autonomy goal [and its] proposed solutions, is rather questionable in terms of ethics and sustainability: it propagates a Europe unaffected by the eventual future misery of its neighbours, and fosters the illusion of independency on a rather small and crowded planet.
>
> (DLR, 2006, p. 33)

In addition to environmental benefits, the system imaginers promised a variety of societal benefits for the EU and for North Africa. In contrast to the *Limits to Growth* report, which acknowledges the trade-offs inherent in technological development, DLR and TREC framed the vision as a win-win solution. System imaginers reasoned that since there were surpluses of oranges and tomatoes in Europe, European farmers would lose from imports of fruit from North Africa while North Africans won (DLR, 2006). In contrast, there was not a surplus of renewable generation capacity in Europe, so Europeans and North Africans would both win from electricity exports to the EU. They also claimed that another benefit of the envisioned system would essentially be to link people living in peripheral spaces to "premium networked spaces." An interviewee argued that although the vision had been criticized for its centralized nature, the existence of infrastructural peripheries leads to centralization through urbanization trends. He pointed out that 25% of Mexico's population lives in Mexico City because of its infrastructure, whereas people remain in the rural but fully networked Black Forest of Germany instead of leaving for Stuttgart.

Design I also framed CSP as a technology that could legislate societal benefits. Multipurpose power plants could become "the nucleus of a totally new social paradigm" to prevent desertification, desalinate ocean water, grow food by sheltering crops in the shadow of the mirrors, and facilitate foreign cooperation (DLR, 2006, p. 83). In the loftiest of the *White Book's* promises, a solar power and water project for Gaza sited in Egypt was envisioned as bringing peace to the Middle East. Also in the *White Book*, Professor Bennouna from Morocco argued that Europeans could make up for the wrongs committed during the colonial era by helping Morocco with its industrial development. He argued that Morocco and the EU should utilize renewable energy to fight common challenges of poverty, climate change, and North-South discrimination. He predicted that Morocco was "on its way to [becoming] an energy exporting country" (p. 46). Thus, the proffered win-win societal benefits of Design I can be summarized as Europe wins from clean power while North Africa wins from water and development. The vision gained traction in the public discourse, but the discourse did not focus on Bennouna's perspective that it could help Morocco but rather criticized the vision for being neocolonial, subject to sabotage, and infeasible.

Forces that shaped the Desertec vision and its design, 2007 to 2011

In this section, I explore the forces and processes that shaped the vision until Design II was published in 2011, in order to describe how visions are developed and negotiated over time. These forces and processes included the creation of Dii; the neocolonial critique; differing energy security framings; long timescales; and circumstances that scenario planners call wildcards, including the unexpected reduction in the price of solar PV panels and the Arab Spring. These represent the "shapers" in the model of a vision in Table 4.1. I will later explain how Design II differed from Design I as a result of these processes and how the designs came into conflict.

Mediterranean system imaginers: In 2009, the constellation of actors advocating for the Desertec vision changed, which shaped the vision by incorporating new stakeholder agendas. Euro-Medness infected the energy sector as complementary organizations sprang up, including the Union for the Mediterranean's Mediterranean Solar Plan;[1] Medgrid, the French-run equivalent to Dii; RES4MED, Dii's Italian equivalent; and Friends of the Supergrid, a coalition of companies advocating for an EU supergrid. TREC was disbanded, and a nonprofit organization, the Desertec Foundation, was founded and promoted the Desertec brand to spread it to other parts of the globe. In October 2009, Dii was founded in the largest press conference in Munich Re's history. Twelve shareholder companies convened to sign a memorandum of understanding "to establish a planning entity in the form of an incorporated company," or limited-liability company. No North African companies were involved. Over three years, Dii was to create a legislative framework to develop projects faster, to facilitate the construction of a reference project for proof-of-concept, and to conduct a study with a proposed design for 2050 (which became Design II). These steps sought to prove the vision was feasible and to carve out a space in the future for it.

Dii's complexity rivaled the complexity of the Desertec vision itself and represented an unprecedented effort to convene companies from the core value and associated value chains that would be needed to develop the Desertec vision—banks, utility companies, solar PV manufacturers, CSP developers, TSOs, insurance companies, energy companies, and project developers. It included one of the world's largest reinsurance companies (Munich Re), major banks such as Deutsche Bank, and large energy and utility companies such as Siemens and RWE. Partner companies included those that wanted to find a niche in renewable energy, such as IBM, which is working on smart grid software and software to manage wind parks.

I call Dii and related organizations "system imaginers" embarking on a twenty-first century effort to form a network of power. Their role differs greatly from that of the inventor-entrepreneurs that Hughes (1983) described as the system builders of the original electrical power systems in the late 1800s. System imaginers describe themselves as accelerators, enablers, catalyzers, facilitators, coordinators, and conveners rather than inventors or innovators. Dii attempted to frame, colonize, and control the future, not by physically building it or by inventing new technologies but by convening powerful groups, constructing knowledge and markets, and advocating for particular system configurations. The term "innovation" was largely absent from Dii's discourse.

These large organizations and companies are vested in a particular system configuration, whereas in the early 1900s system builders worked to scale up a system around the technology they invented (Hughes, 1983). While the involved solar companies sought to gain power and footing, the powerful incumbents, such as utility companies, realized that, while electricity systems have gained significant momentum, momentum is not maintained without effort—and if disruptive innovation were to occur that completely changes the system configuration, their business model would become obsolete. An interviewee from a long-established energy company stated,

> It's a problem of knowledge, achieved knowledge, which unfortunately doesn't cover all questions, and it's much more difficult to forget what you know than to learn new things. If we believe that the world of electricity is developing [toward systems integration], then we want to be among those that shape this again, also to be a dominant player in that again.

Power systems momentum was being disrupted in multiple ways, and the system imaginers sought at least to guide its direction, if not to affect its mass and velocity.

Dii was headquartered in Munich, Germany and employed about 30 staff members for four to five years. It also had a shareholder committee with five or six company representatives. Each shareholder signed a three-year contract and paid one million euros to Dii per year; associated partners paid 75,000 euros per year. Dii split into technical working groups focused on solar PV, CSP, networks, power transfer, storage, local context, and political frameworks. An interviewee described how as Dii grew, it became an extremely complex, intercontinental public-private partnership with 50 shareholder and partner companies, each with a different agenda and company culture. "It was a great learning experience for me because I never had seen such complexity." Some employees were selected as cosmopolitan agents of cooperation because of their family connections or previous work experience in the MENA region. They described themselves as balancing complex and sometimes conflicting goals to improve the standard of living in North Africa and to make money.

Dii saw its appeal as representing a wide variety of corporate interests, one step removed from a profit motive. An interviewee from Dii stated in 2013,

"[Policymakers] care about a broad spectrum of industry far more than they would a narrower sector, so Dii is a more attractive forum than, for example, the European Photovoltaic Industry Association" because Dii represents a broader group of companies. Yet in the first year, Dii struggled to negotiate the needs of its single-technology focused shareholders, such as solar companies Abengoa Solar (CSP) and First Solar (solar PV), versus those of its shareholders without a stake in a particular technology, such as reinsurance company Munich Re, which focuses on clean energy for climate change mitigation to reduce insurance risk from natural disasters, and utility companies that wanted the cheapest clean technology. A representative from a shareholder company stated, "We tried not to [focus on CSP] because we wouldn't want to be stamped as a German CSP Club." Instead, Dii came to focus on low-cost technologies, rather than advocating for CSP for dispatchability.

In spite of its ambition, Dii's complexity and conflicting interests meant that it struggled to project a single coherent identity. It presented itself as a project accelerator, a creator of markets, a think tank and lobbying organization, an industrial platform and convener, and a technology-neutral advisor. These multiple personalities affected public and stakeholder understandings of and expectations for the system. Expectations play an important role in shaping new technologies, and lofty expectations pose risks if they are later disappointed (Borup et al., 2006; Van Lente, 2010). The shareholder companies decided in 2010 that Dii would not build projects itself because the shareholders did not want to create a competitor for themselves. Dii then became, according to an interviewee from Munich Re, "a know-how company, where the knowledge is bundled, not a company realizing new power plants directly." While some derided this as conducting studies rather than taking action, it represents an effort to shape what Miller, Iles, and Jones (2013) called energy epistemics, or "whose knowledge counts in governing and reshaping energy futures" (p. 137). It also reflects Haas's idea of epistemic communities composed of "knowledge entrepreneurs" that have agency in persuading elites to pursue regional environmental governance (Haas, 2015).

As a project accelerator, Dii framed the end goal of market integration as inevitable in the future on the basis of market logic. An interviewee from a shareholder company illustrated this perspective, stating,

> It's like water pouring down a river—you can be sure it will go down—but sometimes you can accelerate the development. It's not a question of whether Desertec comes at the end of the day, but it's a question of time and whether you can speed up the process.

Framing the future as proximal raises expectations that progress will be made soon, but it also risks raising expectations that will be disappointed. Dii was not alone in its belief in the inevitability of renewable electricity exports. For example, an interviewee from the World Bank stated,

> I'm pretty sure it will happen at some point; the question is when. . . . Because Europe has made a commitment to 85% to 90% reduction in CO_2 emissions

> by 2050. . . . At some point, it would be cheaper to tap into the solar resources of MENA . . . if you look at the supply-demand curve there is a point at which [it] basically points to imports anyway.

In spite of an epistemic adherence to market determinism, Dii existed because the vision was not, in fact, inevitable: Rather, a powerful actor was needed to construct renewable electricity markets between the EU and North Africa. Failure to achieve short-term results in this area ultimately conflicted with Dii's identity as a project accelerator, as some shareholders wanted a much faster transition from PowerPoints to power plants.

Finally, Dii also projected itself as a technology-neutral advisor, meaning an unbiased and objective organization without a stake in a particular technology. According to an interviewee, this technology neutrality was limited to "feasible, industrially stable" technologies. Thus, Dii's vision was pioneering but not technologically innovative, staking out new territory in terms of systems integration and renewable energy scale-up, rather than developing radically new energy technologies. Eventually, Dii's technological neutrality conflicted with Design I's promotion of CSP for its advantageous system properties, such as dispatchability.

Public perceptions of Dii: The media, as well as other stakeholders, struggled to interpret Dii's multiple personalities and its complicated vision. Early on, the media was enthusiastic about Dii. In 2010, *Der Spiegel* referred to Dii as "the disciples of light" and quoted Dii's chief operating officer who said that solar power should be constructed in high elevations in the desert because "the closer to God, the better the rays" (Meyer, 2010). Some news articles framed Dii as seeking to supply 15% of the EU's electricity needs and all of North Africa's electricity needs. More often, the media framed Dii as a powerful German consortium seeking only to supply 15% of the EU's energy needs by 2050. Design I had framed this as the percentage of electricity imports that would be feasible, which the media and Dii later reified as a goal. This framing fostered a perception that Dii was uninterested in providing electricity to MENA, where demand is skyrocketing. The media also disrupted the perception of Desertec's feasibility by framing it as a single megaproject with significant cost, scale, and ambition, as well as a fantasy, dream, or mirage, or even a "white elephant in the desert." Dii reiterated that its vision was for a set of distributed projects in the region, rather than a single megaproject, but this gained little media attention. The media often portrayed Dii shareholders as poised to invest $492 billion to build the Desertec system, when in reality Dii had limited funding. An official from the Union for the Mediterranean stated in an interview that Dii was "about making money. At the end of the day how will they make money?" Moreover, many regional stakeholders saw Dii as German, rather than multinational. An interviewee from the World Bank stated that Dii was about "let's play German," meaning the project reflected Germany's promotion of renewable energy technologies in other countries to achieve German industrial policy goals.

System imaginers framed 2050 as proximal in order to construct the system's feasibility, but if they framed 2050 as too close then expectations could be disappointed if little were achieved in the short term. An interviewee from a Dii shareholder utility company emphasized that 50 years is one, or at most two, investment cycles in the power sector, considering planning, permitting, construction, operation, and decommissioning. A public relations director for a Dii shareholder company explained that at a 2009 press conference his company projected pictures of CSP plants in Spain and reporters misunderstood, thinking that the first power plants had been built in North Africa for electricity export to Europe. He reflected on this stating, "We're talking about realization in 2050. Not yet, not 2040, not 2015, two thousand *fifty*. Five! Zero! Now, four years later . . . they said 'hey, Desertec is dead. Nothing happened.'" He said the media assumed Dii played a role analogous to that of General Motors, portrayed as intentionally sabotaging electric vehicles in the documentary *Who Killed the Electric Car?*

Tainted by neocolonialism: A number of media outlets framed the Desertec vision as a European scheme to exploit North Africa's natural resources. For example, the *Guardian* quoted Daniel Ayuk Mbi Egbe, international coordinator of the African Network for Solar Energy (which focuses mostly on distributed generation),

> Many Africans are sceptical [about Desertec]. [Europeans] make promises, but at the end of the day, they bring their engineers, they bring their equipment, and they go. It's a new form of resource exploitation, just like in the past.
>
> (Hickman, 2011, para. 10)

The media also circulated a comment made by a US graduate student in Arabic and Islamic Studies living in Morocco. She stated, "Haven't we already been here before? Europeans covet Africa's wealth of natural resources . . . so they make economic investments for the benefit of Europeans and the detriment of Africans" (Zeller, 2009, para. 20). She later explained to me that she was unfamiliar with the project's details but wanted to provide relevant historical context.

Many of the people I interviewed who were associated with Desertec viewed the concepts of sovereignty and the nation-state as outdated and viewed sovereignty as akin to isolationism. Moreover, some interviewees saw the history of colonialism as part of the distant past. At a Dii conference, the CEO of Munich Re asserted that the notion that each region should have its own energy production "is the old dream of complete autonomy and independence." Desertec could either bring together a more unified Europe on the basis of common values or "risk" falling back on national solutions. An interviewee from the Desertec Foundation pointed out that climate change is ambivalent to nation-state borders.

The historical context provided in Chapter 2 illustrates why this vision is viewed as neocolonial, and in Chapter 6 I will further discuss how officials in Morocco understand Desertec's relationship to neocolonialism. Here, I simply point out that the neocolonial critique indelibly shaped Design II, which was intended to change the perception that the Desertec system would drain energy out of North Africa solely to benefit Europe.

Geopolitics of renewable energy: Some news coverage questioned whether Europe would be beholden to the "sun kings" of MENA, who could form a solar cartel and use electricity as a weapon. The Desertec Foundation countered that economic incentives would prevent this because electricity storage technologies that would enable stockpiling electricity for sale later on have not been built. The foundation also focused on a framing of "global energy security" in which independence is not viewed as the most desirable end state and energy governance is not seen as zero-sum, but rather opportunities exist for win-win solutions through integration and cooperation to improve energy governance. "Security of energy supply shall no longer be regarded as a purely national matter, but rather be seen in an interregional, intercontinental and global context" (van Son & Ruderer, 2015, p. 1). In an FAQ list, the Desertec Foundation website stated,

> The big difference with regard to the dependency on fossil fuels lies here: Renewable power can be produced just as well in Europe, albeit at higher costs and requiring more input. It is hence in the interest of the power-exporting countries to offer an economical and reliable product; otherwise the demand would decrease and a drop in investments, export revenues, and employment could be expected. In this sense, an energy cartel much like the model of OPEC tends to be self-destructive and makes little sense. Solar power that is not delivered is thus lost and may not be sold later at a higher price as with oil or gas.

Several academic studies also argued that a Desertec system would pose minimal risk of terrorism or political withholding of electricity supply (Lacher & Kumetat, 2011; Lilliestam, 2014; Lilliestam & Ellenbeck, 2011, 2012; Patt et al., 2011). These studies evaluate the Design I scenario, using the DLR scenarios for HVDC corridors rather than the supergrid scenario that Dii proposes in Design II. Lilliestam (2014) argues that the risk of terrorist attacks on electricity infrastructure is relatively low, although higher than it is for natural gas pipelines. Lilliestam and Ellenbeck (2011) find that the risk that the countries would cut off exports for political reasons is low because five countries would have to work in tandem to disrupt supply. I argue elsewhere that there are potential benefits to energy security from renewable electricity interdependence, especially from a "deep energy security" perspective that incorporates sustainable development, but there are also risks stemming from lopsided dependence (Moore, 2017).

There are few empirical examples of electricity withholding worldwide. As discussed in Chapter 3, in 2011, Turkey threatened to withhold electricity and water exports from Syria. It is possible that Mozambique cut off Zimbabwe for nonpayment of $75 million in 2012 (SAPA-AFP, 2012). Other cases reflect a surprising lack of electricity shutoff even with significant nonpayment. For instance, Zimbabwe owes South Africa $44.5 million for electricity imports (Dziurtwe, 2017). Iraq has owed Iran as much as $1 billion in back electricity payments, although Iran temporarily cut off Iraq (Slav, 2017). Nevertheless, countries are not always economic rational actors; they could withhold electricity for political reasons despite the economic disadvantages. In terms of the political feasibility of cross-border electricity trade, policymakers' subjective risk perceptions of electricity withholding are unfortunately more influential than the calculated probability and significance of the threat.

Wildcards: Scenario planners in the energy industry use the term "wildcard" to refer to a low-probability, high-impact, trend-breaking event that is external to a prospective scenario but disrupts an anticipated future (Roubelat, 2006). In general, system imaginers must design their vision within a concrete context in order to construct feasibility, but in doing so they expose potential problems, especially as wildcards occur. Actors used these wildcards as they occurred in real time as empirical evidence of the Desertec vision's infeasibility.

Beginning in Tunisia in 2010, a series of unexpected political uprisings occurred across the MENA region, dubbed the Arab Spring. (The term Arab Spring is problematic, but I use it here because it is widely recognized.) It is not my intent to reduce the Arab Spring to a mere wildcard—particularly since events in each country differed and had a specific context, historicity, and set of politics—but rather to illustrate the role of unexpected events in shaping visions for sociotechnical systems. The Arab Spring arguably shifted Dii's interest in a pilot project from Egypt to Morocco. Moreover, maps that depicted Syria as part of the vision while the country devolved into civil war could be viewed as geopolitically naïve.

Dii reframed the Arab Spring as a social justification for the vision rather than as a disruptive event. As Rothe (2016b) stated, Dii tried to "redefine the current crisis as an opportunity or historic chance" (p. 200). On March 4 and 23, 2011, Dii released statements emphasizing the importance of continuing its work. Dii stated,

> Political transitions in North Africa and Middle East plus a worldwide reevaluation of our energy supply give greater urgency to the questions of where are the large-scale sustainable energy resources of the future.

The company suggested that the Desertec vision would have positive socioeconomic effects and therefore could improve political stability. Michael Köhler, head of the cabinet for EU Energy Commissioner Oettinger, also articulated this view at Dii's 2012 conference stating,

> We will pursue this because our objective—even more so after the Arab Spring—is to create a shared zone, not only of democracy, but of shared prosperity in the Mediterranean between Europe and North Africa, and we see energy cooperation as one of the key elements in this regard.

Dii demonstrated its commitment to the region by hosting its conference in Cairo shortly after the January 2011 revolution. The conference was held just a week before another wave of massive street protests in November 2011 in Tahrir Square, in which protestors called for free elections and an end to military rule. During a 2013 protest in Tahrir Square, mobs (of opportunists, not activists) stormed the Semiramis InterContinental Hotel, where the conference had been held, as the staff attempted to fend them off (Lynch, 2013). In a different example, in January 2013, a bloody hostage crisis at a BP natural gas facility in the Algerian Sahara made headlines throughout Europe, and Dii again released a statement emphasizing the importance of continuing its work.

European investors viewed "government instability" as a risk factor, and integrating the European grid with grids in unstable countries to increase grid stability seemed a shaky proposition at best. Researchers interviewed Desertec-related investors in 2012 and found that they viewed the risks of Desertec as regulatory, political, and force majeure (including terrorism) (Komendantova, Patt, Barras, & Battaglini, 2012). In contrast, the CEO of the Saudi utility company ACWA Power said at the 2013 Dii conference that "the proof is in the pudding. Our independent power producers in Egypt are still being paid regularly for supplying electricity in Egypt." Shortly thereafter, the CEO of the German energy company E.ON Climate & Renewables exclaimed, to a round of applause, that he was not worried about the political situation in North Africa—he was worried about the political situation in Europe.

These European political challenges related partly to the Eurozone crisis, which occurred after Dii was founded. It disrupted the vision, perhaps to a greater extent than the Arab Spring. When Dii formed, European companies observed that MENA countries were experiencing 7% to 9% GDP growth per year, plus skyrocketing energy demand, while economic growth and energy demand in the EU was stagnating. They had surplus capital to spend in North Africa. During the economic crisis, Dii had a difficult time recruiting new partners and lost prominent shareholders. Siemens withdrew from the CSP business altogether in 2012, closing its CSP office in Germany and withdrawing from Dii. Additionally, electricity demand grew rapidly in Morocco, while falling far short of projections made in Spain before the Eurozone crisis. The Desertec vision's original framing of the "load centers," or centers of demand, being in Europe rather than in North Africa was called into question.

In addition to political and economic turmoil in EU and MENA, the falling price of solar PV systems represented another wildcard that shaped Design II. Conforming to its technology-neutral framing, Dii focused on scaling up the cheapest renewable energy technologies rather than advocating for CSP for its system benefits. Utility-scale solar module prices fell dramatically, from $3.50 to $4.00 per watt in 2008, to below $1 per watt in 2013, to 35 cents per watt in 2017 (Bazilian et al., 2013; Fu, Feldman, Margolis, Woodhouse, & Ardani, 2017). The reasons included an increase in polysilicon production and availability, technological advances in manufacturing and cell efficiency, consolidation in the sector, and Chinese industrial policy that flooded the market with cheap panels (Bazilian et al., 2013). As one shareholder representative put it, "no one anticipated this." Another stated,

> Everybody said we will get there with CSP because we can generate electricity for 17–20 hours [per day]. . . and during that time the price for PV modules just fell down. Wow!! This was an effect! And this changed the discussion in the group.

The shift was amplified by solar PV companies joining Dii. A Dii interviewee explained,

> The PV companies who joined as associated partners started to lobby more for their interests. And First Solar managed to apply successfully for an upgrade to shareholder at the start of [2013]. So this has been also a milestone getting one of the leaders worldwide in photovoltaics on board as a shareholder.

The interviewee expressed that since Dii could not predict what the next technological breakthrough would be, it focused on the cheapest technology: solar PV and wind. This shaped Design II significantly.

Design II: a supergrid and the integration miracle, 2011 to 2014

Next, I trace out how the factors described earlier—the composition of Dii, neocolonialism, energy security, and wildcards—shaped Design II. Dii's first report, *Desert Power 2050*, was published in 2011 in conjunction with the Fraunhofer Institute for Systems and Innovation Research. The report did not envision a set of solar electricity pipelines to provide dispatchable generation from CSP plants to Europe, but rather depicted a deeply integrated supergrid and single regional electricity market using a more complex modeling system.

As a group of private companies and a technology-neutral advisor, Dii designed the system to prioritize the *lowest-cost electricity* and to create a pipeline of renewable energy projects on which the shareholder companies could bid. An interviewee stated that Dii's study was "driven by the question [of] whether the economy under the right boundary conditions may find the best solution by itself," rather than Design I's "heuristic" approach of integrating fluctuating capacity with dispatchable generation. Dii argued that an integrated EU-MENA power system would be worth more than the sum of the two parts because the interconnected system would supply *cheaper* renewable electricity than two separate EU and MENA renewable systems. A Dii shareholder interviewee explained that although dispatchability is a system benefit, there is not a sufficient market for it in many cases. Dii also concluded that it would be cheaper to deploy wind and solar PV in a regionally integrated system than to pay for storage. Dii framed CSP as expensive and immature, whereas it framed wind, the winning technology of the report, as mature, cheap, and less dependent on geographical conditions than CSP. Because of the reduction in the price of solar PV systems since Design I, Dii framed solar PV as cheaper and more easily scalable than CSP. Yet the upfront costs were still massive. The follow-on report in 2013, *Desert Power: Getting Started*, estimated the total system capital costs to be 4.5 trillion euros over 40 years.

In Design I, CSP in MENA would meet daytime peak demand in Europe. Since the development of Design I, Germany had begun meeting part of its daytime peak demand with rooftop solar PV systems. At the 2013 Dii conference, a group of panelists argued that both baseload power and dispatchable generation are antiquated constructs. An official from the Moroccan Agency for Solar Energy wondered who would let the utility companies know that the era of dispatchability is over. He bemoaned that the focus on backup, conventional power is "bringing us to the wall." An interviewee stated, "We have to change the idea of baseload power . . . [it does not matter if it is] night [and there is] no sun, or no wind . . . we just have to equalize those differences."

Design II addressed the intermittency of renewable energy through what the report termed "the integration miracle," in which massive systems integration would allow electrons to flow where they are needed, when they are needed, across 42 countries. East-West interconnection would balance out solar power between day and night, and wind would be complementary through North-South integration between North Africa and Europe. In this integrated power market, Dii envisioned three categories of countries: super producers (e.g., Morocco and Norway), which have renewable energy resources that far exceed demand, leaving capacity available for export; importers (e.g., Germany and Turkey), which have renewable energy resources smaller than demand, requiring imports; and balancers, which have resources and demand that are well matched (e.g., Saudi Arabia and Egypt) (Zickfeld, Wieland, Blohmke, Sohm, & Yousef, 2012, p. 81). Design II was sometimes called the "copper plate" because copper is an excellent conductor, and electrical charge is instantly available on any part of a copper plate. This design tacitly addressed the neocolonial critique because it would avoid a situation in which a North African country would export electricity to Europe through a solar-energy pipeline when it could not meet its own increasing domestic demand.

Design II defined sustainability in terms of climate change mitigation and a higher percentage of electricity from renewables (90% versus 80% in Design I). There was no longer a focus on integrated solutions, and neither water desalination nor life cycle assessment were addressed. The vision's paradigm shifted from limits to growth to no limits to what systems integration could provide. Integrated is beautiful, rather than small is beautiful. The report framed deep systems integration as more flexible and efficient than distributed energy systems, which Dii deemed to be outside the study's scope.

Like Design I, Design II framed the societal benefit as win-win, including access to affordable renewable energy, reduced cost of decarbonization, mutual reliance, and "sustainable and affordable power for future generations" (ibid., p. 133). Wins in Design II relate to broad, abstract concepts—national energy policy, security, regional stability, peace, and development—rather than specifying which social stratum or specific stakeholders would benefit. In interviews, many stakeholders saw the vision as a win-win but the social impacts as unknowable because of the uncertainty stemming from lengthy timescales. The only interviewee who mentioned a specific social group had received a phone call from Tuareg people in the Sahara Desert, who were concerned about losing their land to solar power plants. Many interviewees discussed the "old incumbents" as losers, including oil companies and state-run utility companies, but they emphasized that these companies have an opportunity to adapt. An executive I interviewed from BP was unconcerned about the security of demand for oil since oil is mostly used for transportation and industry rather than for electricity.

Dii's framing of energy security in Design II responded indirectly to the solar cartel scenario. In Design II, security of supply was not defined as diversifying fuel type (as it was in Design I) but as diversifying the supplier countries. No single exporter would provide more than 10% of Europe's electricity supply, and the new role of fossil fuels in the system would be to provide 10% backup power for Europe as its "insurance policy" against geopolitical withholding. Security would be achieved not through Mutually Assured Destruction, but through the "mutual reliance" provided by an interdependent system. *Desert Power 2050* stated, "[The countries'] complementary roles lead to a situation of interdependence across the system, in which no single country is dependent on another but instead each country is reliant on the system as a whole" (p. 83). So, in fact, the countries would not be reliant on one another, but rather the technological system itself would stand in as a proxy for collaboration, substituting technological integration for political integration.

Finally, in Design II, a modern energy system is one that implicates the control of time by extending centralized electricity systems across a region to compensate for renewable electricity intermittency. Daily and seasonal variations that affect both electricity generation and demand must be predicted to the minute with "perfect foresight" to maintain reliability (ibid., p. 27). The copper plate design would attempt to flatten out the rhythmicity of nature and the vulnerability posed by the seasons through massive systems integration. Adam (2003) argues that through the project of the Enlightenment, the natural rhythm of time, or *natura naturata*, was controlled by developers of technology who "re-creat[ed] the variable rhythmic times of life in uniform, invariable, decontextualized, quantifiable form . . . the control of time is extended to the seasonal rhythmicity of nature . . ." (pp. 68–69). Socially, electricity systems allowed people to stay up later reading, to travel to amusement parks, and to commute via streetcar, changing their daily rhythms of life (Nye, 1990). In the twenty-first century, system imaginers are seeking to integrate *natura naturata* into the centralized system configuration to allow for high penetration of intermittent renewable energy.

Design III: virtual exports, 2012

The German utility company and Dii shareholder company RWE developed Design III in 2012. RWE proposed building a virtual exports system since the physical transmission infrastructure was limited and additional lines would be difficult to site. Through this program—RWE Desert Power—German consumers could have opted into a surcharge on their electricity bill used to finance clean energy projects in North Africa. In 2012, RWE announced that it was pursuing a large-scale wind and solar PV project in Morocco as part of Design III. (It was not built.) Accusations were then made in the media that RWE was trying to "hijack the vision," which it vehemently denied, saying the project would provide a "blueprint" for Moroccan project development for all shareholders.

Design III reflected a broader trend in liberalized electricity markets that Graham (2000) called "infrastructural consumerism," providing consumers with choices for "tailored infrastructure services within internationalizing niche markets" (p. 192).

In this market-based design, the focus on integrated solutions and the benefits stemming from technological integration were largely lost because the HVDC grid that would provide connectivity, stability, and reliability would not be developed. Sustainability in this design related only to climate mitigation, not to life cycle assessment or the triple bottom line. RWE framed Design III as feasible because it avoided a complex, material reorganization of the Mediterranean electricity system. However, Design III could impede the system-level scale-up of renewable energy if it created a niche consumer market in place of an infrastructural shift. According to interviewees, Moroccan government officials believe virtual exports would limit the future possibility for real exports. Virtual exports would also be dependent on a high price for carbon dioxide. This approach is markedly different from the Article 9 approach that allows for exports from third countries to the EU because Article 9 requires that transmission capacity be allocated and electricity be physically transferred (Trieb et al., 2016). Academics have criticized statistical transfers because they will not significantly increase renewable energy development (Klessmann, Lamers, Ragwitz, & Resch, 2010; Klinge Jacobsen, Pade, Schröder, & Kitzing, 2014).

Design III also reflects the theme from Chapter 3 about technological integration and immigration politics in the Mediterranean region. The Desertec vision was initially founded on integration goals. Some stakeholders suggested that the vision would be more socially just if citizens, instead of corporations, funded the power plants as ratepayers. However, would there be sociocultural integration benefits without physical integration? Would consumers know where their virtual electricity comes from? While Design III is easier to achieve from an infrastructural standpoint, its social effects are as murky as those of Design II.

Table 4.2 Comparison of the three designs

Property or technology	*Design I: Integrated solutions for a crowded planet*	*Design II: the integration miracle*	*Design III: virtual exports*
Technology advocacy/ neutrality	Technology advocacy, based on system properties	Technology neutrality	Technology neutrality
CSP technology	Mature, dispatchable, multipurpose, desalination	Expensive, immature	N/A—focused on cheaper renewables
Solar PV	Expensive, fluctuating	Cheap and scalable	Cheap and scalable
Wind	Expensive, fluctuating	Cheapest technology	Cheapest technology
Fossil fuels	Reliable, mature. Needed for peak power	Volatile, limited natural gas needed for security and backup	Carbon emitters
Percentage renewables	80%	90%	N/A (niche market for renewables)

(*Continued*)

Table 4.2 (Continued)

Property or technology	*Design I: Integrated solutions for a crowded planet*	*Design II: the integration miracle*	*Design III: virtual exports*
Transmission system	Single HVDC corridors: efficient, safe, stable, connective, invisible for Europe	Deep transmission integration: mutual interdependence	None
Market	Separate electricity markets	Integrated regional electricity market	Market for carbon, rather than electricity
Overall system benefits	Clean power for Europe (and some for MENA), water for MENA, peace for Gaza	Cheap low-carbon electricity for all	Electricity for MENA, carbon credits for Europe
Export type	Firm power on demand	Fluctuating supply for export	Carbon credits
Sustainability	Triple bottom line (affordability, compatibility, security), using the least materials possible	Least cost possible, mitigating climate change	Mitigating climate change
Cost	EU will pay more for an important system service. The focus is on overall affordability.	The first priority is cost optimization.	Wealthy German consumers will pay for renewables.
Social benefit framing	Win-win. Premium networked spaces expansion, EU should help its neighbors develop.	Win-win. Mutual reliance, clean affordable power for current and future generations	Carbon credits for Europe, renewable energy for MENA
Modern energy system imaginary	A modern energy system prioritizes energy farming.	A modern energy system controls natural, temporal variation.	A modern energy system prioritizes market liberalization.
Energy security	Diversify the number of fuel sources in the portfolio, plus network stability	Limit dependence on a single exporter, plus lowest cost system	Dispense with physical integration and its risks
Paradigm	Limits to growth	Integration miracle	Neoliberalism

The politics of the copper plate

In order for the copper plate to function, the flow of electrons across borders must be unimpeded by politics. Diplomats from Morocco, Spain, France, Italy, Malta, and Luxembourg were scheduled to convene at Dii's third annual conference in

November 2012 to sign a memorandum of understanding with Germany for the first Desertec cooperation project between EU member states and Morocco. However, Spanish officials did not arrive and the negotiations crumbled, disrupting Design II's feasibility. Advocates for the vision still hoped to convince Spain the project would be a win-win. An interviewee from a development bank said, "So, you know, we're trying to seduce them—what can I say?" An interviewee from the German government stated, "We are trying to get Spain in the boat . . . so Spain sees itself as more of a winner in the whole project." But a Spanish energy analyst stated in a 2013 interview, "Exports are science fiction now." Representatives from the Spanish TSO Red Eléctrica de España (REE) dismissed the copper plate design in an interview in 2013 stating, "Desertec, the copper plate, is a dream." They explained that although REE was a shareholder of both Dii and Medgrid and had introduced the concept of electricity exports from North Africa via the existing transmission lines under the Strait of Gibraltar, they believed that Spain had sufficient domestic renewable resources.

Article 9, which Spain accepted before the Eurozone crisis, allows EU countries to meet their emissions reduction targets with renewable electricity imports from countries outside the EU. As Spain's GDP contracted and youth unemployment exceeded 50% in 2013, renewable energy subsidies became politically unpopular, with the feed-in tariff reframed in the public discourse as the feed-in debt (see Chapter 2). Tensions between Spain and Germany grew. Interviewees in Spain thought that Germany was "throwing around its money and power," and some suspected that the Germans would be willing to "break" the Spanish grid by allowing Moroccan green electrons to languish there, causing systems management problems, while Germany took credit for the carbon reductions. Spain has had surplus electricity since demand dwindled during the economic crisis, and Spain's utilization of its power plants was at only 50% in 2012. There was significant curtailment of wind power, as it is easier to shut off than base-load nuclear or coal plants. Therefore, it did not make sense for Spain to import renewable electricity at a higher price than domestic prices. Moreover, Spain is interested in selling its own surplus renewable electricity to other European countries.

An interviewee from REE expressed that if the Maghreb builds significant renewable power capacity for export, then REE will build a new transmission line; furthermore, he blamed Desertec's integration problems on a lack of North African transmission integration. A European grid expert stated, "You are in a real chicken-or-egg problem because . . . without transmission no one would invest [in generation] and without anybody intending to invest [in generation], nobody would start to build transmission." This chicken-or-egg discourse appears in other discussions of major systems-level changes in energy systems. For example, Cifor, Denholm, Ela, Hodge, and Reed (2015) have also noted a chicken-or-egg situation with developing regional electricity markets in the US West. Sovacool (2007) found that policymakers would not consider investing in hydrogen fuel cells without any infrastructure to distribute the hydrogen, but no incentives exist to build the infrastructure without operational fuel cells. The same can be observed with electric vehicle adoption and the development of charging stations; the charging stations are needed for potential users to see electric vehicles as feasible, but the investment in charging scale-up likely will not be made without demand. As these decisions are not only economic but also sociopolitical, this essentially results in a game of chicken in which neither actor agrees to move until the other one moves, and the end result is gridlock. As of 2017, there is no sign that this situation with Spain will change, and Morocco is now pursuing integration with Portugal.

The copper plate attempts to flatten the region, but Europe must still confront contentious transmission siting challenges, even if European countries shift their electricity generation siting challenges to Europe. The existing transmission across the Pyrenees Mountains (on the imagined future superhighway of electricity) is the most congested in Europe. Spain imports electricity from France when French electricity is cheaper, but in 2012, there was only a 1,200 MW capacity connection between Spain and France. This relatively limited interconnection is said to leave the Iberian Peninsula "islanded" (see Chapter 3). A new double-circuit 400 kV line was proposed to increase the transfer capacity to 2,700 MW as early as the 1980s, and again in 2003 and in 2008. Activists protested it because of concerns about construction noise, the environmental impacts on sensitive regions, and the possible disruption to a hot spring that attracts tourists (Ciupuliga & Cuppen, 2013). In 2015, after years of delay, REE and the French TSO RTE finished a new 320 kV DC line that more than doubled the exchange capacity to 2,800 MW. The companies addressed the siting conflict by running the transmission line through a tunnel under the Pyrenees, parallel to high-speed rail, at a cost of 700 million euros (funded partly by the EU and the European Investment Bank). The promotional video for the project uses a similar discourse to Desertec's of "electricity without borders" and "a more connected Europe" with "security of supply."

The actors' lack of imagination of EU-MENA as a single region also impedes the development of an EU-MENA copper plate grid. Spain likely does not want to be seen as African; as Slavin (2001) stated, "Cultural trends had extirpated Spain's Moorish legacy to assert [its] Europeanness and repudiate the adage that 'Africa begins at the Pyrenees'" (p. 90). Additionally, the new Spanish-French transmission line crosses the Catalan Pyrenees, and Catalonia attempted to secede from Spain in late 2017, as Europe itself is not evenly integrated. Spain and France are still short of the 10% EU electricity integration standard, even after a 35-year siting debate resulted in one new line. The copper plate is envisioned to even out both fluctuations of time and space through a massive, centralized grid, but it would have to overcome transmission bottlenecks caused by sociopolitical conflict. The Spanish example illustrates the complex and changing sociopolitical and sociotechnical context in which an electricity system undergoing a disruption in momentum is reshaped.

Visions in conflict

According to an interviewee, "firm power on demand" in Design I shifted to "fluctuating supply for export" in Design II, but trading surpluses of renewable energy among countries put Dii "up against a wall." "We opened the door with our studies. And since 2006 we are standing here and keeping the door open.

And everybody's busy running against the wall beside [our door]." Trieb et al., advocates of Design I, published a paper in 2016 that suggested that there were no Article 9 projects or progress on building Desertec because of the "distracted concept," or Design II. They argued instead for "solar electricity pipelines" analogous to natural gas imports from the Nord Stream offshore gas pipeline, to prioritize CSP for its dispatchability and contribution to diversification of supply. A mostly university-led study and stakeholder consultation process called the BETTER project then rearticulated this concept and called for (yet) another prefeasibility study to be conducted (CIEMAT, 2015). Backhaus et al. (2015) recommend scaling back the vision to omit countries that investors perceive as risky, focusing on several smaller projects instead, with CSP plants in North Africa and several HVDC cables to Europe. Trieb et al. (2016) refer to Backhaus et al. (2015) and other advocates of Design II as having "finally given up" (p. 456).

The debate between Designs I and II also exposed differences of opinion about what type of system configuration would be most just. Design II proposed a fully integrated electricity market based on mutual reliance to avoid perceptions of neocolonialism. In contrast, in 2013, an interviewee associated with Design I argued that it is more just,

> But [Design II] ended in colonialism. . . . With [Design I], you build a solar power plant in Morocco and connect it to an HVDC line, and you send this electricity to Germany where it is needed on demand and fills a gap in the grid. The other concept is, you have power plants here [in North Africa], and you connect the Moroccan grid via supergrid—I don't know how it works but it's a supergrid—to Europe and any fluctuations here, any surpluses, they go to Germany or they go through Spain and France to Germany and backwards. And now imagine that there is some failure. What will Germany do? They will suck energy and the lights will go out in Morocco. If the grids are connected, those who win have more power or more money to pay. . . . So the infrastructure here to cover the load in Germany must be completely separated from the infrastructure in Morocco. And then there is no risk of this type of colonialism.

In contrast, in 2014, Dii released a statement entitled "Europeans do not understand Desertec" that emphasized the importance of integration for both sides of the Mediterranean.

> Many people believed [Desertec] was all about a couple of huge power plants in North Africa with a big exclusive power line to Europe. This will ultimately not happen and makes absolutely no sense. And it was not the main point. The people and the local industry in North Africa and the Middle East need the electricity. . . . Of course, the MENA region and Europe will both benefit from interconnections in the middle and long term. But many people in Europe are thinking mainly in European or national categories.
>
> (Schwarz, 2014, para. 6–7)

In a third actor construction of what arrangement would be most just, the Desertec Foundation reiterated the normative, limits-to-growth approach by developing sustainability criteria for labeling power plants with a Desertec seal of approval. The criteria include the following: The power plant should provide either local electricity for communities and industry or *dispatchable* generation for Europe, excess electricity capacity should be avoided where possible, grid stability should be prioritized, power plants should consume the least possible water, power plants should provide local education and training, the local public should have a say in the project's development, environmental impacts and rare species should be considered, waste should be minimized, and the plant should be properly decommissioned. The development of these criteria extends the concept of feasibility from simply "is it possible" to "under what circumstances should it be done and what benefits should result?"

Hughes (1986) advocated that STS scholars use a network-based approach to eliminate categories and explore networks, or "seamless webs" among science, technology, and institutions. The networks working on electrical power systems integration in the diverse Mediterranean region hardly form a seamless web, but rather constitute a disjointed and at times conflicting jumble. The "internal intrigue" of Dii, as one interviewee termed it, came to a head. Managing Director Aglaia Wieland was fired for stirring up public conflict through an alleged disagreement with CEO Paul van Son over whether Desertec should focus only on North Africa's energy needs or on electricity exports. Following this, in July 2013, the Desertec Foundation announced that Dii was no longer allowed to use the Desertec brand name. Dii then became the "Desert Power Industry Initiative" or simply Dii. As attention grew for the needs of local populations in MENA, the framing of the empty Sahara Desert could no longer be substantiated, but the same shareholders were in conflict about whether Desertec still would be Desertec without cross-continental transmission lines. The actors agreed neither on how to prove the vision's feasibility nor on what they were trying to prove. Not only did the stakeholders break apart, the values and purpose of the vision came apart at the seams.

In 2015, the headquarters of Dii moved from Munich to Dubai to demonstrate Dii's commitment to electricity for MENA. It hired eight staff members from three corporate supporters—ACWA Power, the company developing CSP plants in Morocco; Innogy SE, a subsidiary of RWE focused on renewables; and the State Grid Corporation of China. Dii remains a small, but important, behind-the-scenes epistemic organization working on large-scale clean energy planning in deserts and on grid integration. Dii renamed itself Dii Desert Power MENA. According to its website, it has a new mission to "facilitate the rapid deployment of utility-scale renewable energy projects in MENA desert areas and to integrate them in the interconnected power systems and to make local people benefit." It works on advancing the development of wind and solar PV, as well as grid integration, for the MENA region through a proprietary GIS tool and project database. It also hosts an annual Dii Desert Energy Leadership Summit in Dubai to establish the "voice of the renewable energy industry in the MENA region." Dii hopes eventually to expand to other "desert markets," including Pakistan, South Africa, Namibia, and Chile. Dii-MENA is trying to avoid a conflict with advocates of distributed generation: Van Son stated, "These developments have led to a world [where] small- and large-scale electricity generation will not merely co-exist, but strongly complement each other" (van Son & Ruderer, 2015, p. 1).

The future of visions for regional and global clean power pools

While plans for a deeply integrated EU-MENA power system have stalled, there are multiple, tangible projects and sub-visions for cross-border clean energy development in North Africa and the Middle East, as well as specific projects between North African and EU countries. Footprints of the Desertec concept, broadly construed, and of Dii's organizational structure can be seen in many of these projects. For example, Nur Energie is pursuing the TuNur project in Tunisia, consisting of a 250 MW CSP plant, to be followed by a massive 2.5 GW plant, producing electricity for export through Italy. TuNur is similar to the solar pipeline model of Design I. This long-planned project has gained some traction since a submarine transmission cable between Malta and Sicily was completed in 2015 (Casey, 2017). It remains unclear whether a Tunisia-Malta HVDC link will be built. Additionally, between MENA and the EU, Morocco is pursuing shallow integration with Portugal.

In the Middle East, Dii is working on a grid interconnection study for the Gulf Cooperation Council Interconnection Authority (GCCIA), which includes Kuwait, Saudi Arabia, Bahrain, Qatar, the United Arab Emirates, and Oman. Shallow integration among these countries was first achieved in 2011 through 400 kV interconnections with 400 MW to 1.2 GW capacity. The goal of this system was limited to voltage control, system reliability, and cost avoidance rather than substantial export (Beides, 2013). GCCIA is now working to develop a deeply integrated common electricity market; interconnections beyond GCCIA; and higher integration of intermittent renewable energy, including energy from Saudi Arabia's solar power plan. Between North Africa and the Middle East, Saudi Arabia and Egypt are constructing a large 3 GW HVDC back-to-back interconnection.

Also in the Middle East, an environmental organization called EcoPeace, a partnership among Palestinian, Israeli, and Jordanian environmental activists focused on environmental peacebuilding, advocates for a water-energy community in the Levant to foster "healthy regional interdependence." EcoPeace envisions Jordan as a solar energy powerhouse for the region, with water desalination plants built in Israel and Palestine. It bears resemblance to the original idea from the Desertec *White Book* for a solar desalination plant on the Gaza Strip, an idea that was excluded from Design II and then revived within countries where water scarcity is a pressing problem. Jordan was initially a strong supporter of Desertec, but the allure of nuclear power has been strong, partly for reasons of national pride (Ramana & Ahmad, 2016). EcoPeace frames energy and water interdependence not as being impeded by state instability but as promoting "regional strategic stability" (Mehyar, Bromberg, Khateb, Waxman, & Milner, 2014). EcoPeace's report contrasts mutual interdependence with win-win outcomes, arguing that win-win scenarios that do not promote interdependence are insecure.

Another emerging leader in electricity integration is the State Grid Corporation of China (SGCC). SGCC is listed second on the 2017 Fortune 500 list, with over 900,000 employees and $315 billion in 2017 revenue. SGCC is a strong player in the utility sector and is becoming increasingly important. It has been purchasing shares in utility companies in various countries (e.g., Philippines, Brazil, Australia, Greece, and Portugal) and financing transmission lines in many developing countries. It supports grid integration, including the Asia Super Grid vision. A think tank called the Renewable Energy Institute, located in Japan, is promoting the vision and bears a resemblance to Dii. SGCC is also supporting the Gobitec vision to build CSP plants in the Gobi Desert in Mongolia for electricity export to China (Sovacool & Cooper, 2013). SGCC has begun work on renewable energy development and grid integration in the Mercosur region in South America, as well as engaging in controversial energy investment in Africa, which increases oil supply but raises human rights concerns.

The large-scale Desertec vision is a mere sub-vision within the Chinese vision for a global grid, resembling Buckminster Fuller's vision for a global grid in the 1960s ("Global Energy Grid—The Details," n.d.). China's President Xi Jinping presented the global grid vision at the UN Sustainable Development Summit in 2015. The Global Energy Interconnection Development and Cooperation Organization (GEIDCO) is now promoting this vision, led by former SGCC Chairman Zhenya Liu. In 2015, Liu published *Global Energy Interconnection*, calling for "two replacements," or replacing fossil fuel generation with renewables and replacing internal combustion engines with electric vehicles; "one increase," or achieving total global access to electricity; and "one restore," or restoring the use of fossil fuels as a basic industrial material only. As of 2017, GEIDCO has participation from 265 member organizations, including Dii-MENA.

The plan envisions a copper plate for the world, calling globally for 90% clean electricity, with 10% coal and natural gas to improve affordability for "underdeveloped regions." While the Desertec vision originally framed the "sunbelt" as the new energy frontier, the GEIDCO vision frames both the Equator and the Arctic as new renewable energy frontiers, proposing to build clean energy bases in the Arctic and at the Equator, integrating Africa–Europe, Asia–Europe, and Asia–Africa by 2030, integrating North and South America, Oceania–Asia, and Asia–North America by 2040, and integrating Europe and North America by 2050, with a "backbone grid" and intercontinental interconnection channel between various "renewable energy bases" (Liu, 2015, p. 196). However, GEIDCO geographically shifts what Desertec imaginers called the "technology belt" in Europe to China because of China's work on ultra-high voltage transmission (UHV). In this vision, UHV transmission is framed as innovative, since China is working on improving and expanding it, whereas in the 2006 Desertec feasibility studies HVDC was framed as mature. This further illustrates the subjective framing of technological maturity. GEIDCO envisions a substantial carve-out for distributed generation, subsumed under a broader networking logic.

Liu's book makes similar arguments to the Desertec discourse about the need for interdependence and global energy security. He argues that energy development, energy distribution, energy security, and environmental impacts are all global in nature. In terms of institutions, GEIDCO envisions a global dispatch center that "will play an important role in safeguarding global power security and the global allocation of energy sources" (p. 228). Liu also recommends developing a "cooperative alliance" on global energy interconnection within the United Nations. Again, the win-win discourse is utilized, as Liu calls for the creation of a "geopolitical landscape for win-win cooperation." Liu also argues for a "locals first principle" to avoid situations in which a nation exports electricity without meeting its domestic needs (p. 157). This reflects China's policy of noninterference in other countries' affairs, and the lenses of sovereignty and territorial integrity through which it has viewed its energy and resource extraction deals in Africa (Nötzold, 2012). Yet it is doubtful whether deep electricity integration could occur without countries ceding their nationally focused, autarkic energy policies to liberal conceptions of global energy security.

Liu envisions that the global grid will rely on 1,000 GB/s (gigabyte per second) information transfers for instantaneous global electricity trading, incorporating the global grid into the Internet and the Internet of Things to develop a "robust and intelligent" grid. Liu's book does not address privacy or cybersecurity. The Orwellian privacy implications of smart meters, already heavily debated in the United States and Europe (Stephens, Wilson, & Peterson, 2015), apply to the vision, which calls for "urban administrators to perform real-time analysis"; "health sensors [to] remain in contact on a real-time basis," even performing emergency surgery remotely; and "intelligent eyewear, watches, and other wearable devices [to] have powerful functions," with "sensors located everywhere" (Liu, 2015, p. 352).

The final chapter of Liu's book, "Global Energy Interconnection Changes the World," is techno-utopian and technologically deterministic in its treatment of social effects, arguing that the global grid will result in various environmental, economic, and societal benefits. Moreover, those involved in the upfront stages of crafting a vision are the most important players in determining its later impacts on energy justice. The end of Liu's book calls for more inclusivity, but at the time of my book's publication there are no women on the GEIDCO board, nor are there any Africans, even though the Southern African Power Pool dates back to the 1950s, offering long-standing experience.

Innovation beyond the technical: anticipating sociopolitical issues in transnational grid integration

While electricity integration offers a variety of benefits, methodologies of foresight are necessary in order to anticipate potential problems and plan for them in the upfront stages. While grid integration offers numerous benefits, some benefits coincide with new risks that should be mitigated if visions for shallow or deep grid integration move forward. I conclude this chapter by discussing generalizable issues important to the future of renewable electricity grid integration, including energy security, economics and affordability, and the design of institutions for regulation and management.

Security: The BETTER project, which reexamined Design I in 2015, argued that European decision makers must determine what version of energy security is desirable (CIEMAT, 2015). Stated differently, energy security is polysemic, meaning different things to different people, and could be conceptualized across multiple dimensions (Trombetta & Dyer, 2013). To institute policies that foster interdependence, decision makers would need to adopt a framing of energy security as a collective global problem with potential win-win outcomes, rather than viewing it as a zero-sum game (see Goldthau & Witte, 2010).

Adopting a regional framing of energy security is necessary but insufficient. EcoPeace argued that win-win outcomes are insecure without intentionally planned interdependence among countries. Achieving interdependence hinges on developing the social, technological, and institutional means of achieving this interdependence and getting decision makers to agree to it. Two considerations have been largely overlooked in feasibility studies for regional and global grids. First, even where a mutually interdependent system is the end goal, in the interim stages as the building blocks are put in place, the system will likely have lopsided dependencies that pose geopolitical risks. Second, energy storage is being developed that could allow for the future short-term stockpiling of electricity, removing the economic disincentive for withholding electricity.

Cybersecurity risks also need to be accounted for and planned for in transnational grid integration (Zeniewski, Brancucci Martinez-Anido, & Pearson, 2013). Smarter, more connected grids with increasing Internet dependence have greater cyber-vulnerabilities than conventional grids (Onyeji, Bazilian, & Bronk, 2014). Yet issues of cybersecurity did not arise in my empirical study, and the literature on smart grid cybersecurity affords little consideration to regional power pools. Deeply interconnected grids could lead to cascading blackouts, such the 2006 EU blackout that originated in Germany and cascaded to Morocco (van der Vleuten & Lagendijk, 2010). It is easier to plan for cybersecurity risks up front than to retrofit systems later (Pearson, 2011). Thus, cybersecurity risks should be anticipated and accounted for from the start of grid integration to the extent economically and technologically feasible.

Although cybersecurity should not be overlooked, the risks and hazards of climate change stemming from energy emissions should be weighed against cybersecurity risks and hazards. In current grids, physical risks are greater than cyber risks, stemming mainly from weather and damage by squirrels (Burke & Schneider, 2015). Cyberattacks have been limited thus far, although a 2015 cyberattack on Ukraine's grid by Russian hackers left 225,000 people without power for six hours (Freeman, 2016).

Energy economics: From a practical standpoint, this chapter illustrated that the cost of renewables in comparison to fossil fuels is not the sole reason renewable energy has not been scaled up. Dii convened some of the world's most powerful companies and illustrated how difficult it is even to institute a supposed techno-fix that substitutes clean electricity into the existing centralized system configuration.

Dii argued in *Desert Power 2050* that a 90% regional renewable electricity system would be cheaper than a system dominated by fossil fuels. But it was difficult to predict electricity prices for such a large system, and to account for new, unique costs relative to the cost of infrastructure that must be replaced nevertheless because of old age. Different inputs into Desertec models and very different scopes of the models resulted in different cost estimates (Backhaus et al., 2015). Additionally, framings of technological maturity versus immaturity are socially constructed but then used to substantiate investment decisions as though they are objective facts.

New market mechanisms are needed to facilitate regional energy transitions, and they are not inevitable. Rather, regulators and industry actors must develop them. A lack of a broader systems management plan has exposed challenges to traditional electricity pricing. Red Eléctrica runs an advanced renewable energy control center. However, it avoids shutting down baseload nuclear and coal-fired power plants and curtails renewable energy generation instead. According to interviewees, the control center has the means to prioritize green electrons generated by wind and solar plants, but operators do not disconnect the conventional generators because of the high cost entailed. One exception occurred on Easter 2012, a very sunny and windy day with low demand, when Red Eléctrica cut Spain's nuclear generation for only the third time in history. Germany gives priority to green electrons but has experienced negative electricity pricing on windy days, as conventional power plant operators decide whether it is cheaper to pay consumers to take their electricity or to temporarily shut down the power plants. In discussions about energy systems transitions, a lack of price for system services, such as dispatchability, has become visible. In a system that still requires firm power on demand, price mechanisms could be developed for clean dispatchable generation, from sources such as CSP, and ancillary services. An alternative to storage would be to construct large amounts of surplus fluctuating renewables to cover demand and integrate the Mediterranean's grid, as proposed in Design II. However, this strategy is questionable from a life cycle sustainability standpoint, even if it would be cheaper.

Even when actors prioritized low cost at the systems level, regional politics, systems transformation, systems management, and sociopolitical conflict impeded the vision. Dii's arguments about overall lower system cost did not convince all stakeholders and shareholders and were not powerful enough to motivate the system's construction. Representatives from Dii shareholder company M+W Group thought that the added 2 eurocents per kWh cost from the HVDC cables would make imported electricity too expensive. Moreover, as the price of solar PV fell, some shareholders argued that it would be sufficient to build solar PV installations in southern Spain, Greece, or southern Italy. Focusing on multiple goals and priorities in addition to cost could help to enroll additional stakeholders and to improve system services and life cycle benefits overall.

Regulatory and institutional innovation: A fully renewable electricity system for EU-MENA has not been constructed because it is extraordinarily complex across a variety of sociotechnical dimensions. Regulatory and institutional innovations are at least equally important to technological innovation in the process of achieving fully renewable energy systems and will be important to achieving energy security and affordability. When the European Coal and Steel Community was developed in the 1950s, the president of the High Authority, Jean Monnet, thought that a dose of *dirigisme*, which the High Authority provided, would be needed "to ensure genuinely free competition, to eliminate discrimination, to provide investment guidance and to prevent the formation of cartels" (Spierenburg & Poidevin, 1994, pp. 3–4). The High Authority enforced the rules and objectives of the treaty that created the European Coal and Steel Community, crafted common policies, and created a common market for coal (ibid.).

Actors disagree about whether an institution such as the High Authority would be needed to determine common regulations and standards for a regional grid or whether the market could organically develop a regional grid. The EU-funded MEDRING studies identified the lack of common rules and regulations for electricity operation and markets within the Mediterranean region as a greater impediment to systems integration than the absence of cross-border transmission lines (Med-EMIP, 2010). An unprecedented Mediterranean-wide regulatory institution would need to implement rules and standards that facilitate cross-border systems development and investment in renewables. Today, a MEDREG organization exists, but it acts as a think tank and lacks influence (Cambini & Franzi, 2014). Additionally, the discourse around "immature regulatory frameworks" in North Africa is shaped by "institutional orientalism," based on the othering of North Africans rather than the legitimate interests of both sides (de Souza et al., 2018). A just procedure would be needed to develop regulatory frameworks that serve the needs of North Africa as well as of Europe.

An institution or institutions would be needed to manage and maintain the supergrid envisioned in Design II. An advocate for Design I was skeptical that Design II could be adequately managed. It differs from the grids system builders have managed over the past century because it uses a single regional market and grid to balance fluctuating renewable electricity. It could be the responsibility of a private TSO or nonprofit Independent System Operator (ISO) to manage "traffic" on the regional transmission lines and to provide system security, reliability, and resiliency, as well as voltage and frequency control (Med-EMIP, 2010). The literature is mixed on whether a TSO or an ISO would be preferable. Some recommend an ISO to avoid conflicts of interest and to provide nondiscriminatory access to the grid for independent power generators, while others argue this is not a feasible transnational option (Gruber, 2007; Oseni & Pollitt, 2014). The *Desert Power 2050* report (Design II) did not readily acknowledge that a common market would require an ISO or TSO or closely coordinated grouping of TSOs operating across dozens of countries.

This managing organization could possess a significant amount of social and political power; one interviewee argued that in Design II countries with the most money and power will maintain reliable electricity access if a failure occurs. While ISOs appear to be objective, technocratic organizations, the building and rebuilding of an energy system is always a sociotechnical controversy, as Mitchell (2011) argues in his history of the shaping of modern democracy via hydrocarbons. If the idea of the economy was developed on the basis of networks of pipelines and flows of oil as Mitchell describes, then organizations managing regional transmission would be important political entities in a future with deep grid integration. A Mediterranean Transmission System Operator would ensure that the copper plate provides electricity where it is needed and when it is needed and could decide what country would lose power if a failure were to occur. Thus, the Mediterranean TSO could gain unseen geopolitical power in the region.

Empirically, central planning for a new large-scale sociotechnical system has been generally unsuccessful in the Mediterranean region, leading some stakeholders to view an incremental, building-blocks approach as more feasible than a central-planning approach. Tholens (2014) suggests that electricity integration should remain under private-sector control, and should develop through a bottom-up, informal technical cooperation approach rather than through a top-down energy community controlled by policymakers. Yet it is unlikely that a bottom-up approach would result in a sufficient level of integration to address the problem of energy intermittency through regional integration, or to displace coal with clean, dispatchable generation from CSP plants in North Africa. More likely, integration would remain at the same level as in the GCCIA power grid, focused on small electricity exchanges for reliability and cost avoidance, without achieving large-scale climate mitigation goals and spurring massive investment in renewables in North Africa. The various visions for Mediterranean electricity integration are now entangled in a feasibility trap, with chicken-or-egg challenges related to developing generation versus transmission. Policy mechanisms that help to overcome chicken-or-egg infrastructure problems are likely essential to achieving energy transformations at a significant scale. These chicken-or-egg controversies require someone to make the first move, and that first mover is unlikely to be the free market alone, given the investment risk entailed and the scale of infrastructural investment and change that must occur in a short time frame.

Moreover, while this chapter illustrated that a centralized, high-modernist plan for an electricity system in the Mediterranean region largely has failed, systems thinking remains a desirable part of energy policymaking. Energy transitions in Germany and Spain have been rocky, with policymakers dramatically influencing markets while hoping that the markets will adjust and yield positive societal outcomes. An interviewee from the German government explained that the first step in the German energy transformation was reaching 20% renewable power capacity. From 20% to 50%, the challenge will be to transform the system, a step he said that North Africa has not yet reached. This reflects a chicken-or-egg problem for electricity policy. The system plan ought to come before developing individual projects to avoid building a technological juggernaut, but the system plan is so ambitious and politically complex that it does not become a collectively held imaginary. Still, system builders and policymakers ought to consider what options—among distributed generation and storage, centralized storage, surplus renewable capacity, and deep transmission integration—are the most affordable, sustainable, secure, and resilient overall. Moreover, regulations go beyond technical rules and standards and implicate social equity drawbacks that can become locked into the system for decades, even though regulators and TSO managers are rarely elected through a democratic process that would give them a social license to make such decisions. Issues of social equity ought to be considered in a democratic systems-level planning process in the visioning phase.

Regional sociotechnical imaginaries: This chapter illustrated how visions that fail still have effects on the energy sector. Visions often have variegated histories through which change and the drivers of change can be tracked over time. Visions also make contestation visible as actors work to anticipate trends in the energy sector. Epistemic communities, through knowledge organizations like Dii and other analogs in regional electricity governance, play an intangible but significant role in spreading ideas.

Whether visions are materially enacted depends partly on whether they become collectively held at the regional level. STS sociotechnical imaginaries theory would suggest that a sociotechnical imaginary for a regional energy system would need to be shared across nation-states in the region. Currently, visions for regional energy systems are caught in a feasibility trap in which various social constructions of feasibility were fairly objectively robust but failed to result in a regionally held sociotechnical imaginary for a regional renewable electricity system. The project is only proven to be feasible if people in power imagine it to be so, yet the actors continue to see feasibility as an objective measure of the likelihood of success. The next chapter will explore Morocco's energy policy in relationship to a national-level sociotechnical imaginary for renewable energy. Rather than viewing CSP as it is framed in the Desertec vision, Morocco views it as emerging and in need of research and development. Dii sought to accelerate toward an "inevitable" future, to develop technologies today whose time Dii perceived had come. In contrast, Morocco is balancing its past, present, and future within its broader sociotechnical imaginary for renewable energy. An exploration of Morocco's sociotechnical imaginary alludes to challenges of fusing together different nation-state imaginaries at the regional level.

Note

1 According to an interviewee, in 2008, Desertec inspired the Mediterranean Solar Plan, which became the flagship initiative for the Union for the Mediterranean (UfM), the political authority responsible for Mediterranean integration, launched in July 2008 at the Paris Summit. The inclusion of Israel in the UfM led to tension, and German politicians were unhappy that France was leading the UfM. Thus, solar power was selected as an apolitical topic to guide the UfM. The UfM's Mediterranean Solar Plan sought to develop 20 GW of renewable energy. However, the plan has stalled because the involved parties did not reach a consensus on the roadmap in late 2014.

Works cited

Adam, B. (2003). Reflexive Modernization Temporalized. *Theory Culture Society, 20*(59), 57–78. Retrieved from https://doi.org/10.1177/0263276403020002004

Backhaus, K., Gausling, P., & Hildebrand, L. (2015). Comparing the incomparable: Lessons to be learned from models evaluating the feasibility of Desertec. *Energy, 82*, 905–913. Retrieved from https://doi.org/10.1016/j.energy.2015.01.100

Bazilian, M., Onyeji, I., Liebreich, M., MacGill, I., Chase, J., Shah, J.,. . . Zhengrong, S. (2013). Re-considering the economics of photovoltaic power. *Renewable Energy*, *53*, 329–338. Retrieved from https://doi.org/10.1016/j.renene.2012.11.029

Beides, H. (2013, May 28). *Pan-Arab interconnection and development of Arab power markets*. Retrieved October 13, 2017, from www.gccia.com.sa/Data/PressRelease/Press_27.pdf

Beneking, A., Ellenbeck, S., & Battaglini, A. (2016). Renewable energy cooperation between the EU and North Africa: Findings of a SWOT analysis. *International Journal of Energy Sector Management*, *10*(3), 312–336. Retrieved from https://doi.org/10.1108/IJESM-11-2014-0005

Bergman, N., Schwanen, T., & Sovacool, B. K. (2017). Imagined people, behaviour and future mobility: Insights from visions of electric vehicles and car clubs in the United Kingdom. *Transport Policy*, *59*, 165–173. Retrieved from https://doi.org/10.1016/j.tranpol.2017.07.016

Borup, M., Brown, N., Konrad, K., & Van Lente, H. (2006). The sociology of expectations in science and technology. *Technology Analysis & Strategic Management*, *18*(3–4), 285–298. Retrieved from https://doi.org/10.1080/09537320600777002

Brooks, K. (2006). *Public power, private dams: The Hells Canyon high dam controversy*. Seattle, WA: University of Washington Press.

Brown, N., Rappert, B., & Webster, A. (Eds.). (2000). *Contested futures: A sociology of prospective techno-science*. Burlington, VT: Ashgate.

Burke, S., & Schneider, E. (2015). Enemy number one for the electric grid: Mother nature. *SAIS Review of International Affairs*, *35*(1), 73–86. Retrieved from https://doi.org/10.1353/sais.2015.0010

Cambini, C., & Franzi, D. (2014). Assessing the EU pressure for rules change: The perceptions of Southern Mediterranean energy regulators. *Mediterranean Politics*. Retrieved from http://dx.doi.org/10.1080/13629395.2013.809197

Casey, T. (2017, August 11). The Desertec Sahara solar dream didn't die after all—it's Baaaack . . . *CleanTechnica*. Retrieved November 16, 2017, from https://cleantechnica.com/2017/08/11/desertec-sahara-solar-dream-didnt-die-baaaack/

CIEMAT. (2015). *Bringing Europe and third countries closer together through renewable energies (BETTER)*. Intelligent Energy Europe Programme of the European Union. Retrieved from https://ec.europa.eu/energy/intelligent/projects/en/projects/better

Cifor, A., Denholm, P., Ela, E., Hodge, B-M., & Reed, A. (2015). The policy and institutional challenges of grid integration of renewable energy in the western United States. *Utilities Policy*, *33*, 34–41. Retrieved from https://doi.org/10.1016/j.jup.2014.11.001

Ciupuliga, A. R., & Cuppen, E. (2013). The role of dialogue in fostering acceptance of transmission lines: The case of a France- Spain interconnection project. *Energy Policy*, *60*, 224–233. Retrieved from https://doi.org/10.1016/j.enpol.2013.05.028

Cottier, T., & Espa, I. (2017). *International trade in sustainable electricity: Regulatory challenges in international economic law*. Cambridge, MA: Cambridge University Press.

de Souza, L. E. V., Bosco, E. M. G. R. L., Cavalcante, A. G., & da Costa Ferreira, L. (2018). Postcolonial theories meet energy studies: "Institutional orientalism" as a barrier for renewable electricity trade in the Mediterranean region. *Energy Research & Social Science*, *40*, 91–100.

DLR. (2005). *Med-CSP: Concentrating solar power for the Mediterranean Region: Final report*. Stuttgart, Germany. Retrieved from http://www.dlr.de/tt/med-csp

DLR. (2006). *Trans-CSP: Trans-Mediterranean interconnection for concentrating solar power*. Stuttgart, Germany: German Aerospace Center Systems Analysis and Technology Assessment Section. Retrieved from www.dlr.de/tt/trans-csp

Dziurtwe, M. (2017, May 8). Zimbabwe risks power cuts from S. Africa's Eskom over arrears. *Reuters*. Retrieved May 8, 2017, from https://af.reuters.com/article/africaTech/idAFKBN1840TH-OZABS

ESMAP. (2010). *Regional power integration: Lessons from global case studies and literature review* (No. Briefing Note 004/10). Washington, DC: Energy Sector Management Assistance Program, The World Bank. Retrieved from www.esmap.org/node/353

Freeman, M. (2016, July 6). *European Union must integrate cybersecurity efforts to protect energy assets*. Retrieved from www.atlanticcouncil.org/blogs/new-atlanticist/european-union-must-integrate-cybersecurity-efforts-to-protect-energy-assets

Fu, R., Feldman, D., Margolis, R., Woodhouse, M., & Ardani, K. (2017). *U.S. solar photovoltaic system cost benchmark: Q1 2017*. Golden, CO: National Renewable Energy Laboratory (No. NREL/TP/6A20-68925*)*. Retrieved from www.nrel.gov/docs/fy17osti/68925.pdf

Geels, F. W. (2004). From sectoral systems of innovation to socio-technical systems: Insights about dynamics and change from sociology and institutional theory. *Research Policy*, *33*(6–7), 897–920. Retrieved from https://doi.org/10.1016/j.respol.2004.01.015

Geels, F. W., & Smit, W. A. (2000). Failed technology futures: Pitfalls and lessons from a historical survey. *Futures*, *32*(9–10), 867–885.

Global energy grid- the details. (n.d.). Retrieved December 8, 2017, from www.geni.org/globalenergy/issues/overview/grid.shtml

Goldthau, A., & Witte, J. M. (2010). *Global energy governance: The new rules of the game*. Washington, DC: Brookings Institution Press.

Graham, S. (2000). Constructing premium network spaces: Reflections on infrastructure networks and contemporary urban development. *International Journal of Urban & Regional Research*, *24*(1), 183–200.

Grave, K., Paulus, M., & Lindenberger, D. (2012). A method for estimating security of electricity supply from intermittent sources: Scenarios for Germany until 2030. *Energy Policy*, *46*, 193–202. Retrieved from https://doi.org/10.1016/j.enpol.2012.03.050

Greacen, C., Engel, R., & Quetchenbach, T. (2013). *A guidebook on grid interconnection and islanded operation of mini-grid power systems up to 200 kW*. Berkeley, CA: Lawrence Berkeley National Laboratory.

Gruber, T. (2007). Cross-border trade in electricity and gas: Obstacles to effective competition from a regulatory standpoint. *ERA Forum*, *8*(3), 417–426. Retrieved from https://doi.org/10.1007/s12027-007-0028-1

Haas, P. M. (2015). *Epistemic communities, constructivism, and international environmental politics*. London: Routledge.

Hickman, L. (2011, December 11). Could the desert sun power the world? *The Guardian*. Retrieved December 13, 2011, from www.guardian.co.uk/environment/2011/dec/11/sahara-solar-panels-green-electricity

Hughes, J. (2000). *Ecology and historical materialism*. Cambridge: Cambridge University Press.

Hughes, T. (1983). *Networks of power*. Baltimore, MD: Johns Hopkins University Press.

Hughes, T. (1986). The seamless web: Technology, science, etcetera, etcetera. *Social Studies of Science*, *16*(2), 281–292. Retrieved from https://doi.org/10.1177/0306312786016002004

International Energy Agency. (2017). *Electricity information 2017*. OECD. Retrieved from www.iea.org/bookshop/755-Electricity_Information_2017

Klessmann, C., Lamers, P., Ragwitz, M., & Resch, G. (2010). Design options for cooperation mechanisms under the new European renewable energy directive. *Energy Policy*, *38*(8), 4679–4691. Retrieved from https://doi.org/10.1016/j.enpol.2010.04.027

Klinge Jacobsen, H., Pade, L. L., Schröder, S. T., & Kitzing, L. (2014). Cooperation mechanisms to achieve EU renewable targets. *Renewable Energy*, *63*, 345–352. Retrieved from https://doi.org/10.1016/j.renene.2013.09.035

Komendantova, N., Patt, A., Barras, L., & Battaglini, A. (2012). Perception of risks in renewable energy projects: The case of concentrated solar power in North Africa. *Energy Policy*, *40*, 103–109. Retrieved from https://doi.org/16/j.enpol.2009.12.008

Lacher, W., & Kumetat, D. (2011). The security of energy infrastructure and supply in North Africa: Hydrocarbons and renewable energies in comparative perspective. *Energy Policy*, *39*(8), 4466–4478. Retrieved from https://doi.org/10.1016/j.enpol.2010.10.026

Lilliestam, J. (2014). Vulnerability to terrorist attacks in European electricity decarbonisation scenarios: Comparing renewable electricity imports to gas imports. *Energy Policy*, *66*, 234–248. Retrieved from https://doi.org/10.1016/j.enpol.2013.10.078

Lilliestam, J., & Ellenbeck, S. (2011). Energy security and renewable electricity trade—Will Desertec make Europe vulnerable to the "energy weapon"? *Energy Policy*, *39*(6), 3380–3391. Retrieved from https://doi.org/10.1016/j.enpol.2011.03.035

Lilliestam, J., & Ellenbeck, S. (2012). Fostering interdependence to minimise political risks in a European-North African renewable electricity supergrid. *Green*, *2*(2–3), 105–109. Retrieved from https://doi.org/10.1515/green-2012-0003

Lilliestam, J., Ellenbeck, S., Karakosta, C., & Caldés, N. (2016). Understanding the absence of renewable electricity imports to the European Union. *International Journal of Energy Sector Management*, *10*(3), 291–311. Retrieved from https://doi.org/10.1108/IJESM-10-2014-0002

Liu, Z. (2015). *Global energy interconnection*. London: Academic Press.

Lynch, S. (2013, January 30). Egypt army chief: "Collapse" a threat; others doubtful. *USA Today*. Retrieved March 19, 2015, from www.usatoday.com/story/news/world/2013/01/29/egypt-army-chief-warns-of-collapse-of-state/1873459/

Meadows, D., Randers, J., & Meadows, D. (2004). *Limits to growth: The 30-year update*. White River Junction, VT: Chelsea Green Publishing. (Original work published in 1972.)

Med-EMIP Euro-Mediterranean Energy Market Integration Project. (2010). *Medring update: Volume-I overview of the power systems of the Mediterranean Basin*. Barcelona: IEMed. Retrieved from at http://www.iemed.org/observatori-en/arees-danalisi/documents/arxius-externs/2011/2010-update-of-the-medring-study-regarding-mediter ranean-electricity-interconnections-overview-of-the-power-systems-of-the-mediterranean-basin

Mehyar, M., Bromberg, G., Khateb, N., Waxman, J., & Milner, M. (2014). *A water and energy nexus as a catalyst for Middle East peace*. Amman, Bethlehem, and Tel Aviv: EcoPeace. Retrieved from http://ecopeaceme.org/uploads/14379199451~%5E$%5E~Water_Energy_Nexus_Web4.pdf

Meyer, C. (2010, May 27). Developing Desertec: European dream of desert energy takes shape. *Spiegel Online*. Retrieved February 18, 2013 from www.spiegel.de/international/world/developing-desertec-european-dream-of-desert-energy-takes-shape-a-695908.html

Miller, C., Iles, A., & Jones, C. (2013). The social dimensions of energy transitions. *Science as Culture*, *22*(2), 135–148. Retrieved from https://doi.org/10.1080/09505431.2013.786989

Mitchell, T. (2011). *Carbon democracy: Political power in the age of oil*. London: Verso.

Moore, S. (2017). Evaluating the energy security of electricity interdependence: Perspectives from Morocco. *Energy Research & Social Science*, *24*, 21–29. Retrieved from https://doi.org/10.1016/j.erss.2016.12.008

Nötzold, A. (2012). Chinese energy policy and its implications for global supply security. *The Journal of East Asian Affairs*, *26*(1), 129–154.

Nye, D. (1990). *Electrifying America : Social meanings of a new technology, 1880–1940*. Cambridge, MA: MIT Press.

Onyeji, I., Bazilian, M., & Bronk, C. (2014). Cyber security and critical energy infrastructure. *The Electricity Journal*, *27*(2), 52–60. Retrieved from https://doi.org/10.1016/j.tej.2014.01.011

Oseni, M., & Pollitt, M. G. (2014). *Institutional arrangements for the promotion of regional integration of electricity markets: International experience*. Retrieved November 30, 2016, from https://openknowledge.worldbank.org/handle/10986/19376

Patt, A., Komendantova, N., Battaglini, A., & Lilliestam, J. (2011). Regional integration to support full renewable power deployment for Europe by 2050. *Environmental Politics*, *20*(5), 727–742. Retrieved from https://doi.org/10.1080/09644016.2011.608537

Pearson, I. L. G. (2011). Smart grid cyber security for Europe. *Energy Policy*, *39*(9), 5211–5218. Retrieved from https://doi.org/10.1016/j.enpol.2011.05.043

Ramana, M. V., & Ahmad, A. (2016). Wishful thinking and real problems: Small modular reactors, planning constraints, and nuclear power in Jordan. *Energy Policy*, *93*, 236–245. Retrieved from https://doi.org/10.1016/j.enpol.2016.03.012

Robinson, D. (2017). Economic and geopolitical determinants of trade in electricity. In T. Cottier & I. Espa (Eds.), *International trade in sustainable electricity* (pp. 54–73). Cambridge, MA: Cambridge University Press.

Rothe, D. (2016a). Energy for the masses? Exploring the political logics behind the Desertec vision. *Journal of International Relations and Development*, *19*(3), 392–419. Retrieved from https://doi.org/10.1057/jird.2014.17

Rothe, D. (2016b). Welcome to the desert: Regional integration and identity change through an EU-MENA energy partnership? In J. Horst & A. Jünemann (Eds.), *Euro-Mediterranean relations after the Arab Spring: Persistence in times of change* (pp. 183–203). London: Routledge.

Roubelat, F. (2006). Scenarios to challenge strategic paradigms: Lessons from 2025. *Futures*, *38*(5), 519–527. Retrieved from https://doi.org/10.1016/j.futures.2005.09.001

SAPA-AFP. (2012, March 15). Mozambique denies cutting power to Zimbabwe. *Times Live*. Retrieved from www.timeslive.co.za/news/africa/2012-03-15-mozambique-denies-cutting-power-to-zimbabwe/

Schwarz, S. (2014, October 11). Europeans do not understand Desertec. *Dii Website*. Retrieved October 16, 2014, from www.dii-eumena.com/press/news/single/article/994.html

Selin, C. (2006). *Volatile visions: Transactions in anticipatory knowledge* (Doctoral dissertation, Copenhagen Business School). Retrieved from http://www.forskningsdatabasen.dk/en/catalog/2389470563

Selin, C. (2008). The sociology of the future: Tracing stories of technology and time. *Sociology Compass*, *2*(6), 1878–1895. Retrieved from https://doi.org/10.1111/j.1751-9020.2008.00147.x

Slav, I. (2017, May 19). Iraq starts paying off $1 billion power debt to Iran. *OilPrice.Com*. Retrieved December 21, 2017, from https://oilprice.com/Latest-Energy-News/World-News/Iraq-Starts-Paying-Off-1-Billion-Power-Debt-To-Iran.html

Slavin, D. H. (2001). *Colonial cinema and imperial France, 1919–1939: White blind spots, male fantasies, settler myths*. Baltimore, MD: Johns Hopkins University Press.

Sovacool, B. K. (2007). Coal and nuclear technologies: Creating a false dichotomy for American energy policy. *Policy Sciences*, *40*(2), 101–122. Retrieved from https://doi.org/10.1007/s11077-007-9038-7

Sovacool, B. K., & Cooper, C. (2013). The Gobitec solar array. In *The governance of energy megaprojects: Politics, hubris and energy security* (pp. 172–206). Cheltenham, UK: Edward Elgar Publishing.

Spierenburg, D., & Poidevin, R. (1994). *The history of the high authority of the European Coal and Steel Community: Supranationality in operation*. London: Weidenfeld and Nicolson.

Stephens, J., Wilson, E., & Peterson, T. (2015). *Smart grid (R)Evolution: Electric power struggles* (1st ed.). Cambridge: Cambridge University Press.

Tholens, S. (2014). An EU-South Mediterranean energy community: The right policy for the right region? *The International Spectator*, *49*(2), 34–49. Retrieved from https://doi.org/10.1080/03932729.2014.906806

TREC Trans-Mediterranean Renewable Energy Cooperation. (2007). *Clean power from deserts: The DESERTEC concept for energy, water and climate security*. White Book (1st Ed.). Hamburg, Germany: TREC and DLR.

Trieb, F., Kern, J., Caldés, N., de la Rua, C., Frieden, D., & Tuerk, A. (2016). Rescuing the concept of solar electricity transfer from North Africa to Europe. *International Journal of Energy Sector Management*, *10*(3), 448–473. Retrieved from https://doi.org/10.1108/IJESM-12-2014-0003

Trombetta, M. J., & Dyer, H. (2013). *International handbook of energy security*. Cheltenham, UK: Edward Elgar Publishing.

Ummel, K., & Wheeler, D. (2008, December). *Desert power: The economics of solar thermal electricity for Europe, North Africa, and the Middle East*. Center For Global Development working paper No. 156. Retrieved from http://ssrn.com/abstract=1321842

van der Vleuten, E., & Lagendijk, V. (2010). Transnational infrastructure vulnerability: The historical shaping of the 2006 European "Blackout." *Energy Policy*, *38*(4), 2042–2052. Retrieved from https://doi.org/10.1016/j.enpol.2009.11.047

Van Lente, H. (2000). Forceful futures: From promise to requirement. In N. Brown, B. Rappert, & A. Webster (Eds.), *Contested futures: A sociology of prospective techno-science* (pp. 43–64). Burlington, VT: Ashgate.

van Lente, H. (2010). Supporting and evaluating emerging technologies: A review of approaches. *International Journal of Technology, Policy and Management*, *10*, 104. Retrieved from https://doi.org/10.1504/IJTPM.2010.032857

van Lente, H., & Rip, A. (1998). Expectations in technological developments: An example of prospective structures to be filled by agency. In C. Disco & B. V. D. Meulen (Eds.), *Getting new technologies together: Studies in making sociotechnical order* (pp. 203–231). Berlin, Germany: Walter de Gruyter.

van Son, P., & Ruderer, D. (2015). *Capturing synergies among the power markets around the Mediterranean* (SSRN Scholarly Paper No. ID 2631088). Rochester, NY: Social Science Research Network. Retrieved from https://papers.ssrn.com/abstract=2631088

Verbong, G., & Geels, F. (2007). The ongoing energy transition: Lessons from a socio-technical, multi-level analysis of the Dutch electricity system (1960–2004). *Energy Policy*, *35*(2), 1025–1037. Retrieved from https://doi.org/10.1016/j.enpol.2006.02.010

Williges, K., Lilliestam, J., & Patt, A. (2010). Making concentrated solar power competitive with coal: The costs of a European feed-in tariff. *Energy Policy*, *38*(6), 3089–3097. Retrieved from https://doi.org/10.1016/j.enpol.2010.01.049

Zeller, T. (2009, June 22). Europe looks to Africa for solar power. *The New York Times*. Retrieved February 19, 2013, from http://www.nytimes.com/2009/06/22/business/energy-environment/22iht-green22.html

Zeniewski, P., Brancucci Martinez-Anido, C., & Pearson, I. (2013). Chapter 3: Framing new threats: The internal security of gas and electricity networks in the European Union. In M. J. Trombetta & H. Dyer (Eds.), *International handbook of energy security* (pp. 40–69). Cheltenham, UK: Edward Elgar Publishing.

Zickfeld, F., Wieland, A., Blohmke, J., Sohm, M., & Yousef, A. (2012). *Desert power 2050: Perspectives on a sustainable power system for EUMENA*. Munich, Germany: Dii GmbH and Fraunhofer ISI.

Part II

Nation-state visions for just and socially sustainable energy development

5 The social pillar of sustainable development in Morocco's solar imaginary

In the summer of 2016, I stepped out of a Jeep near the town of Merzouga in southeastern Morocco, not far from the Algerian border. It was a sweltering day, with temperatures exceeding 100 degrees. The breeze merely served to blow a wall of hot, dry, sandy air into my face. We were surrounded by crumbling mud-brick homes. Some were occupied; one had a small solar panel glimmering on top, providing a bit of electricity from the intense sun. Our guide led us toward a canvas tent, pitched in the sand, and a small hut used for cooking. A down-to-earth woman in a pink robe and turquoise headscarf warmly welcomed us, and invited us to sit next to her as she lit a fire under the tea kettle, her children beside her, one with a soccer ball in hand. I winced as I watched smoke billow up from it, imagining microscopic particulate matter lodged in tender lungs. Such energy poverty is rare in Morocco. But I was cautious about making assumptions, as I was merely a tourist in a long lineage of Western travelers too quick to reduce this complex environment to stereotypes and myths about barren deserts. As the water boiled, our group of 15 people crawled under the canvas tent, removing our shoes to sit comfortably on rugs. Our gracious host informed us that the last few days had been miserable because the family had been forced to hole up in the hut as the wind raged for hours. Today was a better day. We asked whether she had a husband. She had one, but without any elaboration, she stated matter-of-factly that she had decided she was better off without him. She would like her children to go to school, but a daily trek to the nearest town was simply impossible. She survived through a social network—friends brought her goods, as well as visitors like us, and she would soon go into the city, where it was cooler, to fast during the month of Ramadan. As we gathered up change, a coin fell on the floor and one of the boys gleefully scooped it up, oohing at his discovery. She gently punished him for his minor impropriety. Even in this harsh climate, dignity was more important than money.

Markers pointing to the history of this place were visible. Nearby were the remains of a coal mine the French Protectorate had built to fuel colonialism, along with decaying administrative buildings and a cemetery for miners. In the West, these structures would have perhaps been incorporated into a museum, but here they were left as living history, a visual testament to an unforgotten past. The fossils people sold along the roadside suggested the protectorate was, in fact, recent history for this area. Our guide spoke of his childhood when there were thousands of camels and caravans in this place, before the region was shaped by urbanization trends. We remarked that now there are more tourists than camels. He nodded, a wistful look on his face.

As I sipped a glass of tea made with fresh, bitter herbs our host had gathered nearby, my mind drifted to a roundtable discussion on sustainable development just a few days earlier at Cadi Ayyad University in Marrakesh. The Moroccan students had introduced us to elements of sustainable development that are often overlooked in Western-centric sustainability discourses. Our premises for the concept had little in common. The American students talked about improving US education on environmental impacts in order to "correct" aspects of US culture not conducive to environmental sustainability and to encourage people to consume in responsible ways. In contrast, the Moroccan students emphasized the importance of preserving the country's patrimony and heritage for future generations. The Moroccan students agreed that sharing culture so that it does not die is the main idea of sustainability. But they disagreed about the relationship between socioeconomic development and environmental sustainability. One Moroccan student argued, "We haven't reached development, so we can't talk about sustainability." She explained that many people who harm the environment as part of their livelihood did not have the equity of opportunity to go to school, or otherwise to improve their human capabilities, in order to pursue a different livelihood. They should not be expected to live an environmentally sustainable lifestyle without equity of opportunity and inclusion in socioeconomic development. Another student disagreed, saying that environmental sustainability and development must go hand in hand. Then a PhD student implored that our discussion of sustainable development go beyond resources—it is, instead, about humanity, gender, equity, capitalism, and income disparity. She said, "We only need a small amount of resources, but we need a lot more humanity."

These discussions imply that sustainable energy policy is about much more than technology and markets. Sustainable energy development must meet a higher standard than carbon dioxide mitigation alone, by accounting for social inequity. As stated in a Moroccan energy visioning document, "All debates on energy policy become, by definition, a debate on the stakes of the future and returns of the choices of the model of society to which we aspire as a national collectivity" (Haut-Commissariat au Plan, 2006, p. 4).[1] Sitting in a tent in the Sahara Desert, I was left with more questions than answers about the foundations and utility of sustainable development for our host and her two sons, who lacked equity of opportunity. What exactly ought to be sustained through a sustainable development strategy? Is sustainable development possible when not everyone has equity of opportunity to pursue an education and secure a livelihood, when people lack what Amartya Sen and Martha Nussbaum have termed "human capabilities"?[2] What do Moroccans want their future to be, and how will this future correspond to Morocco's past and nation-state identity? What role could sustainable energy initiatives play in sustaining and improving livelihoods in the Sahara, and could low-carbon energy negatively affect livelihoods?

The visible markers of the past in the Sahara call attention to the often overlooked idea that sustainability and sustainable development are fundamentally temporal concepts, relating not only to landscapes and places but also to timescapes and the perception of the passage of time (Adam, 1998). Temporally, sustainable development is most often associated with the future. What future does a nation, its citizens, or the international community wish to build and how can future generations be accounted for in this decision-making process? Often overlooked are the ways in which the past shapes our imaginations, as well as constrains our ability to imagine. Colloquially, "sustaining" something suggests propping up or continuing a practice from the past, implicating the past. The most commonly cited definition of sustainable development from the Brundtland Report—"development that meets the needs of the present without compromising the ability of future generations to meet their own needs"—references equity between current and future generations but does not address the past. In the previous chapter, I detailed how sociotechnical energy systems relate to visions for the future, but respond, inevitably, to the existing system in place and the existing societal context. Examining what a society wishes to sustain for the future also leads one to ask what it is about the past that current generations wish to sustain for the present and future. Moreover, this future is shaped by a history of inequity and injustice. Morocco's process of envisioning its sustainable energy system illuminates the future Morocco wishes to build and what it is about the past that decision makers wish to sustain, as well a history of injustice that oppressed publics, whose imaginaries often go unrecognized, seek to overcome. Therefore, the Moroccan energy imaginary also calls attention to citizens' perceptions of both the past and the future, as well as how environmental development affects them, bypasses them, and relates to previous injustices. Yet sustainable development discourses often focus on what the future ought to become, rather than on sustaining the past or shaping a future that is congruent with a nation's past.

As illustrated by the discussion with students in Marrakesh, the role of the citizen and the social and cultural aspects of sustainable development, along with the term itself, mean different things to different people. As Lélé (1991) argued, sustainable development is interpreted in various, ambiguous, and contradictory ways that are often not conceptually precise, but a society ought to be able to develop its own definition of sustainable development and its own approach to achieving it. Sustainability and sustainable development are sometimes colloquially used as stand-ins for normative beliefs or ethical principles (Parmentier & Moore, 2016); in other cases, sustainability is colloquially used as a mild way of calling something good and unsustainability as a mild way of calling something bad (Thompson, 2007). The social pillar is recognized as the least-understood aspect of sustainability, with various theoretical associations, including social movements, environmental ethics, agricultural ethics, and policy (Thompson, 2010).

Nurse (2006) argues that culture should be the main pillar of sustainable development, chiefly relating to the preservation of cultural heritage. Likewise, the sustainable tourism literature focuses on the inclusion of culture and cultural preservation in sustainable development. Tourism can threaten the cultural patrimony it often seeks to promote to tourists (McCool & Moisey, 2001), and sustainable energy can aid in mitigating this impact (Girard & Nijkamp, 2009). However, the view of sustainable energy as innocuous, neutral, nonpolluting, and static, as it is often conceived of in works that seek not to have an "impact" on culture, fails to see renewable energy as a sociotechnical imaginary, as visions for renewable technology relate to collectively held imaginations of what a society was, is, and could become, and illustrate whose imaginary and whose vision is perceivably collectively held, and whose is not. It overlooks the idea that for a massive renewable energy project well suited for large-scale climate mitigation to be successful, it must become an ordering project of society, a collectively held vision, and a direction for the development of a society. Moreover, renewable energy should be considered as a tool for achieving human development goals, rather than as a means of avoiding making an impact (Goldemberg, Johansson, Reddy, & Williams, 1988).

Viewing culture through the lens of tourism may result in tourism managers preserving what they view as "traditional" for the purpose of entertaining tourists rather than for the purpose of sustaining culture for citizens. For example, from 1925 to 1950, citizens, grain mill operators, and students in Fez's historic religious boarding schools wanted access to electricity, but French Protectorate officials implemented regulations to maintain the traditional look of the medina for tourist purposes (Apelian, 2012), preventing electrification as well as the installation of gas pumps by Shell Oil (Porter, 2000). In some cases, students furtively used electricity to study at night in areas not frequented by tourists (Apelian, 2012). By attempting to make time stand still in the medina, the French Protectorate, in fact, shaped culture on the basis of their perception of what counted as traditional and what counted as modern (Porter, 2000).

This chapter discusses the ways in which the Moroccan state imagines Morocco's energy future, as part of nation-state pride and of socioeconomically sustainable development, theoretically pairing discussions of sustainable development with the STS concept of the sociotechnical imaginary. Jasanoff and Kim (2009) define national sociotechnical imaginaries as "collectively imagined forms of social life and social order reflected in the design and fulfillment of nation-specific scientific and/or technological projects" (p. 120). An imaginary incorporates the past as well as the future. Jasanoff and Kim (2015) state, "The past is prologue, but it is also the site of memory excavated and reinterpreted in the light of a society's understanding of the present and its hopes for what lies ahead" (p. 21). In *Dreamscapes of Modernity*, Jasanoff and Kim (2015) add the importance of civil society participation in developing imaginaries, but they do not see the visions of civil society as "ris[ing] to the status of an imaginary" until they are communally adopted and publicly performed (p. 4). Tidwell and Tidwell (2018) draw attention to the often overlooked ways in which citizens imagine energy futures and the language that they use to do so, as well as the normative importance of examining citizen imaginaries even if they do not rise to the level of a collectively held national imaginary. What their intervention in sociotechnical imaginary theory makes clear is that scholars ought to question whether the visions of members of civil society—visions that the researcher fails to see because of his or her methodological choices—could be collectively held, although not publicly performed on the national stage. These latent visions could aid in anticipating change and redressing injustice. This is perhaps more important in an undemocratic society in which access to the press and the societal discourse is cordoned off, but most research on imaginaries has focused on democratic societies. The role of citizens in the Moroccan energy imaginary is contested and unmasks distributive and capabilities injustices, as the latter part of this chapter will elucidate.

As solar power is often viewed as the epitome of a sustainable development strategy, and among the most lauded techno-fixes for a changing climate, it is a strategic empirical domain for examining sustainable development and sociotechnical imaginaries. This chapter focuses on the Moroccan Solar Plan as a national project, whereas Chapter 6 will discuss Morocco's role in regional energy development, and Chapter 7 will focus on the local effects of the first solar power plant under construction in rural Morocco. In this chapter, I tell three different stories about solar power and sustainable development in Morocco that highlight different aspects of the social pillar of sustainability, and the imaginary of sustainable energy, illustrating how sustainable development strategies can be read through different lenses depending on the scale and perspective of the analysis. By telling these variegated stories together, a broader narrative emerges about the sociotechnical imaginary for renewable energy in Morocco.

First, I discuss the Moroccan Solar Plan in terms of Morocco's quest for a charismatic, centralized technology that will bring it national pride, examining the nation-state-based imaginary for renewable energy development. Second, I overview the more specific socioeconomic challenges on which the Moroccan state is concentrating and their relationship to Moroccan solar energy policy, further drawing out the national imaginary for solar power and how it relates to a vision for socially sustainable development. This story also explains that Morocco is pursuing CSP—which is among the most expensive of commercially available renewable energy technologies—on the basis of Morocco's broader vision for socioeconomic development. By discussing this story, I shed light on a plan that the media has largely misunderstood and presented in overly simplistic ways. Third, I tell the story of the gruesome death of a fishmonger in Morocco, a story that is seemingly unrelated to the analysis of the sociotechnical system on which I focus. But if this story remained untold, my examination of collective imaginaries of Moroccan sustainable energy systems would fail to incorporate the views of members of civil society who are unrecognized and bypassed by solar power development and would fail to scrutinize "win-win" sustainable development discourses. I conclude by arguing that, without recognition and distributive justice for present generations, the natural capital left to future generations will continue to be unevenly distributed, even if equivalent amounts are sustained between generations.

Morocco: geography and economy in brief

Al Maghreb Al Aqsa, as Morocco is sometimes transliterated from Arabic, means the country of the far west (in the Arab world). In Chapter 2, I discussed how the Sahara Desert has long been enrolled in visions of the Western frontier—as Miller (2013) states, "The oft-invoked and misguided image of a Moroccan 'Far West'—an empty land, an unutilized space, ripe for exploitation by the pioneering spirit" (p. 113). In Chapter 3, I discussed how the depiction of Morocco as a nation characterized by sand dunes and camels has been deployed to substantiate the vision of North Africa as an ideal solar energy powerhouse. This narrative fails to capture Morocco's geographic or societal diversity.

Zeigler (1997) compares Morocco to California in terms of land area, length and width, shape, continental location, highest peak, climate, vegetation, and offshore current, as well as population. In the United States, California also was enrolled in visions of the frontier, but few Americans today would frame all of California as a barren desert. Morocco's landscape is diverse, including 1,835 kilometers of coastline; the High Atlas, Middle Atlas, Anti-Atlas, and Rif Mountains; maquis shrubland; short grassland; forests teaming with monkeys; canyons; and gorges, in addition to sand dunes. It is not an unpopulated desert but rather a country with a rapidly growing population of 35.19 million people that is projected to reach 43.56 million people by 2050 (Haut-Commissariat au plan, n.d.-b). Morocco is part of Africa as well as the MENA region, and it is just nine miles from Spain across the Strait of Gibraltar (Figure 5.1). Algeria borders Morocco to the east. From roughly 1912 to 1956, Morocco was under French and Spanish control. South of Morocco proper is a territory the UN calls Western Sahara, the sovereignty of which has been contested since Spain withdrew in 1975. In brief, the Moroccan government views the territory as part of its nation-state, dating back centuries, but the UN does not recognize it as part of Morocco. Not including Western Sahara, Morocco's land area is 172,000 square miles, with a length of 750 miles and a width of 250 miles (Zeigler, 1997).

Of course, geography does not determine development. California's GDP per capita is around $56,365. Morocco's gross national income per capita is roughly $2,850, and the World Bank classifies it as a lower-middle-income country (World Bank, 2018). Moreover, Morocco faces significant domestic socioeconomic inequality. "Per capita household consumption in rural areas is only 54% of that in Morocco's urban areas" (World Bank, 2011). Morocco's urbanization rate in 2018 was 62.4%, as people, especially youth, leave rural areas (with limited infrastructure) for cities (Haut-Commissariat au Plan, n.d.-d).

Today, Morocco is a largely agrarian economy, with agriculture (including forestry and fishing) employing roughly 40% of the population and constituting 40% of exports and 21% of GDP (World Bank, 2012; ESMAP, 2011). In rural areas, this agricultural focus is more pronounced, with 73.7% of the rural population employed in agriculture, forestry, and fishing in 2017 (Haut-Commissariat au Plan, n.d.-a). Tourism is Morocco's second-largest sector; it employed 41% of the population in the fourth quarter of 2013 (Rivlin, 2012). Tourism has suffered

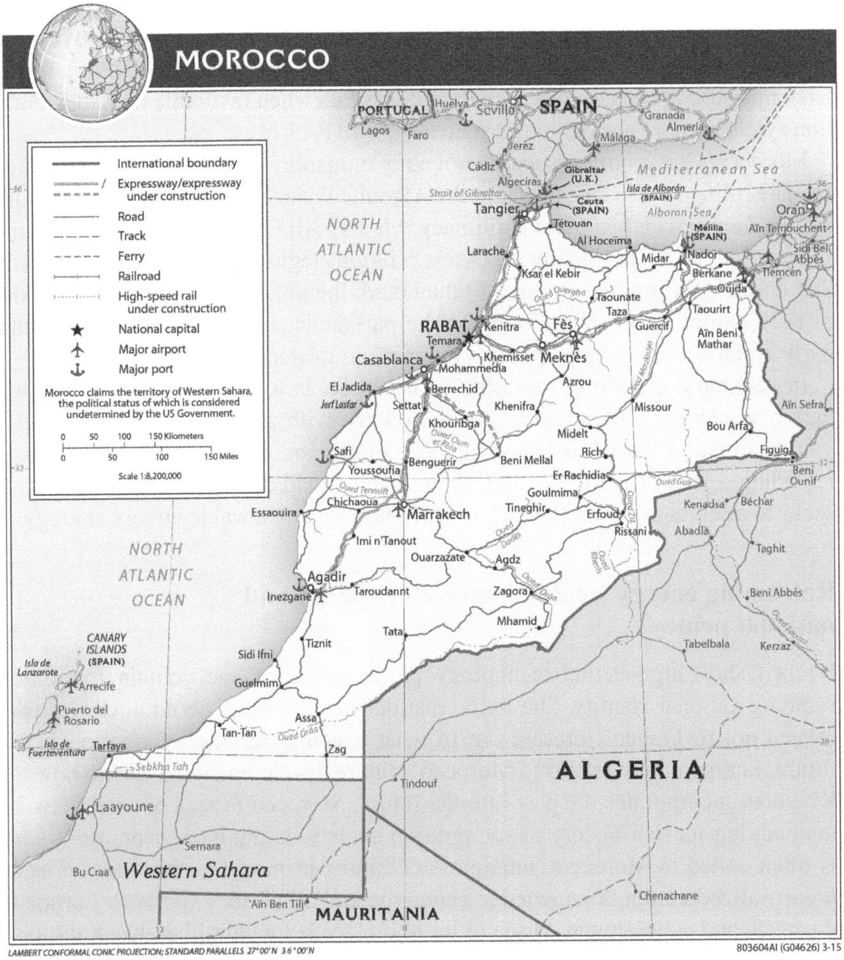

Figure 5.1 Map of Morocco

Source: CIA World Factbook

because of lack of economic growth in Europe, as well as fear of terrorism. Industry (including building) employs around 20% of the population, and manufacturing capacities are not well developed (Rivlin, 2012; Vidican et al., 2013). Morocco lacks the fossil fuel wealth of many other countries in the MENA region. Its most significant natural resource is phosphates, which are added to fertilizers, food and drink, and detergents and are used as a pipe corrosion inhibitor. Morocco holds 75% of the world's estimated phosphate reserves and is one of the top three exporters worldwide (Oxford Business Group, 2012). Despite the magnitude of this resource, in some years remittances sent home by the 4.5

million Moroccans living abroad added more to the economy than phosphates (Maghraoui, 2009; (Boukharouaa et al., 2014)). Overall, Morocco's GDP growth has been relatively strong since 2000, averaging 5% per year (SIE, 2010): the highest growth rate in North Africa.[3] However, these increases in GDP are somewhat misleading because they largely reflect years when favorable climatic conditions yielded more agricultural products (ESMAP, 2011).

Morocco's government is a parliamentary monarchy ruled by King Mohammed IV. The ruling family dates back to the Alaouite Dynasty of the Maghreb, providing the family with historical legitimacy. Moulay Ali Cherif founded the Alaouite Dynasty in 1631. The Alaouites possess religious legitimacy as a *sharifian* family that descended from the Prophet Muhammad; the king is called the Commander of the Faithful (*Amir al-Mu'minin*). The parliament is elected and has a multiparty system, and the king appoints the prime minister.

In summary, despite the framing of Morocco as a desert frontier, it is not a barren wasteland but rather a diverse country with a unique geography and a long history as a state. It is facing multiple sustainable development challenges—including a lack of natural resources, inequality, and agricultural dependence. It seeks to address these challenges, in part, through a renewable energy strategy.

Reframing energy wealth: renewable energy and national pride

Hecht (2009) argued that technology policy is a strategic domain for understanding national identity. She found that debates in France about nuclear power related not to France's present but to what it could become as a nation in the future. Energy policy relates to Morocco's future development but also to how the Kingdom incorporates the past into the future. Morocco faces a tension between maintaining its rich history as an agrarian society dating back centuries—what is often called in Morocco "intangible cultural patrimony"—and a future as an industrialized nation, a knowledge economy, and a hub for exports to Europe to the north and sub-Saharan Africa to the south, traded within liberalized, regional markets. The Moroccan government's policies seek to spur industrialization—attracting foreign direct investment (FDI), building technology parks and call centers, improving training opportunities, and developing new industries. But these policies also reflect historical continuity and an appreciation for sustaining culture. Below I explore how Morocco's sociotechnical imaginary for energy encompasses its past and its agricultural identity, as well as its national pride and future industrialization. Gruber et al. (2017) loosely refer to this as a "Green Growth" imaginary; I write about it more specifically here.

Understanding the relationship among electricity, agriculture, and water, as well as the centralized focus of Morocco's solar energy plans, requires understanding the historical trajectory of Morocco's electricity system. At the start of the French Protectorate in 1912, there was no electricity infrastructure in Morocco (D'Angio, 2002). The French colonists hoped that, after their difficulties colonizing desiccated Algeria, Morocco would provide an "El Dorado of water resources"

(Davis, 2007; Miller, 2013; Swearingen, 1988). They viewed the Moroccan High Atlas Mountains as "a chateau of water," with three large rivers fed by snowmelt: l'oum er Rbia, le Sebou, and la Moulouya (Saul, 2002). They sought to "valorize" Morocco's electrical resources to mine phosphates and transport them from mines to the port of Casablanca using electrified rail (Bouneau, 2002). Because the exploitation of hydrologic and hydroelectric resources required a centralized state and national plan to amortize large investments (Cladé, Beltran, Marseille, Moussa, & Couvreux, 2002)—and centralized electricity was needed for extracting and exporting phosphates—the French Protectorate envisioned the development of electrical power systems at a large scale, with the centralization of generation and interconnection from the start (Cladé et al., 2002; Saul, 2002). A single state-controlled company, Énergie électrique du Maroc, or Electrical Energy of Morocco, was established in 1924 to "rationally develop" hydroelectricity and irrigation (Saul, 2002). This centralized focus was unusual in Africa (Frémeaux & d'Almeida-Topor, 2002) where most electricity systems evolved piecemeal (Saul, 2002), and it illustrates the historical roots of Morocco's contemporary emphasis on centralized, state-run electricity systems.

It was also under the French Protectorate that Morocco's agricultural policy became intertwined with electricity policy. The protectorate developed a policy for large-scale irrigation "to allow not a single drop of fresh water to flow into the sea," and initiated a dam-building plan formative to both Morocco's agricultural and its electrical power systems (Swearingen, 1988, pp. 38–39). The protectorate's grandiose plans for renewable energy (i.e., hydropower) development and irrigation were a source of pride used to legitimate the colonial enterprise, although the protectorate also developed coal-fired power plants, such as the Roches-Noires plant in Casablanca (Swearingen, 1988; Saul, 2002). Between 1929 and the end of the protectorate in 1956, the protectorate built 14 dams in Morocco, 10 of which included hydroelectricity capacity (Swearingen, 1988). The dams' irrigable area totaled 221,450 hectares with 1,920,560,000 cubic meters of reservoir capacity (figures based on data from Swearingen, 1988). By 1954, Morocco's installed capacity was composed of 58% hydroelectric facilities, 37% coal-fired power plants, and 5% diesel-fired power plants (Saul, 2002, p. 505). Saul (2002) argued that, from a development perspective, Morocco was not electrified under the protectorate. Rather, electrification was of "colonial character," meant to serve extractive industries and an agricultural export industry rather than to power Moroccan households (ibid.).

After Morocco's independence, King Hassan II—in power from 1961 to 1999—solidified the centralized configuration of Morocco's electricity infrastructure. He sought to maintain Morocco's traditional past—important to its sovereignty and identity and to the legitimacy of the ruling family dating back to the 1600s—while modernizing the state and the economy. Roberson (2014) encapsulated this "reconciliation of past, present and future" by quoting King Hassan II, who stated, "In this way, we shall graft the past onto the present to prepare for the future" (pp. 57, 59). Roberson (2014) states, "A key question [for King Hassan II] was how to embrace aspects of the modern era while not abandoning significant traditions that served as a source of identity" (p. 59). In the area of energy and water infrastructure, King

Hassan II sought to modernize the country by continuing the French Protectorate's centralized dam-building policies in earnest under his 1968 dam policy—*la politique des barrages*—that renewed the goal to build one dam per year in order to irrigate one million hectares (Swearingen, 1988). *La politique des barrages* interwove electricity policy with agricultural policy to meet the nutritional needs of the country, to develop agricultural exports, "to earn foreign currency," and to increase the standard of living to slow rural exodus (Swearingen, 1988). These are similar to today's goals for CSP (discussed further in Chapter 7).

Under King Hassan II, dams represented a charismatic, centralized technology important to Moroccan national pride. There is a long tradition in Morocco of viewing the "king as [a] builder," who leaves a lasting imprint on the country's infrastructure by transforming a place from quaint to high modernist (Elsheshtawy, 2008, pp. 17, 108–109). The king officiated over the groundbreaking for new projects, giving them national recognition as the king's projects, to build a national identity for Morocco through a new, sovereign version of what was originally a colonial-era quest. Dam building therefore became not the work of Moroccan or foreign engineers but the work of the king. The state also printed Moroccan buildings, dams, and other technologies on currency to foster recognition for these nation-building projects (Roberson, 2014).

By the beginning of the twenty-first century, much of Morocco's hydroelectricity capacity had been exploited, and the country suffered from hydroelectricity intermittency problems because of capricious rainfall and persistent drought. The Moroccan state searched for another charismatic centralized technology, vacillating between nuclear power plants and renewable power plants to fulfill this role. According to interviewees, in the 1970s, King Hassan II announced that he would pursue nuclear power, using the naturally occurring uranium in Morocco's phosphates, as part of his strategy for modernizing the country. The Moroccan state has viewed nuclear power as a means of achieving energy independence and the development of economic clusters, as an engine of economic growth, and later as an opportunity to mitigate carbon dioxide emissions. When the *Maroc 2030* scenario study was completed in 2006, charting a new vision for Morocco's energy development, the Office National de l'Electricité et de l'Eau Potable (ONEE) was planning to construct a 900 MW nuclear power plant by 2017. In 2007, Morocco completed a 2 MW research reactor—for which it had received permission from the United States under a 2001 agreement—associated with the Moroccan National Center for Nuclear Energy, Science and Technology founded in 1986 (American Presidency Project, 2001). In 2010, Morocco signed an agreement with France to advance Morocco's nuclear power ambitions ("France to help Morocco's nuclear energy drive," 2010). France had a financial incentive to sell its nuclear technology to Morocco (Marktanner & Salman, 2011). That same year, Morocco presented plans for two 1 GW nuclear power plants under its submission to the Copenhagen Accord from COP 15 (Conference of the Parties to the United Nations Framework Convention on Climate Change), although it also included a 2 GW solar power plant (World Nuclear News, 2010).

However, geopolitical and geoeconomic roadblocks stood in the way of the development of nuclear power in North Africa (Marktanner & Salman, 2011).

When Spain balked at Morocco's nuclear power ambitions, the plans were largely shelved. The Fukushima nuclear accident in 2011 further soured the public's opinion of nuclear power, and a social movement called Maroc Solaire, Maroc sans Nucléaire (Solar Morocco, Morocco without Nuclear Power) formed to oppose it. In interviews that I conducted in 2012 and 2013, government officials expressed that nuclear power was no longer a viable option.

Morocco's search for a charismatic centralized technology mostly turned toward large-scale solar power. Momentum was already building for this strategy. On March 6, 2009, at the Assises de l'Energie conference, King Mohammed VI presented an updated energy strategy focused on renewables. The Moroccan Strategy for the Development of Renewable Energy and Energy Efficiency sought to increase installed generation capacity to 42% from renewable energy by 2020, split among 14% solar (2 GW), 14% wind (2 GW), and 14% hydroelectricity, at a cost of $9 billion. The strategy also sought to develop domestic hydrocarbons and to achieve electricity market integration within the Maghreb and with Europe. Later, the Kingdom launched a strategy for energy efficiency with the goal of reducing electricity consumption by 12% by 2020 and 15% by 2030, especially in transportation and buildings. Additionally, Law 309 implemented liberal reforms in the electricity sector. As a result of these ambitious plans, half of the 15 large-scale renewable energy projects underway in the MENA region as of 2013 were under development in Morocco (Samborsky, Myrsalieva, & Mahmoud, 2013).

While nuclear power has been part of a development strategy for resource-poor countries, such as Japan and South Korea (Valentine & Sovacool, 2010), large-scale solar power offers more development opportunities for Morocco (Marktanner & Salman, 2011). Yet in October 2017, it became clear that Morocco had not entirely given up on the development of nuclear power as a strategy for increasing nation-state pride, when the government signed a bilateral agreement with Russia to develop "peaceful nuclear energy" ("Russian Prime Minister Strikes Energy Deals with Morocco," 2017). Russia has made similar agreements with other countries in Africa and the Middle East (Ramana & Ahmad, 2016). It is unlikely that this deal will result in the construction of a nuclear power plant in Morocco because of financial and geopolitical constraints.

In the interim, Morocco continues to pursue renewable energy while reimagining itself as what Hecht (2009) termed "a technological nation" and an industrialized economy. Renewable energy has become a shared national project that is part of capacity-building efforts to put Morocco on a level playing field within the region. In *The Radiance of France*, Hecht (2009) explores the "strength and flexibility of the association between technological prowess and national identity" (p. 4). The term *rayonnement* in French means both radiation and radiance. Similarly, *Massolia News* stated that "economic development will not be complete without its radiance (rayonnement) on all of the [Moroccan] national territory. ("Deuxièmes Assises de l'Energie," 2011). This quotation portrays Morocco's renewable energy strategy as a socioeconomic development project designed to radiate benefits to the disadvantaged rural areas of the country. In 2016, the Moroccan Agency for Solar Energy (MASEN) adopted the slogan *"pour un Maroc rayonnant!"*—for a radiant Morocco!—referring to the commercialization of Morocco's sun and wind resources. Large-scale solar power in Morocco could represent what Masco (2006) in the US context has called a "nuclear nation-building technology."

Morocco lacks the hydrocarbon wealth of many of its neighbors but is reframing itself as a country rich in energy resources rather than as a fossil-fuel poor country. Moroccan government agencies use solar radiation maps, such as Figure 5.2, to depict Morocco as a nation awash in solar gold that is waiting to be harvested. Mehdi Alaoui M'Daghri, an event planner for the Morocco Solar Festival in Ouarzazate stated, "We organized this festival to show one idea: gold is in the sky. The sun is a treasure for our country through three main areas: . . . arts and

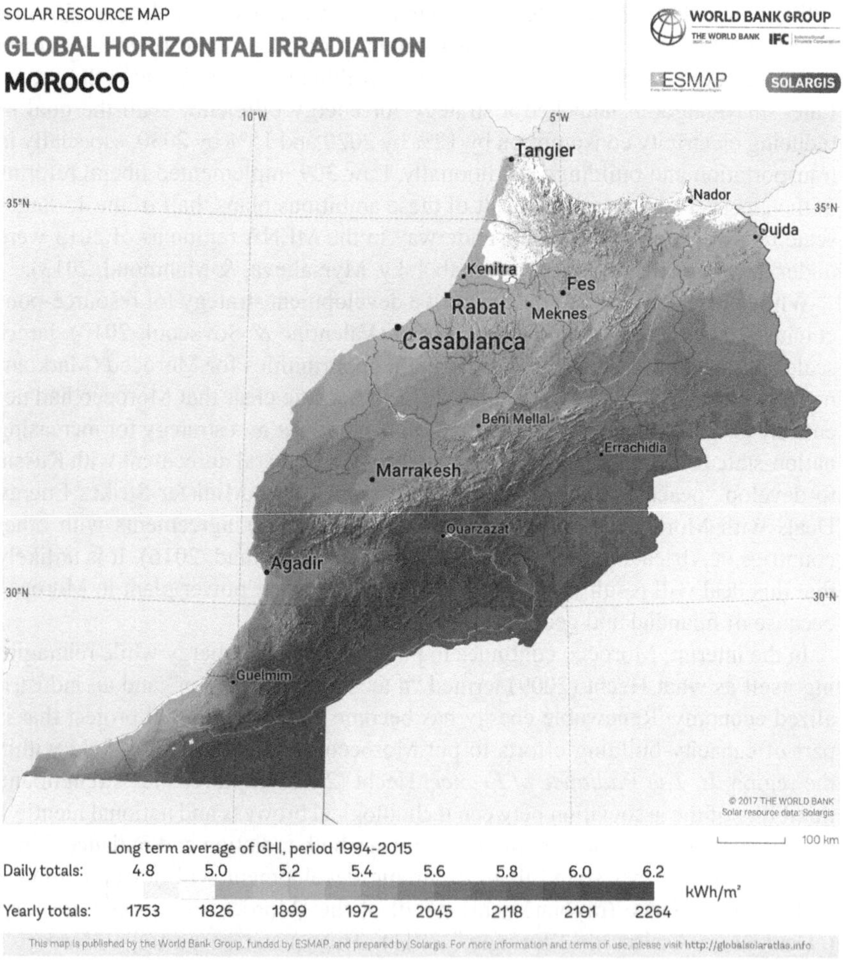

Figure 5.2 Solar irradiation in Morocco measured in kilowatt-hours per square meter

Source: SolarGIS, Creative Commons license

© 2015 GeoModel Solar

culture, science and technology, knowledge and discovery" (Hocke, 2014, para. 8). Some interviewees referred to Morocco's "*gisement solaire*," which literally means Morocco's solar deposits, using terminology that would usually be applied to fossil fuels to describe renewable energy. These renewable energy deposits would be harvested from what in Chapter 3 I termed "renewable energy terroir." The discourse is similar for wind energy. The king furthered this emphasis on the development of Morocco's wealth of sunlight when he bestowed an Arabic name, Noor, which means "light," upon Morocco's CSP plants. The Moroccan state is constructing alternative energy wealth to complement alternative energy technology.

This discourse also discusses renewable resources as having latent societal benefits in need of "valorization." To valorize a resource in this context means to unlock the socioeconomic benefits of the resource through commercialization and technological development. These benefits include fostering national pride, developing rural areas, and profiting from exports. In 2013, the new Moroccan minister of energy said at the Dii conference that the ministry's objective is "to valorize our solar energy resources." In 2016, MASEN cited the "valorization of solar resources" as its top goal, stating on its website that "MASEN was founded in 2010 with the mission of valorizing all of the aspects of solar resources." Moreover, the *Maroc 2030* energy scenario study refers to the "valorization of the privileged geographic position of the country for regional integration" through the Maghreb-Europe Gas pipeline and the Morocco–Spain–Algeria grid (p. 50). Energy efficiency is also viewed as a resource to be valorized, although one that is valued less because it cannot be exported. Other interviewees discussed the valorization of Morocco's human resources through developing entrepreneurship initiatives that spur innovative energy solutions and increasing employment opportunities.

The Moroccan Foundation for Advanced Science, Innovation and Research (MAScIR), a research organization developed as part of efforts to reform Morocco's research and development (R&D) sector, employs a director of business development and valorization. According to MAScIR's former director, the trade winds that brought Morocco phosphates also created plump algae that can be valorized to produce biofuels. He explained that the process for making biofuels from Moroccan algae is an example of a "Champions League" technology, or a technology that produces a product that will compete abroad.[4] These technologies have their own technological specifications, producing fungible fuels from algae that could power any car; they also have social specifications, avoiding genetically engineering the algae so the fuel will be socially acceptable in Europe where genetically modified crops have faced stiff resistance. MAScIR is also designing "local competitor" technologies, such as at-home breast cancer and leukemia testing for rural areas and quality-control testing that will allow small producers to certify their products for export. In this context, "local" means dedicated to the Moroccan market or similar markets in sub-Saharan Africa. The director stated, "It's big science but designed to be cheap and designed to be easy to use by cooperatives." This reflects the dual goals behind valorization: to develop a high technology to facilitate industrialization and to develop appropriate technologies to spur socioeconomic development.

As efforts grew to frame renewable energy resources as energy wealth, renewable energy development became a symbol of national pride in Morocco. Examples of this appear throughout the country. Centers of transportation advertise Morocco's commitment to renewable energy to tourists and citizens alike. Travelers waiting in the airport security line in Rabat and Casablanca gaze at signs that report how much of the airport's energy is coming from solar PV panels at a given moment, as well as the avoided carbon dioxide emissions (see Figure 5.3). While waiting in the customs line in the airport in Casablanca, tourists and business travelers watch a video that shows a man holding the sun in the palm of his hand and then a young girl, with the dawn behind her, saying, "We need new sources of energy to improve quality of life." In another example, there is a free solar charger in front of the train station in Marrakesh where travelers can charge their electronic devices (see Figure 5.4).

Renewable energy policy in Morocco is about visions of a high-technology industrial future but it also shows deference to Morocco's rich past. The documents that lay out Morocco's sustainable development strategy emphasize the protection and valorization of Morocco's natural and cultural patrimony. Renewable energy is becoming enrolled in Moroccan culture in symbolic ways that weave renewable energy technology into traditional economic and religious activities. For instance, SIE, Morocco's energy investment fund, has worked with

Figure 5.3 Solar counter in the Casablanca Airport

Photo by author

Figure 5.4 Free solar charger in Marrakesh
Photo by author

the Moroccan Ministry of Islamic Affairs to promote energy efficiency measures in 15,000 of the country's roughly 50,000 mosques. Morocco is the first Islamic country with a "green mosque" label. It aims to give 600 mosques this label, which would include efficient lighting and/or solar energy and heating, cutting energy use by 40% in each mosque (Nelson, 2016). Integrating renewable energy into mosque architecture and *Khutbah* (sermons) offers an effective communication tool to educate the public about the importance of renewable energy and energy conservation. The government has also improved the efficiency of 2,000 hammam fires (Haut Commissariat au Plan, 2006). The hammam is a public bath, traditionally heated with firewood, which serves important social functions, including as a meeting place for women, a place for women to find wives for their sons, and a place to become spiritually and hygienically clean. It also serves a practical function because some households lack showers. Morocco has also improved the efficiency of 100 pottery kilns—pottery is a traditional economic activity (ibid.). Tying renewable energy and energy efficiency to important aspects of religion and traditional cultural activities sends a powerful educational and symbolic message about the importance of renewable energy and energy efficiency. This pairs energy sustainability with the cultural sustainability that the students at the university in Marrakesh emphasized.

New currency has also promoted Morocco's renewable energy efforts. "Images printed on money support the production and maintenance of national narratives," and currency "creates a link between the state's political identity project and its citizens" (Raento, Hämäläinen, Ikonen, & Mikkonen, 2004, pp. 929–930). The design of currency is one of many processes of "banal nationalism" that influence the ways in which the nation-state is imagined (ibid., p. 930). While previous Moroccan currency featured pictures of dams—then Morocco's charismatic centralized technology—the new Moroccan 100 dirham bill pictures wind turbines in the desert. A Moroccan interviewee who runs the Sahara Wind company stated that

> the previous bill showed a hydropower dam, agricultural activities, and women sorting through oranges, which represented Morocco's main economy. . . . So it's just to tell you this is essentially the building of a nation, the building of a new economy. And Moroccans will see that every day; . . . they'll do business with that.

This quotation emphasizes that the state's renewable energy strategy will be imprinted on the psyche of citizens because they see currency every day.

Although wind turbines have replaced traditional agricultural activities on the 100 dirham bill, the foreground of the drawing depicts one of Morocco's traditional activities: desert caravans. These wind turbines are not sited in a desert devoid of people but are grafted onto a desert caravan and a desert nomad tent. The caravans are in the foreground, and the technology is in the background. The caption on the bill refers to the Moussem de Tan-Tan, which is a festival celebrating the culture of the nomadic tribes of the Sahara. UNESCO has recognized the festival for its role in celebrating the immaterial cultural patrimony of humanity. Urbanization and rural exodus have disrupted these nomadic populations' way of life. According to UNESCO, the Moussem de Tan-Tan is held for the nomadic populations "to protect their way of life, . . . their know-how and traditions" ("Moussem of Tan-Tan," n.d.). On the 100 dirham bill, the Moroccan central bank also alludes to Morocco's long history as a state by referencing this celebration of the past and how it relates to the present.

The building of a nation through this currency gestures toward a past to be sustained as much as it gestures toward a renewable energy future to be built. While scholarship on tradition and modernity in Morocco often treats the categories as temporally fixed (e.g., Sater, 2010), it is unclear what is traditional and what is modern in the picture on the bill. The emerging sociotechnical imaginary of energy calls into question what the slippery concept of modernity means for Morocco. Do the caravans and the nomads represent Morocco's past and wind its future? Or does wind represent a technology that is more compatible with the rhythms of nature and is therefore more compatible with Morocco's patrimony—a technology that is more traditional than modern, in its reflection of *natura naturans*, or the natural temporal rhythms of nature? Could wind power help to sustain traditional nomadic livelihoods in Morocco by reducing environmental degradation?

Or does it risk degrading the landscape in which nomadic populations have lived for centuries? These questions are for Moroccans to answer, but I call attention to them because they risk being lost in discussions about sustainable energy that are limited to technology, capitalism, and carbon inventories.

Morocco's energy imaginary does not necessitate that national pride and wealth come from renewable energy to the exclusion of other sources. Moreover, the coherence of the imaginary is not founded upon the exclusive pursuit of green energy as a climate change mitigation strategy but rather upon the pursuit of socioeconomic development and national energy wealth. Morocco continues to explore for oil off the coast of Western Sahara. It has discovered large reserves of shale oil near Timahdit and Tarfaya, which the government has indicated will be part of its long-term energy strategy (presumably for transportation, not electricity). As discussed earlier, the nuclear option is still on the table. Fossil fuels have always been a part of the Moroccan energy strategy, but they have not received the attention in the national performance of energy policymaking that renewable energy has, first through dam building and now through large-scale solar and wind power. For all its ambiguities, the blending of the past, present, and future is an integral part of Morocco's national imaginary for sustainable energy development, and this has been overlooked in regional-scale visions for solar power. The imaginary highlights what renewable energy could offer Morocco that an offshore oil discovery, a shale play, a continuation of a colonial dam-building policy, or a French nuclear plant could not. What it could offer is not energy independence alone, as shale oil does, but the energy sovereignty necessary to grapple with questions of development and an opportunity to incorporate social considerations into a sustainable development conversation that is too often dominated by natural resource exploitation.

The socioeconomic aspects of Morocco's renewable energy imaginary

When King Mohammed VI inherited the throne from his father in 1999, his ascension fostered hope for a better future in Morocco. The new king was nicknamed the King of the Poor for his focus on social inequality. He renewed the country's focus on human development, improved public services, and instituted a Truth and Reconciliation Commission to investigate and to rectify the human rights abuses of the past. He furthered economic liberalization under the theme of "new public management" in order to increase the state's efficiency and rationalization and its provision of public goods (Catusse, Vairel, & Bono, 2010). His slogans—"a new concept of authority" (*un nouveau concept de l'autorité*) (ibid.) and "the New Morocco" (*le nouveau Maroc*)—reflected both economic liberalism and an increased focus on marginalized populations and women's rights (Lamrini, 2006).

Seven years after King Mohammed VI took the throne, Moroccan author Rida Lamrini cynically asked whether there is a future for Morocco, or rather whether citizens can imagine a future for Morocco. Lamrini stated that this is literally a question that Moroccans inhabit, because the dream of young people often is to leave the country rather than to stay and imagine and construct a better future. His conversation with a young Moroccan woman in the book captured the anomie of youth subjected to inequity of opportunity. He criticized the country's elites stating, "Politicians are discredited. Short of ideas. With neither a project for society nor a mobilizing workforce project. The elites are conspicuously silent, indigent" (ibid., p. 18).[5] On February 20, 2011, thousands of young people marched in the streets of Morocco demanding governmental reform in what was called the February 20th Movement, part of the "Arab Spring" across the MENA region. The government soon adopted a new constitution and held parliamentary elections. It also adopted a "National Charter for the Environment and Sustainable Development," and Article 152 of the new constitution extended the purview of the Economic and Social Council to environmental affairs (Kingdom of Morocco, 2012). The constitution increased the power of the prime minister as well as the separation of powers between the Moroccan Parliament and the king, but the "inviolable" right of the king remained.

Those guiding Morocco's renewable energy policy hope renewable energy will become the mobilizing project enabling society to imagine a bright future for Morocco. Previously, I discussed how renewable energy has become a project of national pride threading together Morocco's future and its past. In this section, I will explain more concretely how solar energy policy has been framed in relation to Morocco's development challenges. Examining energy policy and energy systems makes visible Morocco's core societal challenges and goals, as Moroccans continue to transform their electricity system from one focused on unsustainable colonial resource extraction to one offering sustainable fulfillment of state and local needs. A sustainable energy strategy must address the social pillar of sustainability, not just the environmental one—a requirement that is sometimes obscured in international discussions of global climate change mitigation.

The entity guiding sustainable energy development in Morocco is MASEN, which was founded in March 2010 as a limited-liability company to put solar technologies on a level playing field with other energy technologies. MASEN is an example of what Victor and Heller (2006) call a "dual-firm," which is a company managed as a private enterprise but partly state-owned. This state ownership provides the political connections necessary to develop viable projects within a centralized state system, although Komendantova (2016) has argued that public-private partnerships do not alleviate all governance risks associated with renewable energy scale-up in the MENA region. MASEN is owned equally by the state utility company, ONEE; the Hassan II Fund; and the state-run energy investment company, SIE.[6] MASEN's initial, ambitious primary objective was to build 2 GW of solar generation capacity by 2020, resulting in 14% of total generation capacity from solar. Morocco's renewable energy plan would develop 42% renewable generation capacity overall. As plans advanced for pursuing renewable energy as a sustainable development strategy for Morocco, MASEN took over wind development from ONEE. It changed its name to the Moroccan Agency for Sustainable Energy (rather than the Moroccan Agency for Solar Energy); the Ministry of Energy, Mines, Water, and the Environment (MEMEE) meanwhile changed its name to the Ministry of Energy, Mines, and

Sustainable Development (MEM). After Morocco was selected to host the COP 22 international climate change conference in 2016, MASEN increased its goal to 52% of electricity capacity from renewable energy by 2030, including 10 GW of solar and wind power.[7]

Although MASEN will test some large-scale solar PV panels, its main focus is on CSP, which it plans to develop at five sites: Ouarzazate, Laayoune, Boujdour, Midelt, and Tata.[8] MASEN runs a competitive bidding process for private companies; these companies must provide 20% of the project's capital (e.g., 130 million euros or US$177 million for the first phase, Noor I). The first three solar contracts have been awarded to NOMAC, a subsidiary of ACWA Power. ACWA Power is a Saudi power and water utility company, founded in 2004, which works in 12 countries; in 2016 it had a net income of 214.4 million (ACWA Power, 2016). It is also a shareholder of Dii-MENA. Construction has started on a 580 MW solar power plant in Ouarzazate, set to be the largest in the world. The first phase of the Ouarzazate power plant, called Noor I, came online in January 2016. It is a parabolic trough power plant with three hours of molten salt storage to allow for electricity generation after sunset. Noor II is also a parabolic trough plant, but with seven to eight hours of storage. Noor III will be a CSP trough plant, also with seven to eight hours of storage, followed by Noor IV, a small demonstration of large-scale solar PV panels. While CSP is more expensive than solar PV, it is more compatible with Morocco's overall electricity system, as it generates electricity during Morocco's peak hours of demand in the evening. More importantly, I argue that CSP more closely matches Morocco's socioeconomic goals within its renewable energy imaginary than does solar PV.

The goals of Morocco's renewable energy strategy are socioeconomic more than environmental. An interviewee from the Moroccan electricity sector stated, "Energy is not a good in and of itself but rather a means of developing a national industrial fabric through renewable energy development." In addition to meeting increasing energy demand, MASEN's "integrated solar development" strategy incorporates ambitious ancillary goals to promote the social and economic sustainability of solar power. These ancillary goals are rarely part of the conversation about regional energy integration and renewable energy scale-up but are essential to Morocco's energy imaginary. They include increasing FDI, advancing industrialization, addressing rural-urban inequality, improving equity of opportunity and jobs for young people, and growing an R&D system.

Morocco does expect to reduce its CO_2 emissions by 3.7 million tons per year through the Moroccan Solar Plan, which MASEN emphasizes is a noble and voluntary effort made when some nations responsible for the problem are implementing fewer mitigation initiatives. MASEN is proud that it is developing the technology to do the right thing whereas some other nation-states are not. Moroccan electricity agencies largely view climate change as a strategic opportunity for attracting finance and demonstrating Morocco's prowess in renewable energy technologies. For example, the Moroccan minister of energy stated at the 2013 Dii conference that "the carbon emissions market is an opportunity for Morocco to finance clean projects."[9] Additionally, Morocco gained recognition and prestige as the host of the 2016 COP 22 conference. However, the outcome—in terms of reduced carbon dioxide emissions—of a solar power strategy focused on socioeconomic gains is the same as it would be for a strategy focused only on environmental benefit. This is an important observation for decision makers who seek to spur low-carbon energy development in lower- and middle-income countries. National pride and socioeconomic development are carrots, whereas many low-carbon development policies, such as a carbon tax, are sticks. The following sections discuss Morocco's socioeconomic goals.

Meeting increasing demand

Morocco needs to increase its electricity generation capacity to improve quality of life for a burgeoning population and to meet the needs of industry. Morocco's self-described energy crisis is characterized by its lack of fossil fuel wealth coupled with skyrocketing energy demand. According to an interviewee from MASEN, Morocco imported 95% of its total energy in 2013 and 99% of its fossil fuels. As electricity access increases and Morocco's GDP grows, demand projections indicate that Morocco must at least double its installed generation capacity by 2020 and double it again by 2030. Electricity consumption has increased between 6% and 8% per year over the past 10 years and is projected to continue to do so. Growth in the residential sector has caused a greater increase in evening peak demand, which can be supplied by CSP molten salt storage.

Morocco's rural electrification rate, which is among the highest in Africa, has improved quality of life and increased electricity demand. It grew from 65% in 2002 to 86% in 2006 to 95% in 2008 (World Bank, 2009) to over 99% in 2017 (World Bank, 2017).[10] Morocco's total installed electricity generation capacity was 6.2 GW in 2009 and 8.2 GW in 2016[11] (compared to 75 GW of installed capacity in California and 177 GW in Germany) (ESMAP, 2011; Hamane, 2016).

Until recently, electricity was produced mostly from fuel oil and coal, with some hydroelectricity and natural gas. In addition to implementing its solar and wind goal of 10 GW, Morocco is quietly building a 700 MW extension of its Jorf Lasfar coal-fired power plant and a new 1.38 GW coal-fired power plant in Safi to be finished in 2018 (Hamane, 2016). Both locations manufacture fertilizer that contains Moroccan phosphates. It built two 300 MW (.3 GW) oil-fired power plants, one in Kenitra and one in Mohammedia that will be transitioned to .4 GW natural gas combined cycle power plants (ESMAP, 2011). Additionally, Morocco plans to develop 520 MW of hydroelectricity, 54 MW of microhydroelectricity and 950 MW of pumped hydroelectric storage by 2020, although the percentage of hydroelectricity generation capacity will go down because of an overall increase in capacity (Choukri, Naddami, & Hayani, 2017). It is largely generation capacity from fuel oil that will be retired, while more renewables, natural gas, and coal will be added as the installed capacity increases to 25 GW by 2030, with a little over half of that capacity from renewables. Figures 5.5 and 5.6 compare 2009 capacity to the 2030 goal. The consumption of renewable energy will be smaller than the capacity. If Morocco's targets are met, its renewable electricity proportion will exceed the renewable energy goal under the EU's 20-20-20 targets.

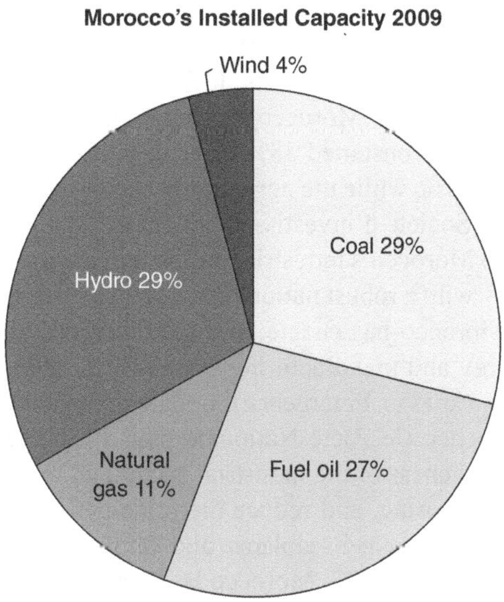

Figure 5.5 Morocco's installed generation capacity, 2009, 6.2 GW capacity

Data source: MEMEE, chart by author

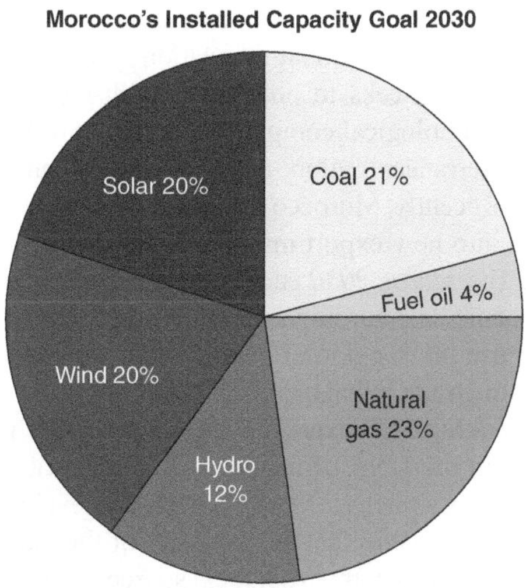

Figure 5.6 Morocco's installed generation capacity goal, 2030, 25 GW capacity

Data source: ONEE employee; (Hamane, 2016); chart by author

Energy as part of industrial policy

Despite Morocco's advantageous geographical position between Europe and Africa, it has not yet developed significant export industries except for phosphates, which suffer from price volatility. France and Spain remain Morocco's largest commercial partners, with 53% of exports going to the EU (Conseil Economique, Social et Environnemental du Maroc, 2014; Zisenwine, 2012). Morocco's trade deficit increased from $1.2 billion in 2001 to $15.5 billion in 2013, owing partly to increasingly costly fossil fuel imports. It fell to $10.5 billion in 2016, as fossil fuel prices bottomed out ("Morocco Trade Summary Data," n.d.; Zisenwine, 2012). In 2010, industry consumed 38% of electricity, households 33%, and the transportation sector 18%, while the agricultural sector used only 6% and the service sector used 5% (Société d'investissement énérgetiques, 2010).

Recent changes to Morocco's industrial policy reflect its ambitions to become a knowledge economy with a robust national innovation system. To grow and diversify its economy, Morocco has enacted a comprehensive industrial policy, particularly in the energy and manufacturing domains. In 2005, Morocco launched the Emergence Plan (le Plan Emergence), updated in 2009 to the National Pact for Industrial Emergence (le Pacte Nationale pour l'Emergence Industrielle) to advance the competitiveness of the industrial sector and to increase GDP growth, add jobs, increase offshoring, and reduce the trade deficit. These policies led to the Renault vehicle factories in Casablanca and Tangier and the Tanger-Med container port, which opened in 2007. Morocco later updated the policy to the Industrial Acceleration Plan 2014–2020 (le Plan d'Accélération Industrielle) to increase the industrial sector's contribution to GDP from 14% to 23%. Along with the industrial emergence plans, the Moroccan Innovation Strategy, led by the Ministry of Industry, Investment, Commerce and the Digital Economy (Ministère de l'Industrie, de l'Investissement, du Commerce et de l'Economie Numérique, or MCINet), seeks to improve its national innovation system. According to an MCINet presentation, MCINet seeks to bolster research-industry collaborations, to improve Morocco's technological competitiveness, to turn Morocco into a center for R&D in the Mediterranean region, and to provide financial mechanisms that promote innovation. Recently, Morocco has sought to become an economic leader in Africa and to tap into new export markets in sub-Saharan Africa as it rejoins the African Union. The *Maroc 2030* energy scenarios depicted a future in which Morocco would become an emerging economic power similar to China or India, projecting it would use the "engine of export" and comparative advantage from low wages to offset the trade imbalance.

An interviewee from MASEN explained that creating new industrial opportunities for Morocco is "in the DNA of the Moroccan renewable program." As part of its "integrated projects strategy," MASEN seeks to develop renewables-related industries in Morocco to ensure that the majority of the materials (e.g., mirrors, steel, turbines) for CSP plants come from local sources. For the first phase of Noor in Ouarzazate, the government required that 30% of the plant's materials come from local content. From a private sector perspective, this increases the projects' risk because they are dependent on fledgling enterprises, which could result in schedule slippage and cost overruns. However, at the 2013 Dii conference, the CEO of ACWA Power argued that using local suppliers allowed the company to decrease its costs and its bid on Noor-Ouarzazate. He emphasized his stake in Morocco's successful industrialization stating, "I need to pray every day for the wealth and success of Morocco so I get my money back" from investment in the Noor-Ouarzazate plants. ACWA Power's bid for Noor III increased local content to 35%. Although this model of development is not without drawbacks, it differs from the neocolonial model in which foreign companies build power plants without providing local benefits or technology transfer.

Capital needs, foreign direct investment, and subsidies

To expand its electricity sector, Morocco requires an infusion of at least US$30 billion in its electricity system by 2030 (Hamane, 2016).[12] Local banks could provide part of the necessary financing, but they would concentrate all of their risk in the energy sector (Haut-Commissariat au Plan, 2006). Therefore, Morocco requires FDI in energy, but historically it has had limited success in attracting FDI (ibid.). Much of the necessary capital is in Asia (ibid.). Investors' perceptions of risks stemming from corruption and government inefficiency have impeded renewable energy project investment in North Africa (Komendantova, Patt, & Williges, 2011). In the World Bank's Ease of Doing Business rankings, Morocco rose from 130th in 2009 to 71st in 2015 to 68th in 2016, but in the Corruption Perceptions Index it fell from 80th in 2014 to 90th in 2016 (of 176 countries). In April 2014, Morocco called on Japan to follow through on its 2013 promise to provide $2 billion in financing for African energy projects (Alternative Energy Africa, 2014). Morocco has secured investment for coal-fired power plants from China, Japan, and South Korea from 2013 onwards (Zawya, 2014; "Organization: China Electric Power Equipment and Technology Company," n.d.).

Thus far, Moroccan CSP plants have been mostly financed by grants and loans from multilateral investment banks and ACWA Power. The Moroccan government, and, in small part, private banks provided the rest. Development banks provide concessional financing at an interest rate of 3.5% rather than the 8% rate charged by private financiers; according to MASEN, this is expected to reduce the price of the electricity by 20% in the long term. According to an interviewee from MASEN, electricity from Noor I costs 18 cents per kWh; this is expected to come down to 14 cents per kWh by the second phase owing to learning effects, despite the cost increase of switching to air cooling to reduce water usage, as well as expanding the amount of storage. The Moroccan government will not enact a feed-in tariff, but it will subsidize the difference between the price of electricity from CSP and the grid price. Wind, which is much cheaper than CSP, has received more private financing. France's EDF (a utility company) and Japan's Mitsui are financing the wind power plant under construction in Taza in northern Morocco.[13] Additionally, Siemens has built a wind turbine manufacturing facility in Morocco. In order for renewable energy to be economically sustainable, future investment in CSP plants by private industry will be necessary.

The need for capital coincides with financial stress in Morocco's energy sector. According to interviewees, some newly electrified villages cannot pay their electricity bills, leaving ONEE insolvent and unable to pay all of its contractors. By the summer of 2014, ONEE was in debt over US$6.2 billion (51 billion Moroccan dirhams, or MAD), and the government signed a plan to raise US$4.6 million (45 million MAD) in capital by 2017 for ONEE (Agueniou, 2014).

Moroccan energy policymakers must also reduce fossil fuel subsidies. The Moroccan government has been under pressure from development banks to eliminate the subsidies to reduce public debt (60% of GDP in 2012) and to maintain Morocco's credit rating (Verme, El-Massnaoui, & Abdelkrim, 2014). Macroeconomists argue that energy subsidies are regressive and inequitable because they largely benefit the middle and upper classes in Morocco that use more energy, particularly given that few Moroccan households can afford a vehicle (ibid.) However, poor households depend on the subsidies to afford energy, and eliminating subsidies would disproportionately affect their livelihoods. For years, citizens protested when the Moroccan government announced decreases in subsidies. For example, in April 2014, trade unions took to the streets of Casablanca to protest the reduction in energy subsidies and the end of the wage increases that had followed the Arab Spring ("Morocco's unions protest austerity plans," 2014). During the summer of 2014, the Moroccan government eliminated industrial fuel, gasoline, and butane subsidies while providing 14 billion MAD to subsidize electricity rates over four years (Agueniou, 2014). Electricity rates were not projected to change for households that consume 100 kWh or less of electricity per month, but were projected to increase for those that use more: presumably wealthier households (ibid.). Because of changes in subsidies and a cash infusion from the government, ONEE's financial health has improved, with profits reported in late 2017 (Telquel, 2017).

Bolstering research and development for social outcomes

While in the 1990s energy development was left mostly to multinational firms, such as former Dii shareholder ABB, today Morocco seeks to advance its domestic R&D capacities. One of MASEN's goals is to conduct R&D to create "disruptive technologies" that reduce the cost of solar power. Morocco could currently build cheaper solar PV plants, but government agencies see more opportunities to become a global leader in CSP R&D than in solar PV. According to an interviewee, energy policymakers aim to increase Morocco's R&D capacities to foster a competitive sector once Morocco's low-cost labor advantage is expended through economic development.

A number of systemic challenges in Morocco's R&D sector must be addressed in order to advance Morocco's national innovation system. The head of the country's energy investment fund stated that the country's R&D sector was "sleeping" until recently. In 2008, Morocco's R&D budget was only 800 million MAD, or US$85 million.

According to interviewees, the structure of Morocco's higher education system reflects the French system's focus on basic research rather than the German system's focus on applied vocational skills. The Institute for Research on Solar Energy and New Energy Technologies (l'Institut de Recherche en Energie Solaire et Energies Nouvelles or IRESEN) was founded in 2011 to increase applied energy R&D. Its head explained in an interview that basic research is a luxury that the French can afford since France's industrial sector conducts applied research. Since Morocco does not have a well-developed industrial sector, focusing only on basic research in universities does not serve Moroccan interests. "Morocco is above all today a technology consumer," he said. "We always have a tendency to copy-apply, copy-apply. And that is dangerous."[14]

IRESEN has built a Green Energy Park in Ben Guerir, a phosphate mining town near Marrakesh, to study the heat-transfer fluid used in CSP, the linear Fresnel reflector, and solar desalination. The researchers are also conducting niche research on solar PV panels. Researchers I interviewed at the park explained that they aim to design a Moroccan, or Sahara Desert, solar panel, with anti-dust and anti-soiling properties that would work well in Morocco's climate and would be marketable across the region. Additionally, by developing different kinds of solar technologies in Ouarzazate, MASEN will identify Morocco's specific CSP R&D needs. Such efforts begin to move Morocco away from the "copy-apply" model of innovation and toward capacity building to become a true collaborator in international R&D projects and a competitor in the international solar energy industry.

Renewable energy policymakers also seek to increase R&D capacities in universities and to develop more vocational and industry-specific training programs. Morocco has 15 public universities with science and technology departments, with 6,000 professors working in science and technology. These universities host 31 educational programs in renewable energy, encompassing 150 researchers active in renewable energy, with 65% working on solar energy and 25% working on wind, biomass, or hydroelectricity (Barhdadi, 2013).[15] As of 2013, these research units had produced 600 publications and were working on 35 international R&D projects (ibid.). New opportunities are under development, including the energy and sustainability program at the International University of Rabat, founded in 2010 by a professor who returned from living abroad. Additionally, the Moroccan government has launched the Institute for Vocational Training in Renewable Energies and Energy Efficiency in Oujda, Tangier, and Ouarzazate with support from the German government.

Despite these efforts, one interviewee lamented that foreigners have been hired to teach courses in new programs because of insufficient local expertise. Morocco suffers from a brain drain as hundreds of thousands of Moroccans with graduate degrees currently live abroad, in what has been called the "Moroccan research diaspora" (Boukharouaa et al., 2014). According to interviewees, there are numerous disincentives to return home as an academic researcher. For instance, any professor, no matter how accomplished, would have to start as an assistant professor at the lowest salary of 12,000 MAD per month (US$1,449 per month). The renewable energy sector has recruited experts home, such as the head of Morocco's energy investment fund, who was working in the space and telecommunications sector in France before the king asked him to return. MASEN has also recruited engineers living abroad back to Morocco to work on solar energy.

Another challenge in the R&D sector is the lack of competitive research funding and university autonomy. According to interviewees, until recently, research funding was shared evenly among the existing research centers and was sufficient for conferences but not for systematic research. Moreover, research teams often do not receive their funds until the end of the project. Universities lack autonomy. The Ministry of Economy and Finance evaluates each line item public universities spend, making it difficult to gain approval for purchasing laboratory equipment. In 2000, Law 0100 gave universities greater autonomy on paper, but it has not been enforced. To address this, IRESEN funds competitive energy research proposals from research teams that include international collaborators. In summary, these efforts in industrial policy and R&D amount to major capacity-building efforts to put Morocco on a level playing field with its neighbors to the north, and these efforts could increase Morocco's capacity to achieve positive outcomes from regional energy integration projects.

Rural development

MASEN's renewable energy development targets stimulating economic development in marginalized areas and reducing rural unemployment and the urbanization rate. Morocco's overall unemployment rate was 10.6% at the end of 2017. In fact, the unemployment rate is much higher in cities (14.9%) than in rural areas (4.6%) (Haut-Commissariat au Plan, n.d.-c; African Development Bank, 2016). However, rural employment is more tenuous, and the poverty rate is higher in rural areas. Additionally, women are disproportionately affected by unemployment (African Development Bank, 2016). People are leaving rural areas in droves in search of employment, resulting in an urbanization rate of 62.4% in 2018 as well as the growth of urban slums (Haut-Commissariat au Plan, n.d.-d). A gulf separates the Moroccan elite living in large cities from the poor living in villages, as well as in urban slums where people pirate electricity to gain access (Zaki, 2010). Thus, renewable energy policy in Morocco risks exacerbating uneven development.

Similar to Dii, MASEN sees itself as an "animator, enabler, and catalyst." In contrast to Dii, it sees itself as a catalyst not just in accelerating technology development but also in spurring socioeconomic development in Ouarzazate and the surrounding areas. MASEN is dramatically altering land use in Ouarzazate by consuming a plot of land the size of the city of Rabat in a relatively isolated rural area. Socioeconomic development initiatives are necessary to gain public trust. A MASEN official stated,

> One of the most important aspects of the Moroccan Solar Plan, is the development of . . . the region where we are implementing [the project]. It's a fact that the most suitable parts [of the country] to develop solar energy in Morocco are the places where you have no other industry—no other economic sector that is developed and the region is very poor. . . . We drop a very huge investment in this region, and the local population can be stressed by that.

MASEN built a new road for the villages adjacent to the Ouarzazate CSP plants. Some of the low-skilled renewable energy construction jobs are earmarked for the uneducated workforce living in the Ouarzazate region. MASEN has also undertaken "charity work" in the region, such as providing bicycles, glasses, dentistry, and school buses for local children, as well as supporting the boarding school for girls in Ouarzazate. According to interviewees, MASEN is particularly focused on women and girls in the region who have less access to education and entrepreneurial activities than men. Chapter 7 will explore the implementation of these national goals at the local level.

Youth unemployment and lack of access to education

The MENA region's youth unemployment rates are among the highest in the world, coupled with a rapidly growing youth demographic (ILO News, 2015). Morocco's overall youth unemployment rate was 20.8% in 2015 (African Development Bank Group, 2016). Many of these unemployed youths are educated. While 9.7% of the overall workforce was unemployed in 2015, 24.4% of the educated workforce lacked jobs (ibid.). Morocco's spending on education is high and increasing, at 21.3% of the public budget and 5.9% of GDP in 2014 (Blancher, Fayad & Auclair, 2016). Morocco has not achieved the expected outcomes from this investment; one example of this is Morocco's adult literacy rate of only 69.4% (World Bank, 2012).[16] Additionally, while public university is free in Morocco, a high-quality education and advanced degrees are often out of reach for lower- and middle-class Moroccans.

The Moroccan government hopes renewable energy will improve educational opportunities for young people that lead to jobs in a sector in which young Moroccans will find work and meaning. A young Moroccan participating in frequent protests related to unemployment told a *New York Times* reporter in 2013 that she dreamed of finding a job in Morocco's expanding renewable energy sector (Alami, 2013). Yet many young people are left with uninspiring jobs at call centers in Rabat and Casablanca.

At a 2013 forum hosted by the Moroccan Society for the Development of Renewable Energy (SMADER) on developing training and educational programs in renewable energy, a young woman in the audience expressed frustration that such training opportunities would likely only be available to the wealthy, perpetuating existing disparities. I saw examples of this lack of equity of opportunity throughout my travels in rural Morocco. In the rural area in which the second phase of the Moroccan Solar Plan is underway, I spoke with a farmer who promotes organic methods of agriculture. He saw his college chemistry degree as having been a waste of time, as there were no opportunities for him to apply it. When staying at the girls' boarding school in Ouarzazate, some young women looked to me for information on Morocco's energy system—one stated (in English), "I am a Moroccan girl, but I don't know about energy in Morocco." While this quotation reflects a lack of public awareness of green energy in Morocco, as discussed by Komendantova and Yazdanpanah (2017), it more importantly reflects a lack of equity of opportunity. At the girls' school the year before, two young women studying at a nearby university had asked me what they could do with their educational background in physics and energy. I began to tell them about government agencies and companies that work on renewable energy, and it quickly became clear I had no opportunities to list that they imagined as being attainable for young women from a rural area.

Socioeconomic development as part of Morocco's renewable energy imaginary

The Moroccan Solar Plan exemplifies what Hjerpe and Linnér (2009) refer to as the "Brundtlandian idea that development and environmental protection are mutually reinforcing and are thus regarded as prerequisites for each other" (p. 238). Overall, Morocco's goals for renewable energy reflect a focus on solving socioeconomic problems through a sustainable development strategy, rather than achieving development first through fossil-fuel-intensive strategies and later through environmental sustainability initiatives.

Table 5.1 summarizes the relationship between socioeconomic goals and MASEN's solar development goals. Morocco's level of success in renewable energy development will be measured by its achievement of these socioeconomic goals, rather than by increased electricity generation or environmental benefits alone. Several interviewees explained that without the achievement

Table 5.1 Socioeconomic goals underpinning the Moroccan Solar Plan

Societal goals	*MASEN's role*
Providing reliable energy access for increasing demand and improving quality of life →	Increasing the electricity generation capacity by building 10 GW of CSP, with molten salt storage providing for evening peak consumption
Raising capital for energy development →	Managing a competitive bidding process, securing investment from Saudi Arabia, securing multilateral development bank funding, reducing subsidies
Industrialization →	Increasing local content, R&D, and technology transfer
Improving equity of opportunity, stemming rural exodus →	Instituting renewable energy education programs, siting plants in rural areas and providing socioeconomic development opportunities there, developing fulfilling jobs
Instituting an R&D system for societal benefit →	Developing a knowledge economy
Fostering national pride, securing financing →	Building low carbon dioxide energy systems
Generating green electrons →	Developing an export commodity to Europe, strengthening Morocco's standing in the Mediterranean region (*discussed in Chapter 6*)
Improving energy independence →	Valorizing solar energy resources to frame alternative energy wealth, foster national pride

of these ancillary goals, Morocco may as well generate electricity by means with which it has more experience, namely coal developed by transnational corporations.

CSP's payoff in local jobs is uncertain (Patt, Komendantova, Battaglini, & Lilliestam, 2011), and MASEN's ancillary goals increase the demands on it. The idea that CSP can legislate socioeconomic change is arguably techno-utopian. However, the Moroccan Solar Plan is part of a sociotechnical energy imaginary that explicitly views development and sustainability as mutually reinforcing and interdependent, developing a local vision of global climate change mitigation that is historically and locally situated in a Moroccan conception of "the good life." In the remainder of this chapter, I will further explore the relationship between Morocco's sociotechnical imaginary and citizens' visions of the good life and citizens' space to imagine.

The role of the current generation of citizens in inclusive sustainable development

Sustainable energy policy requires a higher standard than generating electrons for the sake of electrons. Instead, it requires an improvement in people's capacity to secure their human rights, to sustain what they value, and to live out fulfilling lives, fostering what Karlsson-Vinkhuyzen and Jollands (2013) call "deep energy security." A Moroccan energy scholar said in an interview that Morocco's greatest resource is people and its second greatest resource is solar energy. If the country does not valorize its greatest wealth—people—through human development and democracy, then its other resources cannot be valorized. Another interviewee argued that energy development ought to improve citizenship. He stated,

> What is important is to create the conditions for all these people to become citizens. And citizenship for me is critical—very important—to make people aware of that and to all people to offer them the chance to be citizens.

Similarly, the students at Cadi Ayyad University said that sustainable development requires humanity even more than resources in order to make a green energy plan a socially sustainable plan.

In 2016, tensions related to uneven development boiled over in Al Hoceima, in the Rif Mountains in northern Morocco, where people are protesting social exclusion, lack of education, injustice, and humiliation. The death of Mouhcine Fikri, a fishmonger from a village outside Al Hoceima, catalyzed the protests. In October, authorities confiscated Fikri's illegally caught swordfish and tossed the catch in a garbage truck. Fikri crawled into the back of the garbage truck along with his business partners to retrieve the costly fish. According to social media accounts, cell phone videos recorded police officers saying "grind them." The trash compactor was turned on and Fikri's business partners escaped, but Fikri was crushed to death ("Fishmonger's gruesome death sparks protests," 2016).[17] An uproar occurred on social media and a drawing was circulated of Fikri being crushed by the Moroccan flag. Commentators on Twitter asserted that Moroccans were not, in fact, garbage.

Reporters claimed that the Arab Spring had finally begun in Morocco. They compared Fikri's death to the self-immolation of a fruit vendor in Tunisia, which marked the beginning of the Arab Spring, and drew connections to the use of cell phone video and social media in Arab Spring movements. Rallying cries of "we are all Mouhcine!" were compared to "we are all Bouazizi!" in Tunisia and "we are all Khaled Said!" in Egypt. Social media provided a means for citizens to visually attest to a human being's death in a garbage truck, and Fikri's death became a symbol of social exclusion, inequality, and indignity in the Rif and across Morocco as protests spread. The ongoing protest movement, called Hirak (which means "the popular movement") argues for greater equity of opportunity and access to infrastructure, clean water, education, jobs, and health care, as well as decriminalization of small cannabis production (Calvo, 2017; Sakthivel, 2017). Protestors are also demanding the release of people jailed during the protests (Oumlil, 2017). The protest movement's controversial leader is not formally well educated (ibid.), distinguishing Hirak from the main protests in Morocco in 2011, which Robinson and Parmentier (2014) argue did not resonate with the wider society because of these movements' abstract messages about free speech and civil liberties.

Historical context is necessary for understanding Hirak and provides a more accurate framing than the "Arab Spring," or "civil unrest," which is how protest movements for greater democracy, equity, and inclusion are often framed in European discourses on electricity integration (see, for example, CIEMAT, 2015). The Rif is ethnically Amazigh (or Berber—plural *Imazighen*), and has a long history of resistance to centralized rule. The Abd al-Karim al Khattabi family from the Aith Waryaghar tribe in the region of Al Hoceima led a rebellion against the Spanish from 1921 to 1926, a movement that was later violently crushed with French assistance (Miller, 2013). Uprisings occurred from 1958 to 1959 in Al Hoceima as Morocco gained its independence, to protest unemployment that resulted from closing the border with Algeria, where some people from Al Hoceima previously went to work ("Rural and tribal uprisings in post-colonial Morocco," 1999). In addition, citizens demanded the extension of government agricultural development programs to the area, greater access to education, and greater representation in government posts that were being filled by Arabs rather than Imazighen (ibid.). Uprisings occurred again in 1984 in Al Hoceima (ibid.). Protestors in the Rif on February 20, 2011 and afterward have called for greater regional governance, or even autonomy, and protested the marginalization of the periphery (Collado, 2015). Young protestors idolized Al Khattabi, suggesting that there is historical continuity to Riffian resistance (ibid.). While the current king has made efforts to reconcile with the Rif, policymakers in Rabat often frame the Rif as an ungovernable area with high rates of production of hashish and illicit goods. There are no solar projects being planned for the Rif Mountains; there is some wind development in this area near the cities of Tangier and Taza but not near Al Hoceima. Moreover, the Tanger-Med port and Renault vehicle factories have benefited Tangier, rather than surrounding rural areas.

Put differently, the radiance of the Moroccan energy imaginary shines unevenly across the country. Protestors argue that the benefits of infrastructure development have bypassed their region. They do not say "why me?" but rather "me as well!" because sustainable development ought to be inclusive. Being bypassed is not a benign or accidental act, but rather the direct result of a lack of recognition justice and distributive justice. In October 2017, the king fired the director general of ONEE, Ali Fassi Fihri, the architect of early versions of the Moroccan Solar Plan, as well as the ministers of education, health, housing, and professional development, for failing to implement development projects under the Al Hoceima Lighthouse of the Mediterranean program, including desalination and clean drinking water access. *Telquel* magazine provocatively referred to this incident as the short-circuiting of Mr. Energy by Hirak (Telquel, 2017).

Another aspect of this story is the fact that the EU has pressured Morocco to follow environmental regulations that are reflective of sustainability rather than sustainable development. For more than a decade and a half, the Moroccan government has been under pressure by the EU and Western environmental groups, such as Oceana, to prevent driftnet fishing and overfishing of swordfish to preserve these resources for current and future generations. Swordfish are often caught using driftnet fishing, which results in enormous bycatch in the southwestern Mediterranean Sea each year, including thousands to tens of thousands of sharks, dolphins, and sea turtles (Benhardouze, Aksissou, & Tiwari, 2012; Casale, 2011; Tudela, Kai Kai, Maynou, El Andalossi, & Guglielmi, 2005). The Moroccan government banned driftnet fishing in 2003, but this ban was not enforced. In 2010, the Moroccan government promised it would enforce the ban and phase out driftnets. The Moroccan government also agreed to a quota and a ban on catching swordfish from mid-February to mid-March as well as in October and November, which was when police confiscated Fikri's catch (Mokhliss, 2016). The state-run daily newspaper *Le Matin* ran a story shortly after Fikri's death, stating that the Moroccan government had rejected a recent EU request to further reduce the swordfish quota, projecting an image of Moroccan independence from European pressure (ibid.).

Morocco's sustainable development charter and supporting documents emphasize the importance of public participation in sustainable development. A government document on sustainable development states, "[The 2011 constitution] develops an inclusive society where everyone enjoys security, liberty, equal opportunity, dignity and social justice under the principle of correlation between rights and citizenship duties" (Kingdom of Morocco, 2012, p. 8). Three different conceptions of the place of the citizen in sustainable development emerged in 2017. First, a week after Fikri's death, delegates gathered in Marrakesh for the COP 22 climate change conference. A protestor was pictured on *Al Jazeera* and other news outlets with a sign exclaiming, "Welcome to COP 22; we grind people here!" Protestors called for the international climate change delegates to support human rights and sustainable development for citizens, rather than focusing only on greenhouse gas mitigation. This conceptualizes the role of the citizen in sustainable development as protestors who pressure the Moroccan government and the international community to pursue equitable sustainable development.

It illustrates the importance of citizen participation and procedural justice in the framing of problems. Second, prior to COP 22, several news outlets ran stories about environmental volunteer groups—of which there are approximately 2,000 in Morocco (Kingdom of Morocco, 2012)—picking up trash, recycling, and planting trees (Hackwill, 2016). This idea that citizens' role in sustainable development is to be volunteers is similar to the conception of the appropriate role of citizens in sustainable development espoused by the American students I discussed in the introduction to this chapter—that educated citizens will clean up the environment. Third, in February 2017, the Moroccan government held a ceremony to announce the development of an environmental police force that will fine people who violate environmental laws, such as illegal dumping or possession of contraband or dangerous goods (El Masaiti, 2017). This conceptualizes citizens as people in need of discipline for violating environmental rules. It also enforces good environmental behavior before ensuring human development. In part, these different conceptions of how citizenship ought to be enacted in sustainable development and environmental protection call into question how general calls for civil society to be engaged in the process of transnational grid integration could be achieved (CIEMAT, 2015). Komendantova and Yazdanpanah (2017) argue that citizens in Morocco should be better informed about renewable energy so they are "responsible and active citizens" rather than passive consumers. But how responsible and active citizenship is conceived is far from straightforward.

The incongruity between the protection of the marine ecosystem in the Mediterranean Sea and the lack of equity of opportunity to secure livelihoods in Al Hoceima could be interpreted using the perspective of the Moroccan student quoted at the beginning of this chapter, who argued that development must precede environmental protection. Alternatively, it could provide a rationale for a conception of sustainable development that fully intertwines inclusive human development and citizenship rights with environmental sustainability from the start. Amartya Sen's capabilities approach focuses on putting humans first in sustainable development to improve the capabilities of individuals to secure rights and to live fulfilling and enriching lives (Sen, 2009). Drèze and Sen (2015) state,

> If development is about the expansion of freedom, it has to embrace the removal of poverty as well as paying attention to ecology as integral parts of a unified concern, aimed ultimately at the security and advancement of human freedom.
>
> (p. 43)

Human development and environmental sustainability should be approached in a unified fashion, but human development and freedom should be at the core and environmental protection should serve this anthropocentric core.

In contrast to either of these approaches, ecocentrists argue for valuing and protecting the intrinsic value of nature through a non-anthropocentric lens, and they think that sustainable development is merely about development, not environmental sustainability and that the interests of humans and nature diverge.

In contrast, Norton's (1991) pragmatic convergence hypothesis postulates that non-anthropocentric and long-sighted anthropocentric viewpoints converge on the same policy goals because human and nonhuman interests converge in the long run. It is reasonable to assume that the needs of future fishing communities depend on ecosystem integrity and the availability of fish in the Mediterranean Sea, but without putting human development and capabilities first is it possible to reach a point in which people in Al Hoceima can imagine and enact a different future than unsustainable fishing? Without recognition, distributive, and capabilities justice, lack of development seems to be ensured for both current and future generations in Al Hoceima. Protection of nature will then be imposed through top-down authoritarian forms of governance, while inequity of opportunity guarantees that development will not result for those who are most in need of it. By pointing this out, I am not arguing for a strong anthropocentrism, in which natural kinds are valued purely economically and the focus is only on human well-being. Instead, I agree that natural capital ought to be maintained from one generation to the next (see Jamieson, 1998). However, the distribution of access to natural capital for both current and future generations must be considered and cannot be understood without addressing power dynamics in the present. I argue that the interests of disadvantaged social groups will not eventually converge with overall human and nonhuman interests without recognition, distributive, and capabilities justice for currently disadvantaged social groups to chart new paths for their progeny. Short-term (but not short-sighted) anthropocentrism is necessary to chart a future for these populations that differs from their present. Additionally, a conception of sustainable development that focuses on the ability of technological advancements to generate welfare (see Hjerpe & Linnér, 2009) must recognize that new technologies are scaled up within sociotechnical imaginaries that implicate particular views of what sustainable development ought to look like for a nation-state and that exclude particular populations.

Alternative imaginaries, spaces to imagine, as well as time and space to chart new paths that level the playing field for disadvantaged populations in the current generation to be able to create new futures are necessary, and redressing past injustice for these groups is equally important to considering future generations writ large. Otherwise, Lamrini (2006) is right to question whether there is a future for Morocco for social groups that lack inclusion in sustainable development in the present. What Jasanoff and Kim (2015) refer to as a "communal sense of justice" that is produced within a sociotechnical imaginary is observed in Morocco as the idea that everyone ought to benefit from a national sociotechnical project of climate change mitigation that spurs socioeconomic development (p. 10). Yet the benefits of this project are out of the reach of disadvantaged populations in the absence of efforts to redress past and current injustice. Moreover, the effects of distributive injustice stemming from the past and affecting the present are often overlooked in theories of sustainable development and intergenerational justice that reduce future populations to homogenous wholes. An understanding of inequity among social groups was also lacking in the EU's push for sustainable fishing in the Mediterranean Sea.

Citizens in the Rif arguing for inclusion in infrastructure development are demanding to be included in the benefits of the energy imaginary. It seems that it is not technological progress that they are contesting, or even that they are contesting being the losers of technological development—as Chapter 7 will explore in rural Ouarzazate—but rather they are contesting being excluded from the vision of socioeconomic development implicated in Morocco's sustainable development imaginary. In other words, they seem to be demanding inclusion rather than a change of agenda. This is a significant consideration for win-win sustainable development discourses that exclude from the frame not only losers but also those who are bypassed by technological progress. Considering this, the achievement of distributive justice for marginalized social groups in the current generation should be incorporated more clearly into sustainable development goals. Additionally, it is important to acknowledge that it is unclear how the citizens of Al Hoceima would imagine their energy futures if their capability to imagine were bolstered. Ironically, what has been called an imaginary can constrain citizens' ability to imagine. What has been termed an imaginary could implicate a lack of opportunity to imagine and a failure to "open up" debates in the way that Leach, Scoones, and Stirling (2010) advocate in their pathways to sustainability approach.

Conclusion

This chapter covered three different narratives related to Morocco's sociotechnical imaginary: one about how the Moroccan government's imaginary for renewable energy constructed renewable energy wealth and national pride, one about the imaginary's socioeconomic development goals, and one about the importance of inclusive and equitable sustainable development that improves the capabilities of marginalized populations in the current generation so that future generations can chart a course for more environmentally sustainable livelihoods that ultimately benefit both humans and the environment. In the first story, I discussed how renewable energy is being enrolled in fostering national pride, within the context of an energy imaginary that incorporates the country's heritage and historical continuity. More broadly, this story illustrates that the mitigation of climate change by scaling up renewable energy is more likely to succeed on the basis of a collectively held belief that renewables will spur socially sustainable development, than on the basis of a claim that renewable energy technologies are benign and without impacts. In this sense, Morocco provides a model for other countries to emulate. Without the social and cultural shaping of a sustainable energy development strategy, as well as the fostering of national pride, the allure of hydrocarbons, particularly shale gas, is overwhelming. The establishment of renewable energy as a national project illustrates the importance of socioeconomic development to spur the adoption of renewable energy in developing countries, a factor that is often overlooked in discussions of renewable energy development as a global climate change mitigation strategy. The Moroccan sociotechnical imaginary for renewable energy illustrates one case in which solar and wind energy are taken seriously as part of a national project in a developing country. Now that ACWA Power has won the bids for Noor I–III, the Shams I CSP plant has been completed in the United Arab Emirates, and Saudi Arabia is discussing building CSP plants, this sense of pride that is related to CSP technology could extend to other parts of the fossil-fuel-rich MENA region. Several other researchers have noticed this relationship between clean energy and nation-state pride in other countries. Uddin, Taplin, and Yu (2007) argue that environmental protection and renewable energy in Bhutan relate to the preservation of rich traditions and national heritage. Bhutan also lacks energy resources, it is also a monarchy, and its main export is hydroelectricity, to India. Barr (2012) highlights sustainable development and energy as part of China's efforts to develop an image at home and abroad as orderly and prosperous.

The second story examined the sociotechnical imaginary of renewable energy as encompassing a socioeconomic development strategy and industrialization strategy for Morocco, viewing solar power first as a means of addressing socioeconomic challenges and only second as a climate mitigation strategy. The relationship of CSP to national plans for developing a national innovation system, improving R&D, creating jobs, meeting demand, balancing intermittency, and improving industrial policy illustrates the reasons for pursuing an expensive renewable energy technology in a lower-middle-income country. While some critics dismiss this as green capitalism, there is much to admire in Morocco's intentional treatment of energy as socioeconomic. Global climate change mitigation processes that fail to recognize the importance of social aspects and sociotechnical imaginaries may succeed in mitigating greenhouse gas emissions but will not succeed in achieving sustainable development. Morocco is an exemplar of a nation susceptible to climate injustice (Gardiner, 2004). Morocco emits less than 1% of global greenhouse gas emissions but is already suffering from the effects of desertification and water scarcity. In the Moroccan Solar Plan, Morocco has developed a project for climate change mitigation and environmental protection that is designed in a way that puts socioeconomic development at its core.

For actors pursuing regional grid integration as a climate change mitigation strategy, this chapter illustrates that large-scale renewable energy is part of a Moroccan sociotechnical imaginary with locally specific development goals. The sociotechnical imaginary as a concept illustrates the locally specific and idiomatic aspects of a scaled-up nation-state strategy for renewable energy. What would it take to develop a regionally held sociotechnical imaginary for a renewable-energy-based supergrid, or for all individual nation-states to incorporate regional electricity into their national energy imaginaries? It is likely that the energy particularities, needs, and goals of dozens of nation-states would need to be fused to develop a deeply integrated renewable energy supergrid across nation-state lines.

The third story discussed the history of protests in the Rif, which were recently inflamed by the gruesome death of a fishmonger who dove into the back of a garbage truck in an attempt to rescue swordfish that—according to regional environmental regulations—was caught out of season. From the nomad woman in the Sahara Desert to the citizens in the Rif, the lack of recognition of marginalized populations illustrates the importance of inclusivity in sustainable energy development, particularly recognition, distributive, and capabilities justice for marginalized populations in the current generation so that future generations can be on an even playing field. Development and environmental sustainability were integrally interwoven under the Moroccan Solar Plan. However, this story highlights the importance in sustainable development of understanding the temporal continuity of injustice, as well as considering distributive justice in sustainable development approaches that seek to preserve natural capital for future generations.

Notes

1 Translated from French.
2 Sen and Nussbaum have written numerous works on the capability approach to human development, discussed further in Chapter 8. For an overview, see: Alkire, S. (2015.) The capability approach and well-being measurement for public policy. OPHI Working Paper. Retrieved from http://www.ophi.org.uk/wp-content/uploads/OPHIWP094.pdf.
3 Morocco weathered the global economic crisis fairly well, in part by spending 2.2% of GDP on a 2010–2020 stimulus package (ESMAP, 2011).
4 This refers to the soccer (football) tournament organized by the Union of European Football Associations.
5 Translated from French.
6 According to an interviewee, MASEN had 2,250,000 Moroccan dirhams in capital as of 2016.
7 Figures for the percentage of consumption are not yet available.
8 Two of these sites are in Western Sahara.
9 Translated from French.
10 One interviewee indicated that occasionally the rate is calculated on the basis of whether a village has any access to electricity rather than whether each individual household has access. However, an interviewee from ONEE said that the company counts each household.
11 In 2006, ONEE contributed 4.4 GW of capacity and independent power producers contributed 1.8 GW.
12 It is unclear whether this figure is for generation alone, or whether it includes transmission and distribution. Some Moroccan media report figures as high as $42 billion.
13 A wind facility outside Tangier is privately owned by ACWA Power, with a loan from the European Bank for Reconstruction and Development; the Clean Technology Fund; and BMCE Bank, a Moroccan investment bank.
14 Translated from French.
15 Owing to lack of data on R&D and public universities in Morocco, for these figures I relied on unpublished research conducted by interviewees and on a forum I attended on renewable energy education in Morocco.
16 Seventy percent is the average literacy rate for lower-middle-income countries. Measures of literacy in Morocco vary.
17 The police later asserted that the information (spread via social media) that the police officer instructed and paid the driver to turn on the trash compactor was false ("La DGSN fait le point sur la mort d'un citoyen à Al-Hoceima," 2016).

Works cited

ACWA Power. (2016). Annual report. Retrieved from http://acwapower.com/en/stakeholder-relations/annual-reports/

Adam, B. (1998). *Timescapes of modernity: The environment and invisible hazards*. London: Routledge.

African Development Bank Group. (2016). *Morocco country strategy paper 2017–2021*. African Development Bank Group. Retrieved from www.afdb.org/fileadmin/uploads/afdb/Documents/Project-and-Operations/Morocco_-_Country_Strategy_Paper_2017-2021.pdf

Agueniou, S. (2014, May 30). La Compétitivité ne se résume pas au coût de l'énergie. *La Vie Éco*. Retrieved March 19, 2015, from http://www.mem.gov.ma/SiteAssets/Interview/Interview2014/Inter30-05--05-06-14.pdf

Alami, A. (2013, September 29). Gaps in graduates' skills confound Morocco. *New York Times*. Retrieved October 2, 2013, from www.nytimes.com/2013/09/30/world/africa/gaps-in-graduates-skills-confound-morocco.html?_r=2&

Alternative Energy Africa. (2014, April 27). Morocco's energy minister asks for Japanese investment. *Alternative Energy Africa*. Retrieved March 19, 2015, from www.ae-africa.com/read_article.php?NID=5550

American Presidency Project. (2001, October 24). George W. Bush: Message to Congress Transmitting a Protocol to the Morocco-United States Agreement on the Peaceful Uses of Nuclear Energy. Retrieved from http://www.presidency.ucsb.edu/ws/index.php?pid=73438

Apelian, C. (2012). Modern mosque lamps: Electricity in the historic monuments and tourist attractions of French colonial Fez, Morocco (1925–1950). *History and Technology*, *28*(2), 177–207. Retrieved from https://doi.org/10.1080/07341512.2012.694208

Barhdadi, A. (2013, September 13). *Solar energy research in Morocco*. Retrieved March 21, 2015 from http://users.ictp.it/~barhdadi/

Barr, M. (2012). Nation branding as nation building: China's image campaign. *East Asia*, *29*(1), 81–94. Retrieved from https://doi.org/10.1007/s12140-011-9159-7

Benhardouze, W., Aksissou, M., & Tiwari, M. (2012). Incidental captures of sea turtles in the driftnet and longline fisheries in northwestern Morocco. *Fisheries Research*, *127–128*, 125–132. Retrieved from https://doi.org/10.1016/j.fishres.2012.04.002

Blancher, N., Fayad, D., Auclair, G. (2016) Education: Investing in Morocco's future. International Monetary Fund. Retrieved from https://www.imf.org/external/np/blog/nafida/050416.pdf/

Boukharouaa, N-E., Berrada, M., Abdelhak, C., Dinia, S., El Ftouh, A., El Maliki, A., Farah, K., Bennani, I., Elyoussoufi, O., Ouardirhi, Y. (2014). Chapter 8: The Moroccan Diaspora and its contribution to the development of innovation in Morocco. The global innovation index 2014. Geneva: The World Intellectual Property Organization. Retrieved from http://www.wipo.int/edocs/pubdocs/en/wipo_pub_gii_2014-chapter8.pdf

Bouneau, C. (2002). Electrification des chemins de fer et empires coloniaux : De l'Afrique du Nord française jusqu'au Second Conflit. *Outre-Mers*, *89*(334), 135–146. Retrieved from https://doi.org/10.3406/outre.2002.3929

Calvo, O. (2017, July 29). Social unrest boils over in Morocco's al-Hoceima. *Al Jazeera*. Retrieved October 23, 2017, from www.aljazeera.com/indepth/inpictures/2017/07/social-unrest-boils-morocco-al-hoceima-170727090556224.html

Casale, P. (2011). Sea turtle by-catch in the Mediterranean. *Fish and Fisheries*, *12*(3), 299–316. Retrieved from https://doi.org/10.1111/j.1467-2979.2010.00394.x

Catusse, M., Vairel, F., & Bono, I. (2010). *Le Maroc de Mohammed VI: Mobilisations et action publique*. Paris: Karthala.

CIEMAT. (2015). *Bringing Europe and third countries closer together through renewable energies* (BETTER). Intelligent Energy Europe Programme of the European Union. Retrieved from https://ec.europa.eu/energy/intelligent/projects/en/projects/better

Choukri, K., Naddami, A., & Hayani, S. (2018). Renewable energy in emergent countries: Lessons from energy transition in Morocco. *Energy, Sustainability and Society*, *7*(25), 1–11.

Cladé, J., Beltran, A., Marseille, J., Moussa, P., Couvreux, J-P. (2002). L'Électricité outre-mer: Un heritage industriel et humain. Table ronde de conclusions. *Outre-Mers*, *89*(334-335), 601–619.

Collado, Á. S. (2015). Territorial stress in Morocco: From democratic to autonomist demands in popular protests in the Rif. *Mediterranean Politics*, *20*(2), 217–234. Retrieved from https://doi.org/10.1080/13629395.2015.1033908

Conseil Economique, Social et Environnemental du Maroc. (2014). Cohérence des politiques sectorielles et accords de libre-echange: Fondements stratégiques pour un développement soutenu et durable. Retrieved from http://www.ces.ma/Documents/PDF/Auto-saisines/AS_16_2014-Coherence-des-Politiques-Sectorielles-et-ALE/rp-as16-vf.pdf

D'Angio, A. (2002). L'électrification du Maroc vue à travers l'action de la société Schneider et Cie (1907–1954). *Outre-Mers*, *89*(334–335), 317–329.

Davis, D. (2007). *Resurrecting the Granary of Rome*. Athens: Ohio University Press.

Deuxièmes Assises de l'Energie: Intégration industrielle au Maroc, développement économique et perspectives de complémentarité Europe-Maghreb. (2011, June 2). Retrieved May 5, 2014, from www.massolia.com/energie-2/deuxiemes-assises-de-l%E2%80%99energie-integration-industrielle-au-maroc-developpement-economique-et-perspectives-de-complementarite-europe-maghreb/

Drèze, J., & Sen, A. (2015). *An uncertain glory: India and its contradictions*. Princeton, NJ: Princeton University Press.

El Masaiti, A. (2017, February 24). Morocco Launches Environmental Police. *Morocco World News*. Retrieved March 14, 2017 from https://www.moroccoworldnews.com/2017/02/209229/environmental-police-to-be-launched-on-february-23/

Elsheshtawy, Y. (2008). *The evolving Arab city: Tradition, modernity and urban development*. London: Routledge.

Energy Sector Management Assistance Program (ESMAP). (2011, January 20). *Power sector financial vulnerability assessment impact of the credit crisis on investments in the power sector: The case of Morocco*. World Bank. Retrieved from www.esmap.org/node/1128

Fishmonger's gruesome death sparks protests in Morocco. (2016, October 31). *Al Jazeera*. Retrieved December 5, 2016, from www.aljazeera.com/news/2016/10/fishmonger-gruesome-death-sparks-protests-morocco-161031131330359.html

France to help Morocco's nuclear energy drive. (2010). *Expatica France*. Retrieved June 28, 2014, from www.expatica.com/fr/news/local_news/france-to-help-morocco-s-nuclear-energy-drive_80826.html

Frémeaux, J., & d'Almeida-Topor, H. (2002). Rapport de synthèse. *Outre-Mers*, *89*(334), 411–414. Retrieved from https://doi.org/10.3406/outre.2002.3947

Gardiner, S. M. (2004). Ethics and global climate change. *Ethics*, *114*(3), 555–600.

Girard, L. F., & Nijkamp, P. (2009). *Cultural tourism and sustainable local development*. Burlington, VT: Ashgate.

Goldemberg, J., Johansson, T. B., Reddy, A. K. N., & Williams, R. (1988). *Energy for a sustainable world*. New York City, NY: John Wiley & Sons Inc.

Gruber, B., Gunay, C., Rizvan, A., Jamea, E. M., Komendantova, N., & Zejli, D. (2017). *Working paper: Green modernization—The political, social and societal setting of Morocco's solar energy policies*. OIIP. Retrieved from www.ethz.ch/content/dam/ethz/special-interest/usys/ied/climate-policy-dam/docs/pub/wp97green-modernization_background-report.pdf

Hackwill, R. (2016, October 31). Morocco cleans up its act for COP 22. *Euronews*. Retrieved December 5, 2016, from www.euronews.com/2016/10/31/morocco-cleans-up-its-act-for-cop-22

Hamane, T. (2016). *A snapshot of Morocco's power sector*. Retrieved October 19, 2017, from www.energynet.co.uk/webfm_send/2025

Haut-Commissariat au Plan (2006). *Prospective Maroc 2030: Energie 2030 Quelles options pour le Maroc ?* Retrieved from www.hcp.ma/file/111440/

Haut-Commissariat au Plan. (n.d.-a). *Emploi par branche d'activité de la population active occupée au milieu rural*. Retrieved March 14, 2018, from www.hcp.ma/Emploi-par-branche-d-activite-de-la-population-active-occupee-au-milieu-rural_a154.html

Haut-Commissariat au plan. (n.d.-b). *Population du Maroc par année civile (en milliers et au milieu de l'année) par milieu de résidence : 1960–2050*. Retrieved March 14, 2018, from www.hcp.ma/Population-du-Maroc-par-annee-civile-en-milliers-et-au-milieu-de-l-annee-par-milieu-de-residence-1960–2050_a677.html

Haut-Commissariat au Plan. (n.d.-c). *Taux de Chômage selon le milieu*. Retrieved March 14, 2018, from www.hcp.ma/Taux-de-Chomage-selon-le-milieu_a255.html

Haut-Commissariat au Plan. (n.d.-d). *Taux d'urbanisation (en %) par année : 1960–2050*. Retrieved March 14, 2015, from www.hcp.ma/Taux-d-urbanisation-en-par-annee-1960-2050_a682.html

Hecht, G. (2009). *The radiance of France: Nuclear power and national identity after World War II* (new ed.). Cambridge, MA: MIT Press.

Hjerpe, M., & Linnér, B-O. (2009). Utopian and dystopian thought in climate change science and policy. *Futures*, *41*(4), 234–245. Retrieved from https://doi.org/10.1016/j.futures.2008.09.007

Hocke, Z. (2014, October 21). Morocco's first solar power plant to start operating in 2015. *Voice of America News*. Retrieved October 21, 2014, from www.voanews.com/content/morocco-first-solar-power-plant-to-start-operating-in-2015/2490557.html

ILO News. MENA region retains highest youth unemployment rate in the world. (2015, October 8). Retrieved October 20, 2017, from www.ilo.org/beirut/media-centre/news/WCMS_412797/lang-en/index.htm

Jamieson, D. (1998). Sustainability and beyond. *Ecological Economics*, *24*, 183–192.

Jasanoff, S., & Kim, S-H. (2009). Containing the atom, sociotechnical imaginaries and nuclear power in the United States and South Korea. *Minerva*, *47*(2), 119–146. Retrieved from https://doi.org/10.1007/s11024-009-9124-4

Jasanoff, S., & Kim, S-H. (2015). *Dreamscapes of modernity: Sociotechnical imaginaries and the fabrication of power*. Chicago, IL: University of Chicago Press.

Karlsson-Vinkhuyzen, S., & Jollands, N. (2013). Human security and energy security: A sustainable energy system as a public good. In M. J. Trombetta & H. Dyer (Eds.), *International handbook of energy security* (pp. 507–525). Cheltenham, UK: Edward Elgar Publishing.

Kingdom of Morocco. (2012). *Sustainable development in Morocco: Achievements and perspectives from Rio to Rio+20*. Retrieved from https://sustainabledevelopment.un.org/content/documents/1010file.pdf

Komendantova, N. (2016). Renewable energies in the Middle East and North African region: Can private-public partnerships address existing barriers and risks? *International Proceedings of Chemical, Biological and Environmental Engineering*, *91*, 31–37.

Komendantova, N., Patt, A., & Williges, K. (2011). Solar power investment in North Africa: Reducing perceived risks. *Renewable and Sustainable Energy Reviews*, *15*(9), 4829–4835. Retrieved from https://doi.org/10.1016/j.rser.2011.07.068

Komendantova, N., & Yazdanpanah, M. (2017). Impacts of human factors on willingness to use renewable energy sources in Iran and Morocco. *Environmental Energy and Economic Research*, *1*(2), 141–152. Retrieved from https://doi.org/10.22097/eeer.2017.47240

La DGSN fait le point sur la mort d'un citoyen à Al-Hoceima. (2016, October 30). *Le Matin*. Retrieved October 23, 2017, from https://lematin.ma/express/2016/la-dgsn-fait-le-point-sur-la-mort-d-un-citoyen-a-al-hoceima/257065.html

Lamrini, R. (2006). *Y a-t-il un avenir au Maroc me demanda Yasmina*. Paris, France: Marsam Editions.

Leach, M., Scoones, I., & Stirling, A. (2010). *Dynamic sustainabilities: Technology, environment, social justice*. London: Earthscan.

Lélé, S. (1991). Sustainable development: A critical review. *World Development*, *19*(6), 607–621.

Maghraoui, D. (2009). The strengths and limits of religious reforms in Morocco. *Mediterranean Politics*, *14*(2), 195–211. Retrieved from https://doi.org/10.1080/13629390902985976

Marktanner, M., & Salman, L. (2011). Economic and geopolitical dimensions of renewable vs. nuclear energy in North Africa. *Energy Policy*, *39*(8), 4479–4489. Retrieved from https://doi.org/16/j.enpol.2010.12.047

Masco, J. (2006). Chapter 1: The enlightened earth. In *The nuclear borderlands: The Manhattan Project in post-cold war New Mexico* (pp. 1–40). Princeton, NJ: Princeton University Press.

McCool, S. F., & Moisey, R. N. (2001). *Tourism, recreation, and sustainability: Linking culture and the environment* (1st Ed.). Oxfordshire, UK: CABI.

Miller, S. G. (2013). *A history of modern Morocco*. Cambridge: Cambridge University Press.

Mokhliss, B. (2016, November 10). Le Maroc rejette une demande de l'UE de réduire son quota de pêche de l'espadon. *Le Matin*. Retrieved October 23, 2017, from https://lematin.ma/journal/2016/le-maroc-rejette-une-demande-de-l-ue-de-reduire-son-quota-de-peche-de-l-espadon/257832.html

Morocco trade summary data. (n.d.). Retrieved October 19, 2017, from https://wits.worldbank.org/CountryProfile/en/Country/MAR/Year/2012/Summary

Morocco's unions protest austerity plans. (2014, April). *Al Jazeera English*. Retrieved April 21, 2014, from www.aljazeera.com/news/africa/2014/04/morocco-unions-protest-austerity-plans-201446131645914944.html

Moussem of Tan-Tan. (n.d.). Retrieved October 24, 2017, from https://ich.unesco.org/en/RL/moussem-of-tan-tan-00168

Nelson, A. (2016, September 5). Morocco to give 600 mosques a green makeover. *The Guardian*. Retrieved November 6, 2016, from www.theguardian.com/environment/2016/sep/05/morocco-to-give-600-mosques-a-green-makeover

Norton, B. G. (1991). *Toward unity among environmentalists* (1st Ed.). New York, NY: Oxford University Press.

Nurse, K. (2006). *Culture as the fourth pillar of sustainable development*. London: Commonwealth Secretariat Malborough House Pall Mall. Retrieved from http://placemakers.wdfiles.com/local–files/theoretical-analysis-examined/Cultureas4thPillarSD.pdf

Organization: China Electric Power Equipment and Technology Company (CET). (n.d.). Retrieved March 7, 2017, from http://china.aiddata.org/organizations/1665

Oumlil, K. (2017, June 4). Making sense of recent protests in Morocco. *Al Jazeera*. Retrieved October 23, 2017, from www.aljazeera.com/indepth/opinion/2017/06/making-sense-protests-morocco-170604092533766.html

Oxford Business Group. (2012). *The report: Morocco 2012*. Oxford: Oxford Business Group. Retrieved May 4, 2014, from Google Books.

Parmentier, M. J., & Moore, S. (2016). 'The camels are unsustainable': Using study abroad as a pedagogical tool for teaching ethics and sustainable development. *Teaching Ethics*, *16*(2), 207–221. Retrieved from https://doi.org/10.5840/tej2016113038

Patt, A., Komendantova, N., Battaglini, A., & Lilliestam, J. (2011). Regional integration to support full renewable power deployment for Europe by 2050. *Environmental Politics*, *20*(5), 727–742. Retrieved from https://doi.org/10.1080/09644016.2011.608537

Porter, G. D. (2000). The city's many uses: Cultural tourism, the sacred monarchy and the preservation of Fez's Medina. *The Journal of North African Studies*, *5*(2), 59–88. Retrieved from https://doi.org/10.1080/13629380008718398

Raento, P., Hämäläinen, A., Ikonen, H., & Mikkonen, N. (2004). Striking stories: A political geography of euro coinage. *Political Geography*, *23*(8), 929–956. Retrieved from https://doi.org/10.1016/j.polgeo.2004.04.007

Ramana, M. V., & Ahmad, A. (2016). Wishful thinking and real problems: Small modular reactors, planning constraints, and nuclear power in Jordan. *Energy Policy*, *93*, 236–245. Retrieved from https://doi.org/10.1016/j.enpol.2016.03.012

Rivlin, P. (2012). Morocco's economy under Mohammed VI. In B. Maddy-Weitzman & D. Zisenwine (Eds.), *Contemporary Morocco: State, politics and society under Mohammed VI* (pp. 82–90). Milton Park, Abingdon: Routledge.

Roberson, J. (2014). The changing face of Morocco under King Hassan II. *Mediterranean Studies*, *22*(1), 57–83.

Robinson, R., & Parmentier, M-J. C. (2014). The Arab Spring in North Africa: Still winter in Morocco? In N. Agarwal, M. Lim, & R. Wigand (Eds.), *Online collective action* (pp. 197–211). New York City, NY: Springer.

Rural and tribal uprisings in post-colonial Morocco, 1957–1960: An overview and a reappraisal. (1999). *The Journal of North African Studies*, *4*(2), 84–102. Retrieved from https://doi.org/10.1080/13629389908718363

Russian prime minister strikes energy deals in Morocco. (2017, October 11). *Washington Post*. Retrieved October 19, 2017, from www.washingtonpost.com/business/russian-prime-minister-strikes-energy-deals-in-morocco/2017/10/11/b661e576-aecb-11e7-9b93-b97043e57a22_story.html

Sakthivel, V. (2017, August 18). Protests erupt after Moroccan fishmonger dies desperately trying to get back his confiscated catch. *The Washington Post*. Retrieved December 5, 2016, from www.washingtonpost.com/news/morning-mix/wp/2016/10/31/protests-erupt-after-moroccan-fishmonger-was-crushed-to-death-in-garbage-truck/?utm_term=.084cb74e513a

Samborsky, B., Myrsalieva, N., & Mahmoud, M. (2013). *Arab future energy index AFEX 2013*. Cairo, Egypt: RCREEE.

Sater, J. N. (2010). *Morocco: Challenges to tradition and modernity* (1st ed.). London: Routledge.

Saul, S. (2002). L'électrification du Maroc à l'époque du protectorat. *Outre-Mers*, *89*(334–335), 491–512.

Sen, A. (2009). *The idea of justice*. Cambridge, MA: Harvard University Press.

Société d'investissements énérgetiques (SIE). (2010). *Etude conjuncturelle sur l'énergie au Maroc: Les Chiffres Clés*. Rabat, Morocco: SIE.

Swearingen, W. D. (1988). *Moroccan mirages: Agrarian dreams and deceptions, 1912–1986*. London: I. B.Tauris.

Telquel. (2017, October 28). Ali Fassi Fihri, le spécialiste de l'énergie court-circuité par le Hirak. *Telquel*. Retrieved November 3, 2017, from http://telquel.ma/2017/10/28/ali-fassi-fihri_1566160

Thompson, P. B. (2007). Agricultural sustainability: What it is and what it is not. *International Journal of Agricultural Sustainability*, *5*(1), 5–16. Retrieved from https://doi.org/10.1080/14735903.2007.9684809

Thompson, P. B. (2010). *The Agrarian vision: Sustainability and environmental ethics*. Lexington, KY: University Press of Kentucky.

Tidwell, J. H., & Tidwell, A. S. D. (2018). Energy ideals, visions, narratives, and rhetoric: Examining sociotechnical imaginaries theory and methodology in energy research. *Energy Research & Social Science, 39*,103–107. Retrieved from https://doi.org/10.1016/j.erss.2017.11.005

Tudela, S., Kai Kai, A., Maynou, F., El Andalossi, M., & Guglielmi, P. (2005). Driftnet fishing and biodiversity conservation: The case study of the large-scale Moroccan driftnet fleet operating in the Alboran Sea (SW Mediterranean). *Biological Conservation, 121*(1), 65–78. Retrieved from https://doi.org/10.1016/j.biocon.2004.04.010

Uddin, S. N., Taplin, R., & Yu, X. (2007). Energy, environment and development in Bhutan. *Renewable and Sustainable Energy Reviews, 11*(9), 2083–2103. Retrieved from https://doi.org/10.1016/j.rser.2006.03.008

Valentine, S. V., & Sovacool, B. K. (2010). The socio-political economy of nuclear power development in Japan and South Korea. *Energy Policy, 38*(12), 7971–7979. Retrieved from https://doi.org/10.1016/j.enpol.2010.09.036

Verme, P., El-Massnaoui, K., & Abdelkrim, A. (2014, February). Reforming subsidies in Morocco. *Economic Premise* (No. 134). The World Bank. Retrieved from http://siteresources.worldbank.org/EXTPREMNET/Resources/EP134.pdf

Victor, D., & Heller, T. (Eds.). (2006). *The political economy of power sector reform: The experiences of five major developing countries*. Cambridge: Cambridge University Press.

Vidican, G., Böhning, M., Burger, G., Regueira, E. de S., Müller, S., & Wendt, S. (2013). *Achieving inclusive competitiveness in the emerging solar energy sector in Morocco*. Bonn: The German Development Institute/Deutsches Institut für Entwicklungspolitik. Retrieved from http://edoc.vifapol.de/opus/volltexte/2015/5611/pdf/Studies_79.pdf

World Bank. (2009, November 30). *Implementation, completion and results report: (Loan-74480-MOR)*. The World Bank and Sustainable Development Department, Middle East and North Africa.

World Bank. (2011). *Poor places, thriving people: How the Middle East and North Africa can rise above spatial disparities*. Washington, DC: World Bank. Retrieved from http://elibrary.worldbank.org/doi/book/10.1596/978-0-8213-8321-6

World Bank. (2012). *Literacy rate, adult total*. Retrieved June 28, 2014, from http://data.worldbank.org/indicator/SE.ADT.LITR.ZS/countries/1W-XN-XQ-ZQ-MA?display=graph

World Bank. (2017). *State of electricity access report 2017*. Washington, DC. Retrieved from http://documents.worldbank.org/curated/en/364571494517675149/pdf/114841-REVISED-JUNE12-FINAL-SEAR-web-REV-optimized.pdf

World Bank. (2018). *Doing business in Morocco*. Retrieved March 14, 2018, from www.doingbusiness.org/data/exploreeconomies/morocco

World Nuclear News. (2010). Nuclear Named for Copenhagen Reductions. Retrieved March 15, 2015, from http://www.world-nuclear-news.org/EE_Nuclear_named_for_Copenhagen_reductions_0202101.html.

Zaki, L. (2010). L'électrification temporaire des bidonvilles Casablancais: Aspects et limites d'une transformation 'par le bas' de l'action publique. Le cas des Carrières Centrales. In M. Catusse, F. Vairel, & I. Bono (Eds.), *Le Maroc de Mohammed VI: Mobilisations et action publique*. Paris, France: Karthala.

Zawya. (2014, November 25). *Morocco power projects light up with funding boost: OBG*. Retrieved October 19, 2017, from www.zawya.com/story/Morocco_power_projects_light_up_with_funding_boost-ZAWYA20141123091325/

Zeigler, D. J. (1997). Morocco and California: Comparing environment and society. *Journal of Geography*, *96*(4), 191–195.

Zisenwine, D. (2012). Mohammed VI and Moroccan foreign policy. In B. Maddy-Weitzman & D. Zisenwine (Eds.), *Contemporary Morocco: State, politics and society under Mohammed VI* (70–81). Milton Park, Abingdon: Routledge.

6 Neocolonial or not?

Evaluating North-South-South partnership on electricity integration

The challenge of North-South-South electricity integration

This chapter discusses the prospect of "deep electricity integration" between the EU and North Africa. Deep integration is characterized by a single market for electricity in the region, which is governed by regional regulatory institutions and transmission system operators (ESMAP, 2010). This chapter explores Moroccan actors' perspectives of electricity integration. In doing so, it fills a gap in the literature, which is dominated by European perspectives. It examines the compatibility of Morocco's sociotechnical imaginary of sustainable energy with plans for regional electricity integration.

While the amount of electricity traded across borders is currently small, dozens of transnational and regional electricity integration projects are emerging across the globe, which suggests that there is a growing integration trend. This trend demands attention from social scientists in order to better understand the implications for international relations, energy security, and energy justice. The literature suggests that it may be easier to develop power pools within more homogenous regions than across significant sociopolitical divides. For example, Nord Pool, which originally connected Norway, Finland, Sweden, and Demark, is arguably the most deeply integrated electricity market in the world. Hira & Amaya (2003) argue that the success of integration in this case may be idiosyncratic, as the Nordic countries have similar generation technologies (i.e., hydroelectricity), governments, and cultures. In contrast to the case of Nord Pool, an EU-North Africa electricity system would require North-South cooperation or, more accurately, North-South-South cooperation.

Much of the literature on North-South cooperation focuses on research collaboration or on the effects of geopolitical dynamics on international relations among countries that have differential access to resources. The North Africa-EU electricity integration case study is more complex than other cases because it encompasses R&D partnership, private sector cooperation, nation-state cooperation, public-private partnerships, and sustainable development goals. Such North-South-South cooperation is exactly the sort of cooperation needed for mitigating climate change—which is an inherently transnational and global challenge—but achieving this cooperation is difficult in practice (Hurrell & Sengupta, 2012).

While "North" and "South" are problematic representations (Doty, 1996), from a practical standpoint they refer to existing power and wealth asymmetries between the EU and North Africa.

A common critique of North-South cooperation is that it results in neocolonialism because equal partnership is not achieved. This is partly owing to the North setting priorities that result in the continued extraction of resources from the South rather than resulting in the addition of socioeconomic value to the South (Appadurai, 1996). Moreover, today's perceptions and realities of North-South collaboration are shaped by the history of imperialism and "the geography of empire," which are, in fact, "inaccurate and senseless to repress" (Said, 1994, p. 15).

The literature on North-South collaboration suggests that North-South interactions that reinforce colonial relationships ought to be replaced with interactions that result in "true partnership" (Binka, 2005; Hwang, 2008) or "real cooperation" (Mervis & Normile, 1998). The literature does not clearly define "true" partnership, but it relates to the development of a level playing field between North and South, as well as a cosmopolitan respect for cultural differences (ibid.). Similarly, Hwang (2008) investigates international cooperation in the context of "inequitable international relations" and argues that a true partner is the opposite of a technology "transferee and technology evaluator" who lacks power and who lives in the "periphery" (p. 121). Additionally, it is important to understand actors' motivations for the initiation of a North-South partnership and to develop shared priorities between North and South (Katz & Martin, 1997). Procedural justice requires mutual priority setting and avoidance of countries in the North setting the priorities and then demanding that countries in the South follow these priorities (Rosseel, De Corte, Blommaert, & Verniers, n.d.). "Demand-led research" should respond to the South's development priorities (Hall, 2002; Nair & Menon, 2002; Velho, 2002). "Rich industrialized nations can no longer count on access to raw materials, and consumer markets in Southern countries are increasingly gaining influence. . . . These relations are no longer based solely on development aid but on reciprocal trade as well" (Rosseel et al., n.d., p. 19). Developing countries seek to become economic or research partners with the North rather than to become passive recipients of aid or technology transfer.

"True" partnership cannot be achieved without long-term capacity building efforts in the South that aid in overcoming North-South asymmetries in access to information and resources (Gaillard, 1994). Velho (2002) argues that this capacity-building effort should be a part of a national innovation systems approach that puts "collaboration and policy institutions at the centre of the analysis" (p. 172). The innovation systems approach should not merely establish checkboxes for good collaboration and partnership but should also seek to overall improve the South's innovation and sustainable development capacity (ibid.). The Moroccan government's goals for its renewable energy policy, as described in Chapter 5, reflect this innovation systems approach. Its energy policy is part of its broader industrial policy, encompassing an ensemble of societal goals that seek to increase innovation in Morocco and to develop a knowledge-based economy. Morocco is approaching the development of renewable energy as part of an "integrated projects strategy" that includes the development of a domestic supply chain, North-South R&D cooperation, university degree programs and vocational training, and (eventually) the development of an electricity export market.

Surprisingly, I found in my research that critics who saw the Desertec vision as neocolonial largely came from Europe. Some Moroccan interviewees saw the flawed process of negotiating the vision as neocolonial but did not see the concept of renewable electricity integration as intrinsically neocolonial. Through Morocco's renewable energy strategy, the Moroccan government is bolstering Morocco's capacity to become an equal partner in negotiating the development of a regional energy system that is free from coercion and that improves Morocco's energy security. I found greater support for regional electricity integration among Moroccan policymakers than I expected to uncover. Yet the outcome from regional energy partnerships that the Moroccan government desires exceeds simplistic goals related to security of supply. Furthermore, the challenges of energy partnership go beyond the techno-economic factors and encompass conflicting paradigms between North and South related to the role of the state versus of the market and related to differences in the benefits that the actors expect to achieve from the construction of a renewable electricity system. Claims that electricity integration will result in "win-wins" but that focus only on wins relating to security of supply exclude Morocco's goals for achieving socioeconomic benefits from renewable energy transitions. The North-South collaboration literature has focused largely on the criteria for successful research partnerships; this chapter illustrates that it would be far more complex to achieve true partnership in the development of a regional sociotechnical electricity system.

This chapter begins by discussing the EU and Morocco's shared goal of ensuring the reliability and security of electricity supply, particularly of intermittent supply from renewable energy. I argue that despite this shared goal, different perspectives between North and South on the role of the state versus the market, as well as a lack of procedural justice in the visioning process, suggest that deep electricity integration in the Mediterranean region would be difficult to achieve. Deep North-South electricity integration would require multilateral cooperation, a shared commitment to deregulation—which would require viewing electricity as a commodity rather than as a strategic resource or social necessity—and the development of a just visioning process that does not reinforce neocolonial relationships. A "true partnership" for Mediterranean electricity integration would need to offer value-added for North Africa's national innovation systems. These findings are applicable to other cases of electricity integration that are in the planning phase, such as the CASA-1000 project to export electricity from Central Asia to Afghanistan, US-Mexico electricity integration, African regional power pools, and the GEIDCO vision for a global grid.

Reliability and security of supply as a mutual North-South goal

An important aspect of achieving fruitful North-South collaboration is the setting of mutual priorities. A key mutual priority and potential mutual benefit between

North and South Mediterranean is to ensure the security and reliability of electricity supply in a system with high levels of renewable energy penetration. Proponents of Mediterranean electricity integration frame it as a "win-win solution," in part because of mutual supply security. While security of supply has various definitions (Winzer, 2012), I narrowly define it as the 24-7 availability of electricity supply free of political or technological disruption. Elsewhere, I have argued that Morocco's energy security would be better understood using a polysemic definition of energy security that encompasses affordability and sustainable development in addition to availability and reliability of supply (Moore, 2017). However, in this chapter, I use a narrower definition because it reflects a mutual priority from the perspectives of the actors. I found that from a technological standpoint, improving the reliability of supply from intermittent renewable electricity can be accomplished by integrating grids. But from a political standpoint, a lack of trust and partnership dredges up issues related to the reliability of supplier countries and to the North-South-South politics that impede integration.

Existing electricity integration among Morocco, Spain, and Algeria is already balancing out intermittent hydroelectricity and improving Morocco's reliability of supply. Morocco is connected to Spain and Algeria through "shallow integration" (ESMAP, 2010), meaning that the countries share transmission lines that allow imports and exports but do not share a single grid and liberalized market for electricity. There are currently two AC submarine transmission cables between Morocco and Spain under the Strait of Gibraltar: a 400 kV line with a transfer capacity of 700 MW completed in 1997 and a 400 kV line with a transfer capacity of 900 MW (in the Spain to Morocco direction) completed in 2006 (ENTSO-E, n.d.; IEA, 2016). A third line is under construction from Ceuta in Morocco to Spain, and a new line has been proposed between Morocco and Portugal (ENTSO-E, n.d.). Algeria and Morocco are also shallowly integrated through three transmission lines, including one double-circuit 400 kV overhead line and two single-circuit 220 kV overhead lines (ibid.). Morocco's imports from Spain and Algeria increased from 10% of its overall electricity consumption in 2006 to 18% in 2008, because drought decreased Morocco's share of electricity consumption from hydroelectricity (Energy Sector Management Assistance Program (2011; "Indicateurs Clés," 2014). Imports decreased in early 2009 because of increased water availability for hydroelectricity in Morocco, falling to 14% in 2012 ("Indicateurs Clés," 2014). This hydroelectricity intermittency should be considered in the planning of grid integration projects focused on new hydroelectricity development, including the ASEAN power cooperation in Southeast Asia that would develop hydroelectric facilities on the Mekong River and the Grand Renaissance Dam in Ethiopia that would export electricity to Kenya.

Morocco's renewable energy policy would result in 52% of its electricity portfolio sourced from wind, solar, and hydroelectricity, in order to reduce its high dependence on fossil fuel imports and to increase domestic supply to meet skyrocketing demand. While this strategy would increase energy independence, reduce carbon dioxide emissions, and aid Morocco in its goal to industrialize, the intermittency of renewable energy could technologically impede grid stability and reliability in the short term, which could compromise security of supply. Peaks in electricity consumption during certain time periods make it more difficult to achieve supply reliability. While demand increased 5% on average from 2011 to 2012, it increased 10% during the evening peak hours and increased more than 5% during the hottest and coldest months, which required the construction of expensive power plants that operate mainly during peak hours. As one energy expert in Morocco put it, "7:00 to 10:00pm [are] three hours that ONEE fears like the devil." Moreover, the reliability of supply decreases in Morocco's peripheral areas. When I visited the small desert town of Merzouga, there was load shedding during peak hours, but demand was consistently met while I lived near Rabat. Large and uneven increases in demand could impede reliability for both the domestic and industrial sectors. If rising demand cannot be met, citizens who have come to rely on electricity could lose access; thus, reliability of supply is critical for human development in Morocco.

Electricity imports could improve reliability during peak hours of demand and times of limited availability of renewable electricity. Although new renewable power plants would increase domestic supply, an interviewee from ONEE explained that "the integration of the grid with [intermittent] renewable energy is a major challenge for ONEE." To meet this challenge domestically, ONEE has built 464 MW of pumped hydroelectric storage (STEP) and plans to build an additional 950 MW. STEP stores electricity during off-peak hours by using it to pump water from a basin that is lower in altitude to one that is higher. The water is later released to generate electricity during peak hours. ONEE is also working on consumer load balancing programs, in which consumers are asked to reduce their consumption when wind resources are scarce, and it is developing a pilot smart grid program in Casablanca. However, these measures are insufficient to ensure the future reliability of a system that includes high levels of intermittent energy. Therefore, the systems-level aspects of a renewable electricity transformation provide impetus for increasing regional integration in tandem with Morocco's energy independence initiatives. The interviewee from ONEE explained that achieving energy security in Morocco requires the development of independent, domestic energy resources *in addition* to a favorable regional context for coordination and integration that would balance energy intermittency. He stated, "Regional coordination is necessary for ensuring energy security." Electricity interconnection can help to maintain reliability by importing electricity during peak hours and exporting it during off-peak hours. The systems context of electricity makes achieving total energy independence difficult. In Morocco's case, the solution to fossil fuel dependence has led to a system that would benefit from regional electricity interdependence, pulling Morocco toward neighboring countries despite political tensions with them.

Dii's CEO Paul van Son said that Morocco was the first country to approach Dii when it was established. At the 2013 Dii conference in Rabat, the Moroccan minister of energy stated,

> 'The EU has moved toward renewable energy legislation and creating an institutional environment for renewable energy. The south of the Mediterranean

> has also put ambitious plans for renewable energy in place and for growing an electrical market. Our signature indicates that we support this market. We wish to share between north and south of the Mediterranean.' "With these interconnections with Algeria and Spain, we're geostrategically positioned for playing a major role in the exchange of electricity with the Mediterranean."[1]

This sentiment can also be seen in the open regionalism scenario from the *Maroc 2030* energy forecast study, which viewed the most desirable future for Morocco as one in which the Moroccan grid is deeply integrated within neighboring electricity systems (Haut-Commissariat au Plan, 2006). The scenario included extensive electricity interconnection with Europe and the Maghreb region. The study argued that Morocco has significant potential to export green electricity to Europe, which would foster North-South solidarity and address renewable energy intermittency, and would protect the environment and reduce greenhouse gas emissions. It concluded that the regionalism scenario would be the most positive for sustainable development.

Similarly, in 20 interviews I conducted with Moroccan energy policymakers in 2013, I found strong support for regional energy integration and the export of renewable electricity. For example, MASEN supports Morocco's integration with the EU grid and the export of solar electricity after meeting its 2,000 MW domestic goal. A representative of MASEN stated, "For us, 2,000 MW is only like an appetizer . . . especially if we reach grid parity."[2] Moreover, even though MASEN sought to increase Morocco's energy independence by the development of renewable energy, it adopted a European identity for Morocco's electricity system. Because of the cables connecting Morocco to the Spanish grid, MASEN's view is that "Morocco is already electrically European." The Moroccan grid operates at the same frequency as Europe, and "we produce the same electron as Sweden." An interviewee from ONEE emphasized that the success of the interconnections between Morocco and Spain offered evidence of ONEE's European infrastructural know-how. He stated, "One can consider that Morocco is a country integrated into the European market because it is interconnected to the European market.. . . Morocco is the first country to have the *courage* to construct interconnections with Europe." A decision was made over a decade ago to pursue the "logic" of regional electricity cooperation. He recalled that there initially was significant skepticism among funders, operators, and other stakeholders, but the interconnection has worked well and was even reinforced.

North-South reliability challenges: Although there is a techno-economic logic to North-South electricity integration, without partnership and mutual trust, integration relates to traditional geopolitical concerns about the reliability of supply for Morocco as well as for EU countries. Interviewees from the Spanish TSO explained that they do not see Morocco as electrically European because Morocco's electricity sector has not been fully liberalized. While Morocco is the only country in the MENA region that has attracted private investment in renewable energy, it does not yet have an independent energy regulator. It is therefore only welcome in the Spanish grid as a market player, not as a partner.

This attitude toward the Moroccan electricity sector is also reflected in European concerns about the withholding of electricity exports for political reasons. As mentioned in Chapter 4, analysis of news clippings related to Desertec showed that some critics of the vision argued that it would make the EU vulnerable to electricity sabotage. In contrast, Lilliestam (2014) has argued that the risk of terrorist attacks on regional grids is low, although higher than for natural gas, and that electricity infrastructure is not a popular terrorist target because it lacks symbolism and fails to evoke fear in the way that other terrorist attacks do. Moreover, Lilliestam and Ellenbeck (2011) find that there is low risk that a single supplier country could cut off supply; rather, five supplier countries would have to work in tandem to disrupt supply. Even so, some EU decision-makers are wary of importing electricity—especially given the history of Russia withholding natural gas exports—and this wariness reduces the potential of developing a trusting partnership. Van der Vleuten and Lagendijk (2010) illustrate that risks to electricity supply are not objective threats, but rather various stakeholders interpret and construct these threats. This suggests that stakeholders' perceptions are more important to achieving successful integration than the objective existence of a threat.

What has received less attention is that many Moroccan policy officials viewed Morocco as unfairly excluded from the European electricity market. I found in my fieldwork that many Moroccan policymakers wanted to develop mutually agreed on terms that avoid coercion and that allow Morocco to access the EU electricity market. However, some Moroccan energy policymakers thought that political problems in Europe were impeding Morocco's chances of exporting electricity to the EU. An ONEE interviewee argued that it was necessary for Europe to implement mechanisms that favor the import and export of electricity. He also stated,

> We are convinced by regional cooperation through the development of two components—namely interconnections and renewable energies—to enhance the energy needs. But it is also necessary for the European electricity grid to develop, to avoid having electrical islands in Europe, especially between Spain and France.

A MASEN interviewee viewed the Desertec vision as a "good and simple concept" to generate electricity where it is cheapest and to transmit it using HVDC lines. However, the conditions for export must be right, he explained, as the issues go "way beyond electricity" to encompass geopolitics; security of supply; and the Eurozone crisis, after which 35% of Spain's power generation capacity was shuttered because of lack of domestic demand. An advisor to the Moroccan Ministry of Energy (MEME) emphasized that Morocco and Germany could achieve successful, bilateral cooperation on electricity, but Spain and France are physically and politically "in the way." The interviewee thought that there had been a reduction in European enthusiasm for electricity integration since the Eurozone crisis began, and the ministry decided that it could proceed with a solar plan faster alone instead of waiting for European companies to develop projects. The interviewee emphasized that Desertec is a vision for the distant future and declined to estimate when renewable electricity export might be achieved. He faulted Dii for setting unrealistic expectations, even within the consortium, and cited high staff turnover within Dii as evidence that expectations were not internally met. He thought that Dii had moved away from an emphasis on exports from North Africa to the EU because Dii did not meet the investors' expectations for how quickly exports could be achieved. More proof-of-concept would be required, in the form of political and regulatory frameworks that lay the groundwork for regional integration, but Dii had little success with this category of proof-of-concept.

EU policymakers perceive risks associated with the lack of reliability of supplier countries. However, Morocco has negotiated unfavorable terms in past free trade agreements with the EU (Dawson, 2009), and it also risks negotiating a deal for electricity exports in which it does not secure sufficient local and national benefit. A Moroccan energy professor I interviewed wrote a prescient book in the 1980s about Morocco's potential to export wind electricity to Europe. In it, he argued that robust North-South political collaboration is necessary to achieve electricity export, in which the needs of both sides are satisfied and the social rift between North and South is closed. However, by bolstering its negotiation and R&D capacities through the Moroccan Solar Plan, Morocco will be in a better position to avoid being coerced by neighboring countries in integration processes and to become a true partner in these processes.

South-South reliability challenges: Lack of trust—and the uncertainty that comes with it—is also of concern in cases of South-South integration. Although the authors of the *Maroc 2030* report favored the regional integration scenario, they also expressed skepticism about the scenario's feasibility because it would require the reactivation of Maghrebian multilateral cooperation, possibly through the Arab-Maghreb Union (AMU) that has only met once since 1996. Some shallow integration already exists. In 1972, Morocco, Algeria, and Tunisia's state-run utility companies decided to develop the Maghreb Electricity Committee (Comité Maghrébin de l'Electricité or COMELEC) (ICA, 2016). In 1989, five Maghrebian countries formed the AMU to promote political cooperation in the Maghreb, and Libya's and Mauritania's utility companies joined COMELEC (ibid.). Today, all countries except Mauritania are interconnected through 400 kV or 220 kV connections (ibid.). The lines between Spain and Morocco are frequently used, but only 5% to 16% of interconnection capacity among Morocco, Algeria, and Tunisia is used (ibid.). Significant South-South electricity and other forms of trade are lacking partly because of South-South political disputes and North Africa's predominant export orientation toward Europe. However, there are ongoing plans for integration. For example, Algeria hopes to increase its electricity exports to its neighbors (Alternative Energy Africa, 2018), and Morocco is exploring building a transmission line to Mauritania (Ratcliffe, 2016).

Proponents of the Desertec vision have argued that technological integration can facilitate political integration. Even if this is true, as van der Vleuten and Lagendijk (2010) suggest it could be, it is unclear whether technological integration that precedes political integration will result in security of supply. Morocco's technological integration with its neighbors currently is tenuous for Morocco's security of supply due to a lack of political integration and mutual interdependence. Several interviewees said that Morocco and Algeria cooperate best on energy issues, and the political tensions between the two countries impede deep electricity integration. The border between the nation-states is closed to people even as electrons flow across it. Due to poor relations, Morocco's lopsided dependence on Algeria puts Morocco at risk of Algeria withholding supply from Morocco for political reasons. Electricity integration without political integration and mutual trust lacks the security of supply that Morocco seeks to develop through its renewable energy policy, especially if energy dependence increases and political integration remains the same.

Considering Morocco and Algeria have long been at loggerheads, the interconnection between Morocco and Algeria appears to be an example of technological integration without political accord. On further examination, I found that the French government first established the transmission lines between Morocco and Algeria in the 1940s when it controlled both countries. In 1941, French General Maxime Weygand held a conference in Algiers on connecting the Moroccan, Algerian, and Tunisian grids (Saul, 2002). With an annual production of 220 million kWh (ibid.), French protectorate officials perceived Morocco as the most advanced North African country in electrification, especially since Algeria and Tunisia had less hydroelectric potential (Berthonnet, 2008). At the conference, it was envisioned that Moroccan electricity production would be quadrupled through the construction of new hydroelectric facilities, and Morocco would then export 200 million kWh to Algeria within 8 to 10 years, although energy planners stipulated that Morocco's electricity demand had to be met before exports began (Saul, 2002). In 1945, the export goal was reduced to 100 million kWh from 1953 to 1956, to increase thereafter (ibid.). Morocco exported electricity to Algeria from 1956 until 1974, when Morocco experienced electricity shortages due to the oil crisis (Debbarh, 2006). In 1988, Algeria began exporting electricity to Morocco because Morocco's hydroelectric capacity had declined because of drought (ibid.). Algeria had more energy wealth than Morocco because it had discovered large natural gas reserves after achieving independence from France. This history suggests that the achievement of technological integration without political integration and accord was idiosyncratic in this case and does not provide evidence that technological integration generally works well without political integration. Nor is there yet evidence that technological integration between the countries has improved political integration and political relations.

In this section, I suggested that North-South and South-South political integration and mutual trust must co-evolve with the development of electricity exports in order to ensure security of supply. While security of supply is a shared priority between North and South, achieving it is more difficult than is implied in win-win discourses. Today, Morocco's government appears to be focusing on integration with sub-Saharan Africa, as well as on developing additional lines with Portugal and Spain. King Mohammed VI expressed interest in entering the West Africa Power Pool in the 2017 speech he made when rejoining the African Union, and MASEN has recently shifted its attention to exporting to sub-Saharan Africa, via its new "MASEN partner of Africa" focus.

Challenges associated with North-South grid integration

Much of the conversation on electricity cooperation in the Mediterranean region has focused on supply security and the techno-economic benefits of integration. However, this artificially constrains the discussion by excluding other North-South priorities, the consideration of different energy governance processes in the North and South, and acrimonious histories that shape perceptions of fairness and procedural justice. In Chapter 4, I concluded that energy transformations require institutional and regulatory innovation; this section addresses how difficult it would be to develop mutual institutions and regulations across North-South divides.

Differential role of state versus market: The Spanish TSO's assertion that Morocco's electricity sector must be fully deregulated in order to gain a European identity reflects deep-seated differences in the perceived role of the state and the market in Moroccan and EU energy policy. A global trend toward electricity deregulation in the 1980s and 1990s shaped the electricity markets of at least 70 developing countries (Dubash & Williams, 2006), in part because development banks mandated deregulation as a prerequisite to receiving loans for energy infrastructure (Dubash, 2003, Dubash, 2005). Dubash (2005) critiques this practice as the "orphaning of social policy" because free market governance was privileged over setting explicit social goals for energy systems (p. 3). Peters and Westphal (2013) argue that in some regions energy is again becoming securitized even though promoters of deregulation processes encouraged a commodity-based framing of electricity. While security of supply and geopolitics have received attention in Mediterranean electricity discussions, deep electricity integration is thought to require complete deregulation, and deregulation proponents frame electricity as a commodity not as a strategic resource. Moreover, Dubash's critique remains important in discussions of energy integration focused on security *or* on deregulation because energy is not framed as a social necessity or a public good in either case. In the rest of this section, I compare and contrast perceptions implicit in Morocco's energy policy and in the Desertec vision of the appropriate role of experts in energy policymaking as well as the appropriate role of the state and the market in developing energy systems, which is an important consideration for making social policy. I also discuss the implications of these different perceptions for achieving electricity integration.

Morocco's energy policymaking reflects a technocratic impulse, which has likely been shaped by efforts since the 1980s to liberalize its economy. In her history of the development of French nuclear power, Hecht (2009) discussed the concept of "technocratic pose," which refers to placing technology above and beyond the sphere of politics and out of the reach of politicians (Alder, 1997; Hecht; 2009). Hecht (2009) stated that "technologies gave these elite [technologists] a unique vehicle for political action" (p. 11). These technologists played a role in shaping nuclear power as a technology of national pride in France. Morocco's technocratic impulse is reflected in the framing of energy policy as being rooted in data, called in French *les données* or literally the givens. When Abdelkader Amara was appointed the minister of energy in 2013, MEME's website was overhauled, and "key indicators" and "key statistics" became prominent features of the site. An illustration on the ministry's website depicted the energy policymaking process as a circular iteration of three steps: collecting statistics, analyzing and interpreting statistics, and communicating for a given action. It indicated that data lead to accurate statistics, which are interpreted to form energy indicators and mixes, which lead to evidence-based energy policymaking. Implicitly, this process results in socioeconomic benefits.

This focus on *les données* was also reflected in the founding documents of Morocco's renewable energy policy. A quotation from the *Maroc 2030* study summarized this perspective: "In the field of energy, quantitative data play a primordial (important) role, given the capacity of the structuring of the sector on the rest of the economic and social system" (p. 55). The study explains that errors in the estimation of demand could have future, negative macroeconomic consequences.[3] An evidence-based energy policy supported by *les données* is, in part, a strategy to grapple with uncertainty and to rationalize the energy policymaking process.

Even so, experts appear to be missing from the technocratic equation. Traditionally, Moroccan elites fell into one of three categories: the *makhzen,* or royal family and close advisors; wealthy merchant families; and technocrats, or unelected bureaucrats who work for the government on the basis of their expertise (Ben Ali, 1997). Since neoliberal reforms began in the 1980s, there has been some decoupling of the sphere of public action open to experts and the political sphere controlled by wealthy families and political parties, but this itself has been a political maneuver, serving to carve out boundaries between rational and values-based topics (Catusse, Vairel, & Bono, 2010). Yet Morocco's energy policy has not become what Sarewitz (2004), in a Western context, has called a "scientized" debate, or a debate that has become about scientific data rather than about the societal values at stake. I demonstrated in Chapter 5 that Morocco's renewable energy policy directly relates to social and political goals. Additionally, the *Maroc 2030* study framed electricity planning mainly as a social, political, and development issue, despite its emphasis on *les données*.

While MEME's illustration suggests that there is a technocratic impulse in Morocco, there is no "technocratic pose" because technology and technologists are subordinate to the Morocco state to the extent of being invisible. Experts who use technology as a vehicle for political action are notably absent. One interviewee from a state research agency said that researchers should not be involved in forming policy. It was up for the state to choose a way and for the experts to follow. Not all scientists agreed and, instead, complained that when the government requires technical expertise for the development of national policy strategies, foreign experts are often consulted rather than domestic experts. Some Moroccan solar researchers—including those who were among the first to study CSP before the state adopted it—perceived themselves as disempowered and ignored. Two nonprofit associations provide coordination and networking functions for academics studying renewable energy and promote collaborations between industry and academia: SMADER and MANEREE (Master National en Energie Renouvelable et en Efficacité Energetique). The researchers who had tried to promote solar energy before it became popular in Morocco later developed these organizations to gain a voice once solar became politically popular. Additionally, some Moroccan researchers felt excluded from international events, such as the Dii conferences, which tainted the vision with procedural injustice. An energy researcher stated,

> We have approximately 150 active researchers, and they have worked on this field [solar power] for many years. . . . They have skills, they have expertise because they . . . have given many lectures, they have produced many publications in very prestigious journals and reviews, so they . . . are experts in their fields. But . . . practically, nobody—*no one of them*—has been frequently approached and asked to present and to provide his expertise in the form of the implementation of Desertec or solar energy plants. . . . There are experts also wanting to participate in the implementation of the Moroccan strategy. They want to be involved in the implementation. They don't want to ask to be involved. They want to be invited to be involved. . . . Most of the [Moroccan] authorities are looking for experts from outside . . . and we are here! And we don't ask for money. We are nationalists. We would like to work for our nation, for our country. We like our country, and we want to participate. But . . . we would like to be invited.

The researchers want the deference and dignity typically bestowed upon experts, but the state does not regard them as players in the political process. Therefore, unlike in France, technology does not provide experts in Morocco with a unique license to participate in the political sphere. While there is a technocratic impulse in Morocco, it does not involve rule of experts, particularly not domestic ones.

Moroccan energy policy reflects a bridging of neoliberal and traditional governance in which a policy challenge with social goals is segmented into (1) a rational management process bridging technocratic and political spheres and (2) a state-led, centrally-controlled process. Neoliberalism has not removed the state from technology planning, but rather has changed the role of the state (Catusse et al., 2010). The Moroccan state chooses winning and losing technologies, as one interviewee put it, "setting a direction and going for it," in addition to allowing some voluntary actions, such as enabling an independent wind company to sell electricity to the grid. Vidican (2015) argues that governance accountability is lacking in the development of the Moroccan solar innovation system; here, governance refers to the state's exercise of power. This is true, but it should be acknowledged that this system of governance is culturally and historically specific.

The Moroccan government mitigates the uncertainty of long planning periods by using both *les données* and governance by the *makhzen*. Since the state must shoulder risk by supporting 20 to 40 year energy development projects—beyond the lifetime of some companies—this interviewee saw it as justifiable for the state to put conditions in place, such as requiring local content and requiring foreign companies to partner with a local research institution. Here, the government is simultaneously involved in incentivizing private investment and in placing it under state control. The state is viewed as assuming the risk rather than private investors. The interviewee summarized this stating, "So the condition is, if you want [the state] to buy electricity from you [the foreign company] for the next 30 years, you need to transfer the technology to these [local] companies" to boost Morocco's development. He continued,

> In Morocco we believe that the state has to show the way, because the public involvement is important, because there is not something called the truth and we have to find it. The state is there to show the way, to show *a* way, because we need that.

Vidican (2015) makes an important critique that Moroccan actors from various sectors do not necessarily have the capacity and coalitions required to follow the state's path. Similar to how the Moroccan state acts as a moral arbitrator to exemplify the appropriate actions of Moroccans in society (Ben Ali, 1997; Maghraoui, 2009), the state acts as a technological arbitrator to select appropriate technologies for Morocco. Thus, the Moroccan state's idea of technological neutrality—a term which European interviewees used to emphasize the role of the market in choosing the best technology—is that the state should set the overarching technological agenda rather than allow the invisible hand of the market to randomly select a technology. MASEN calls its approach "technology neutral," but by this it means that the market will choose among solar technologies, not that the market will choose the winning electricity generation technology overall.

In contrast, Dii's vision was both technocratic and neoliberal. Dii framed the vision's outcomes as an inevitable endpoint of techno-economic progress. From this perspective, the free market is the technological arbiter that selects winning technologies. Dii framed this market-based process as objective and unbiased, although in the absence of a belief in market determinism this scenario implies that a private company composed of experts unelected by the public should make decisions on the basis of their understanding of what the market demands. Dii's approach suggests that experts, in fact, play a crucial role in accelerating progress toward the vision's goal, provided that politics do not "get in the way." One Moroccan interviewee argued that this approach is as arbitrary as the state choosing the winning technologies. From the Moroccan government perspective, the state assumes the risk and sets the course, but from Dii's perspective, private investors assume the risk and market logic sets the course.

Dii's model of prioritizing market control was not, in fact, universally accepted in Europe. For example, the director of GIZ AgenZ and a resident fellow at the German Council on Foreign Relations argued that what was at stake in the Desertec project was far broader than clean energy development alone (Gnad & Viëtor, 2011). Instead, a "Mediterranean Solar Union" of "supranational dimensions" could revive EU integration processes and develop positive mutual dependencies between the north and south of the Mediterranean. They argued that what started as a comprehensive visioning process had become insular and needed to be opened back up to political and societal scrutiny. They argued that the Desertec vision was too important to be left in the hands of an industrial initiative alone.

Therefore, they recommended that the vice president of the EU as well as the European External Action Service should take charge of the vision. They stated, "Visions that affect society as a whole—especially a common European vision, which is able to replace financial transfer payments as a substitute for policy—need to receive political priority" (para. 33). Similarly, the *Irish Times* accused politicians of "outsourcing" their duties to Dii instead of addressing crucial questions about the EU's future, North-South collaboration, and energy in the Mediterranean region (Kinsella, 2009). However, van der Vleuten and Lagendijk (2010) illustrated that in Europe a decentralized approach that prioritizes state sovereignty has been favored over a centralized, transnational approach. Understanding the tensions among state, private sector, and transnational governance is important to successful North-South electricity collaboration in the Mediterranean region and elsewhere but is rarely discussed. An inclusive North-South energy policymaking process should be developed in the Mediterranean region that acknowledges different models of state and economic development.

Governments' perceptions of the ideal role of the state versus the market matter because achieving deep electricity integration may require electricity liberalization and deregulation. Puka and Szulecki (2014) attribute slow progress on electricity interconnection between Germany and Poland to the contrast between a commodity-based deregulated electricity market in Germany and a state-run electricity system in Poland. This difference is also likely to impede integration in other cases. Despite the shared goal of security of supply, it may be the case that these different perspectives of the role of the market and state, as well as power asymmetries across North and South, will prevent the level of grid integration and political accord achieved in the case of Nord Pool, which has extended its liberalized market model to 20 European countries since its creation.

Advocates of liberalization promise it will achieve numerous benefits including reducing electricity costs, increasing consumer choice, increasing competition, prioritizing lowest cost generation on an hourly basis, and evening out the intermittency of renewable energy sources (Hoven & Froschauer, 2004). Dii has made similar promises, but these benefits may depend upon achieving full deregulation across the region. For example, de Villemeur and Pineau (2012) argue from an economics perspective that the expected benefits of lower price, lower consumption, and lower emissions are unlikely to be achieved when one jurisdiction adopts market-based marginal cost pricing and the other jurisdiction retains a regulated average price. Deregulation may be necessary for securing maximum techno-economic benefit, but some countries see more benefit associated with a state-run approach, particularly those focused on achieving domestic development and meeting what de Villemeur and Pineau call "native load commitments," which are rapidly increasing (ibid., p. 2).

From a political science and sociology perspective, Hoven and Froschauer (2004) illustrated that France and Canada shaped liberal electricity reforms in order to serve national policy interests. Powerful Western countries that promote the liberalization of electricity markets could use new market configurations to serve their needs in ways that may impede achieving sustainable development in the South. While Morocco is criticized for its state-run model, countries in Europe may manipulate what appears to be a neutral market-based model to meet only their interests. Deep grid integration with full deregulation seems unlikely, especially in the short- or medium-term, although opportunities for shallow integration that enhance reliability of supply remain possible.

I often heard at Dii conferences that renewable energy will thrive in the Mediterranean region only if North Africa's power sectors are liberalized. At the 2013 Dii conference, an official from MEME stated that Morocco was liberalizing its power sector to enable it to export to Europe. Yet renewable electricity in Morocco has achieved tangible success in a "dual-market system" (Victor & Heller, 2006)[4] in which government-established ancillary social goals are the focus of renewable energy development, and the Saudi electricity and water company, ACWA Power, has won all of the contracts to build solar power in Morocco, rather than private European companies.

Neocolonialism, procedural justice, and innovation systems: Appadurai (1996) critiques North-South relations for perpetuating colonial injustices and maintaining power asymmetries between former colonizers and colonized. Using analysis of six case studies of electricity integration projects, Oseni and Pollitt (2016) argue that "trust building around electricity trading is possible even between countries with a history of conflict" (p. 636). The Mediterranean case study suggests that it is important to understand and to acknowledge historical injustice and to develop a just process in order to build trust. However, I also caution that the neocolonial critique of electricity integration largely originated in Europe and often did not include North African voices, which ironically impeded a lack of true partnership between North and South.

The Desertec vision was criticized as a European scheme to exploit North Africa's resources to develop a neocolonial resource colony in North Africa for clean electricity. Dii's CEO Paul van Son disputed this, stating in an interview with the *Tripoli Post*: "Your country [Libya] would never allow any more western industries to colonialize it. I cannot imagine it. There is no way of colonialism. Nothing will happen if a country doesn't like it" ("Renewable Energy is a Win-win Situation for Everyone, Says CEO of the Desertec Industrial Initiative (DII)," 2010, para. 18). Van Son's quotation implies that the neocolonial critique treats Africans as potential victims who lack sovereignty. Van Son further argued that colonization entails extracting something from a country, whereas in Dii's vision, countries would export a commodity that would be lost if it were not utilized (because electricity cannot be effectively stored). Several interviewees associated with Dii described claims of neocolonialism as "bullshit." On one hand, overlooking the problematic history of EU-North Africa collaboration and colonial-era projects could perpetuate colonial-era discourses and stereotypes of North Africa and taint plans for grid integration. On the other hand, it is also problematic to assume that North Africans would not want to export electricity to Europe without listening to their opinions and understanding their visions for energy integration. I found more support for exporting renewable energy from North Africa to Europe than European critics of the "neocolonial Desertec scheme" have assumed, but I also observed discontent related to the lack of procedural justice in the development of Eurocentric visions for integration.

Some interviewees distanced today's collaborations with Europe from the colonial past. For example, the director of a Moroccan research center that facilitates cooperation between European and Moroccan researchers, whose grandfather played an active role in achieving independence from France, stated,

> But generally, Morocco, we have no problem with our colonial past. . . . It was our history. . . . Now it's ok, we cooperate . . . no one said FDI is just French entrepreneurs using Moroccan workers. No, actually, they're bringing them jobs and know-how and opportunities to produce something meaningful.

This quotation illustrates that branding all interactions between North Africa and Europe as neocolonial is problematic because it could preclude fruitful collaborations in which knowledge is exchanged and created and economic value is added.

Other interviewees supported Desertec as a concept or idea but were dissatisfied in practice with the lack of avenues for North African participation in the development of the vision for energy integration. A Moroccan interviewee involved in North-South energy research collaboration encapsulated this point of view stating,

> The initial Desertec idea is a proposition, an idea that might be pertinent . . . so the concept itself is interesting. Now, of course, it was the case with this concept that one plus one must be worth eleven. Because one plus one could also be one . . . in this context it could be a total benefit to the European side. This should not be the case . . . and Dii had as its objective a bad strategy. . . . Dii's strategy was 100% commercial—it was European. And . . . the MENA region was only a tool for doing business, that's all. So, all must be added value (so that one plus one is equal to eleven) and [so] that the region benefits, be it through the creation of jobs, expertise, etc. . . . So personally I attended the meeting with Dii . . . where they said "and here we have a proposition, you will give us the best terrain in Morocco. And there, we will install the central power plants, we will take the electricity." So here is the proposition!. . . It can also be a form of colonization. I went, I took the best spot, I used it, et voila! That is it! And that's why I told you, there was a bad strategy or maybe there was not a strategy . . . and there was an amalgam. An amalgam between the Desertec Foundation, the Desertec concept, and the Desertec Industrial Initiative. . . . This model is (I am sorry for the term, I will say it in English) bullshit. It's the right term . . . because the idea, yes, the idea is good, and it is may be pertinent for both, for everyone etc., but the tool that was created because we thought it would help. On the contrary . . . it was not the right tool with the right strategy.

This interviewee did not see North-South electricity integration as intrinsically neocolonial. However, prioritizing procedural justice—including mutual priority setting, capacity building, and stakeholder participation—is essential in future North-South collaborations, especially given the colonial past. He then explained that global teamwork is needed in place of the attitude: "I decide and you provide." North-South energy integration must be based on a just procedure, mutual benefit, and value-added in order to achieve a true partnership.

One example of a Moroccan organization focused on electricity integration is Sahara Wind, which NATO partly funds. It seeks to develop wind power plants first for Morocco and later for export of surplus electricity to European and Maghrebian countries. In an interview, the head of Sahara Wind contrasted the initial use of the North Atlantic trade winds in the Age of Sail—which brought together the continents of Europe and North America to exploit the continent of Africa through the slave trade—to the positive use of Africa's renewable energy potential.

As I sat in a local café waiting for the interview to start, I watched footage on the local news of sub-Saharan African immigrants who were trying to climb the fence from Morocco into one of the two Spanish enclaves in Morocco while police beat them back. The head of Sahara Wind then discussed the inequity among continents stating,

> You're going to be exploited, through the slaves that are going to be taken and exploited in the US, and whatever resource you're going to have is going to be exploited by colonial powers, which will come thereafter. So with that hindsight, you have to see that those North Atlantic trade winds mean a lot for mankind, and you do have to see that hindsight in order to improve the history, not re-write it but learn from it and say: Are those winds just about two continents making fun of the rest of the world in exploiting its resources and peoples beyond limits? Or is there an opportunity to integrate another continent with a better story with a leadership that can look beyond . . . using perhaps a US president that is Black, and a new ambassador of the US to Morocco that will be Black.. . . And I think that, you know, you cannot go into the twenty-first century with a doomed continent, a continent of people trying to climb a fence. You see that and you go like—oh my God. You know this is the world we live in. And I don't think it's acceptable.

While this interviewee did not see North-South integration as intrinsically xenophobic, it could be xenophobic if people from both North and South Mediterranean are not treated as equals and if injustices are not redressed in the integration process. Another Moroccan energy expert stated that as long as people think that there are two classes of people within society, energy justice cannot be achieved. In a postcolonial world, Morocco should be tied to Europe only through a collaborative, equitable relationship, not through a coercive one. The director of Sahara Wind insisted that there must be Moroccan ownership of the process for this equitable relationship to be achieved.

This is not how the process evolved. After Dii and MASEN signed an MOU, the German Embassy held a press conference claiming that the Ouarzazate CSP plant was a Desertec pilot project. The media regularly conflated Desertec and the

Moroccan Solar Plan, fostering animosity by giving credit for Morocco's project to the EU. Desertec's founder stated that although foreign experts were consulted, Moroccans should receive the credit for the Moroccan Solar Plan. He said, "The less they give credit to others, the more it is their project and their responsibility." While Europeans involved in the process are responsible for having taken credit for projects that were not their own, the Moroccan government bears some responsibility for consistently sourcing experts from abroad rather than from Morocco.

Not all collaboration between North and South Mediterranean is intrinsically neocolonial and assuming as much could impede opportunities for beneficial North-South partnership. However, I have argued that the implementation of North-South energy integration would be more complex than actors initially appreciated. While Moroccan policymakers generally viewed regional electricity integration as a desirable scenario, they did not view it as an entirely feasible one due to the sociopolitical complexity that would have to be negotiated to achieve it. A regional Mediterranean regulatory agency would face the daunting task of navigating various goals, visions, and priorities across North and South. The case of Morocco—just one of the countries in North Africa included in the Desertec vision—illustrates that individual nation-states have specific societal goals for their electricity systems and that full deregulation is unlikely to be achieved. While some common ground exists between North and South, this common ground does not encompass all of the priorities that countries have for their energy systems. The framing of the vision as a win-win may serve to exclude certain definitions of what a loss would be. If grid integration were to proceed, the development of a just planning process would be essential to its success. North Africans should have power and agency in judging whether this process and the outcomes are just.

Morocco's integrated projects model, in its ideal form, could be used to better theorize national innovation systems in the context of North-South collaboration. Viotti (2002) argued that processes of technological change in industrializing countries should be analyzed using the concept of national learning systems rather than of national innovation systems. Learning is the process of absorbing a technology, conducting incremental innovation on it, and diffusing the technology (ibid.). The Moroccan case study illustrates that this depiction of innovation may be overly combative. Technologies are neither absorbed nor transferred. Partnership and collaboration are involved as well as competition. As one interviewee expressed, "No business will come to Morocco and transfer their know-how; this is merely false marketing. Working with foreign research teams is more likely to transfer know-how or to develop new knowledge." In the process of North-South collaboration, the technology could be altered to accommodate domestic needs. The Moroccan integrated projects model is a useful addition to national learning systems theory because it acknowledges the international exchange involved in industrial policy as well as the importance of capacity building and social value-added. It could also aid in thinking through how partnerships could be improved to enhance North-South collaboration and to increase the adaptation of existing technologies and the development of new technologies that, aid in fulfilling the promises of social benefit that governments make for electricity systems development. Capacity building through the expansion of industrial and R&D policy could also help

Morocco to avoid being coerced in North-South negotiations and technology-related partnerships. An integrated projects model would go beyond the "copy-apply" model of technology transfer to consider a technology's societal context and to achieve positive North-South partnership and socioeconomic benefits.

Conclusion

In this chapter, I examined North-South motivations for involvement in regional electricity integration. Recognizing that each of the 44 countries included in the Desertec vision has its own goals for electricity systems development is one of several substantial challenges for multilateral, regional electricity systems planning.

In spite of Morocco's focus on increasing its energy independence, I found that Moroccan government officials largely supported regional electricity integration. I found strong support in my interviews for the eventual export of renewable energy from Morocco to Europe, provided that export offers value-added for Morocco's development of a national innovation system and that there is a just process for developing plans for regional integration. Many Moroccan officials I interviewed saw regional integration as a facet of energy security and of sustainable development. Many interviewees viewed Morocco as unfairly excluded from European electricity markets, especially since Morocco generates electrons that are compatible with the EU's grid. The outcomes of the envisioned regional electricity system related to politics and justice are dependent on the development of a just integration procedure. Capacity building that enables less powerful countries to avoid coercion is a prerequisite to achieving just regional integration. The development of North-South partnership on energy research projects would be an important step toward the initiation of a Mediterranean electricity integration process that is not based on coercion but on procedural justice.

As the Desertec visioning process broke down, Morocco shifted its focus to exporting electricity to sub-Saharan Africa, which could be a desirable outcome for reducing energy poverty, although it is a lost opportunity for Europe. Because of its relative wealth, Morocco would be the regional power if it deepens its integration with and electricity export to sub-Saharan Africa. Thus, Morocco should apply the lessons stemming from its rocky experience with European electricity integration to any future cooperation with its neighbors to the south. Moroccans seek to develop the capacity to have a seat at the table and to be considered true collaborators in technology development. Extending Morocco's integrated solar projects model in its ideal form to other parts of the region could foster North-South partnership to address social goals and to treat Moroccan experts as true contributors in energy innovation instead of technology adopters.

Notes

1 Single quotation marks indicate a non-verbatim quotation and double quotation marks indicate a verbatim quotation.

2 Grid parity is reached when the levelized cost of electricity from CSP plants is reduced to the average wholesale price of electricity from the grid.

3 The problems stemming from errors in estimating demand can be seen in the case of South Africa in the 1990s (Victor & Heller, 2006).

4 Victor and Heller (2006) argued that neoliberal reforms in the power sectors of developing countries resulted in "dual-market" systems "that combine some features of textbook reform with powerful residues of state monopolies" (Kindle location 390–391), which they view as "stable and likely outcomes" rather than as "aberrations" (Kindle location 407).

Works cited

Alder, K. (1997). Engineering the revolution: Arms and enlightenment in France, 1763–1815. Chicago, IL: University of Chicago Press.

Alternative Energy Africa. (2018, March 15). Algeria to export electricity to Libya and Tunisia. Retrieved June 1, 2018, from https://www.ae-africa.com/read_article.php?NID=9005

Appadurai, A. (1996). *Modernity at large: Cultural dimensions of globalization*. Minneapolis, MN: University of Minnesota Press.

Ben Ali, D. (1997). Economic adjustment and political liberalization in Morocco. In H. A. Handoussa (Ed.), *Economic transition in the Middle East: Global challenges and adjustment strategies* (pp. 183–218). Cairo, Egypt: American University in Cairo Press.

Berthonnet, A. (2008). La formation d'une culture économique et technique en Algérie (1830–1962): L'exemple des grandes infrastructures de génie civil. *French Colonial History*, *9*(1), 37–63. Retrieved from https://doi.org/10.1353/fch.0.0000

Binka, F. (2005). Editorial: North-South research collaborations: A move towards a true partnership? *Tropical Medicine & International Health*, *10*(3), 207–209. Retrieved from https://doi.org/10.1111/j.1365-3156.2004.01373.x

Catusse, M., Vairel, F., & Bono, I. (2010). *Le Maroc de Mohammed VI: Mobilisations et action publique*. Paris, France: Karthala.

Dawson, C. (2009). *EU Integration with North Africa: Trade negotiations and democracy deficits in Morocco*. London: I.B.Tauris.

de Villemeur, E. B., & Pineau, P-O. (2012). Regulation and electricity market integration: When trade introduces inefficiencies. *Energy Economics*, *34*(2), 529–535. Retrieved from https://doi.org/10.1016/j.eneco.2011.12.004

Debbarh, A. M. (2006). *L'Énergie : Développement énergétique au Maroc depuis 1955, perspectives 2025*. Association Nationale des Collectivités Locales du Maroc. Retrieved from http://doc.abhatoo.net.ma/DOC/IMG/pdf/energie_devpt_energetique_ma.pdf

Doty, R. L. (1996). *Imperial encounters: The politics of representation in North-South relations*. Minneapolis, MN: University of Minnesota Press.

Dubash, N. K. (2003). Revisiting electricity reform: The case for a sustainable development approach. *Utilities Policy*, *11*(3), 143–154. Retrieved from https://doi.org/10.1016/S0957-1787(03)00044-4

Dubash, N. K. (2005, December). *Regulation as an arena for social policy: Examples from electricity in Asia*. Presented at the Arusha Conference, New Frontiers of Social Policy.

Dubash, N. K., & Williams, J. (2006). The political economy of electricity liberalization. In J. Byrne, N. Toly, & L. Glover (Eds.), *Transforming power: Energy, environment, and society in conflict* (pp. 155–190). Piscataway, NJ: Transaction Publishers.

Energy Sector Management Assistance Program (ESMAP). (2011, January 20). *Power sector financial vulnerability assessment impact of the credit crisis on investments in the power sector: The case of Morocco*. World Bank. Retrieved from www.esmap.org/node/1128

ENTSO-E. (n.d.). ENTSO-E Transmission System Map Retrieved June 1, 2018 from https://www.entsoe.eu/data/map/

ESMAP. (2010). *Regional power integration: Lessons from global case studies and literature review* (No. Briefing Note 004/10). Washington, DC: Energy Sector Management Assistance Program, The World Bank. Retrieved from www.esmap.org/node/353

Gaillard, J. F. (1994). North-South research partnership: Is collaboration possible between unequal partners? *Knowledge and Policy*, *7*(2), 31–63. Retrieved from https://doi.org/10.1007/BF02692761

Gnad, O., & Viëtor, M. (2011, July 1). Desert powered progress. *IP Journal*. Retrieved February 18, 2013, from https://ip-journal.dgap.org/en/ip-journal/topics/desert-powered-progress

Hall, A. (2002). Innovation systems and capacity development: An agenda for North-South research collaboration? *International Journal of Technology Management and Sustainable Development*, *1*(3), 146–152.

Haut-Commissariat au Plan. (2006). *Prospective Maroc 2030: Energie 2030 Quelles options pour le Maroc?* Retrieved from www.hcp.ma/file/111440/

Hecht, G. (2009). *The radiance of France: Nuclear power and national identity after World War II* (new ed.). Cambridge, MA: MIT Press.

Hira, A., & Amaya, L. (2003). Does energy integrate? *Energy Policy*, *31*(2), 185–199. Retrieved from https://doi.org/10.1016/S0301-4215(02)00024-1

Hoven, A. van den, & Froschauer, K. (2004). Limiting regional electricity sector integration and market reform the cases of France in the EU and Canada in the NAFTA region. *Comparative Political Studies*, *37*(9), 1079–1103. Retrieved from https://doi.org/10.1177/0010414004268845

Hurrell, A., & Sengupta, S. (2012). Emerging powers, North-South relations and global climate politics. *International Affairs*, *88*(3), 463–484. Retrieved from https://doi.org/10.1111/j.1468-2346.2012.01084.x

Hwang, K. (2008). International collaboration in multilayered center-periphery in the globalization of science and technology. *Science, Technology, & Human Values*, *33*(1), 101–133. Retrieved from https://doi.org/10.1177/0162243907306196

Indicateurs Clés. (2014). *Ministry of Energy, Mines, Water, and Environment*. Retrieved June 28, 2014, from www.mem.gov.ma/SitePages/IndicateursCles.aspx

IEA International Energy Agency. (2016). Large-scale electricity interconnection: Technology and prospects for cross-regional networks. Paris, France: OECD/IEA. Retrieved from https://www.iea.org/publications/freepublications/publication/Interconnection.pdf

ICA Infrastructure Consortium for Africa. (2016, November.) Updated Regional Status in Africa Power Pools Report. Abidjan, Ivory Coast: ICA Secretariat. Retrieved from https://www.icafrica.org/fileadmin/documents/Publications/Regional_Power_Pools_report_April17.pdf

Katz, J. S., & Martin, B. R. (1997). What is research collaboration? *Research Policy*, *26*(1), 1–18. Retrieved from https://doi.rg/10.1016/S0048-7333(96)00917-1

Kinsella, T. (2009, July 20). Politics abdicating responsibility to experts. *The Irish Times*. Retrieved February 20, 2013, from LexisNexis database.

Lilliestam, J. (2014). Vulnerability to terrorist attacks in European electricity decarbonisation scenarios: Comparing renewable electricity imports to gas imports. *Energy Policy*, *66*, 234–248. Retrieved from https://doi.org/10.1016/j.enpol.2013.10.078

Lilliestam, J., & Ellenbeck, S. (2011). Energy security and renewable electricity trade—Will Desertec make Europe vulnerable to the "energy weapon"? *Energy Policy*, *39*(6), 3380–3391. Retrieved from https://doi.org/10.1016/j.enpol.2011.03.035

Maghraoui, D. (2009). The strengths and limits of religious reforms in Morocco. *Mediterranean Politics*, *14*(2), 195–211. Retrieved from https://doi.org/10.1080/13629390902985976

Mervis, J., & Normile, D. (1998). Lopsided partnerships give way to real collaboration. *Science*, *279*(5356), 1477–1479. Retrieved from https://doi.org/10.1126/science.279.5356.1477

Moore, S. (2017). Evaluating the energy security of electricity interdependence: Perspectives from Morocco. *Energy Research & Social Science*, *24*, 21–29. Retrieved from https://doi.org/10.1016/j.erss.2016.12.008

Nair, K. N., & Menon, V. (2002, November). *Capacity building for demand-led research: Issues and priorities*. Retrieved from http://213.154.74.164/invenio/record/15635/files/Nair-2002-Capacity.pdf

Oseni, M. O., & Pollitt, M. G. (2016). The promotion of regional integration of electricity markets: Lessons for developing countries. *Energy Policy*, *88*, 628–638. Retrieved from https://doi.org/10.1016/j.enpol.2015.09.007

Peters, S., & Westphal, K. (2013). Chapter 5: Global energy supply: Scale, perception and the return to geopolitics. In M. J. Trombetta & H. Dyer (Eds.), *International handbook of energy security* (pp. 92–113). Cheltenham, UK: Edward Elgar Publishing.

Puka, L., & Szulecki, K. (2014). The politics and economics of cross-border electricity infrastructure: A framework for analysis. *Energy Research & Social Science*, *4*, 124–134. Retrieved from https://doi.org/10.1016/j.erss.2014.10.003

Ratcliffe, V. (2016, June 27). Morocco studying connections with neighbors. Interfax Global Energy. Retrieved from http://interfaxenergy.com/gasdaily/article/20813/morocco-studying-connections-with-neighbours

Renewable energy is a win-win situation for everyone, says CEO of the Desertec Industrial Initiative (DII). (2010, October 25). *The Tripoli Post*. Retrieved April 26, 2015, from www.tripolipost.com/articledetail.asp?c=2&i=5034

Rosseel, P., De Corte, E., Blommaert, J., & Verniers, E. (n.d.). *Approaches to North-South, South-South and North-South-South collaboration*. Retrieved from https://lirias.kuleuven.be/bitstream/123456789/229636/1/POLICY_PAPER_VLIR_UWC_NSS.pdf

Said, E. W. (1994). *Culture and imperialism* (Reprint ed.). New York, NY: Vintage Books.

Sarewitz, D. (2004). How science makes environmental controversies worse. *Environmental Science & Policy*, *7*, 385–403.

Saul, S. (2002). L'électrification du Maroc à l'époque du protectorat. *Outre-Mers*, *89*(334–335), 491–512.

van der Vleuten, E., & Lagendijk, V. (2010). Interpreting transnational infrastructure vulnerability: European blackout and the historical dynamics of transnational electricity governance. *Energy Policy*, *38*(4), 2053–2062. Retrieved from https://doi.org/10.1016/j.enpol.2009.11.030

Velho, L. (2002). North-South collaboration and systems of innovation. *International Journal of Technology Management & Sustainable Development*, *1*(3), 171.

Victor, D., & Heller, T. (Eds.). (2006). *The political economy of power sector reform: The experiences of five major developing countries*. Cambridge: Cambridge University Press.

Vidican, G. (2015). The emergence of a solar energy innovation system in Morocco: A governance perspective. *Innovation and Development, 5*(2), 225-240. Retrieved from https://doi.org/10.1080/2157930X.2015.1049852

Viotti, E. B. (2002). National learning systems: A new approach on technological change in late industrializing economies and evidences from the cases of Brazil and South Korea. *Technological Forecasting and Social Change*, *69*(7), 653–680. Retrieved from https://doi.org/10.1016/S0040-1625(01)00167-6

Winzer, C. (2012). Conceptualizing energy security. *Energy Policy*, *46*, 36–48. Retrieved from https://doi.org/10.1016/j.enpol.2012.02.067

7 Socially sustainable solar power development

From national dreams to local outcomes

Introduction: solar power siting and sustainability

Electricity systems are being called on to meet complex sustainability challenges in the twenty-first century that go beyond providing light and heat. The construction of large-scale renewable energy has obvious environmental benefits in terms of reducing greenhouse gas emissions and other types of pollution. However, it would be a mistake to assume that a transition to renewable energy will automatically result in socially sustainable development. Poorly planned clean energy development could perpetuate the inequities of the existing electrical power system as well as produce new configurations of winners and losers. While technologists advocating for concentrating solar power (CSP) argue that a surfeit of land exists to build enough CSP to meet global electricity demand, each plot of land is tied to complex and sometimes conflictual social values and livelihood needs. This makes it difficult to build a solar power plant that is sustainable not only in terms of low greenhouse gas emissions but also in terms of the improvement of social equity and the avoidance of "energy boom town" effects, in which an influx of energy investment in an area fails to provide long-term economic growth that is sustained and that improves livelihoods (e.g., Freudenburg, 1981). Building a socially, environmentally, and economically sustainable energy system is what sustainability practitioners term a "wicked problem": one that is highly complex, systems-level, and addressable only by developing workable solutions rather than optimal ones.

This chapter explores the social sustainability of CSP by examining its relationship to existing social inequities in the region in which the first CSP plant of the Moroccan Solar Plan is under construction, public participation in the siting process, connections between CSP and scarce water resources, and the effects of centralized versus distributed solar power on sustainable development. A focus on the local scale is useful in understanding how macroscale energy transitions shape people's livelihoods in localities, as well as in thinking in generalizable terms about how energy policy choices could be adapted to improve these outcomes. The effects of wind and solar power siting have been largely studied utilizing developed country case studies for example: Devine-Wright's edited volume on public participation on wind, wave, and solar siting in Western Europe (2011); Apostol et al.'s volume on changing renewable energy landscapes in the United States and Europe (2016); and articles focusing on large-scale solar siting conflicts in the deserts of the US Southwest (Carlisle, Kane, Solan, & Joe, 2014; Hunold & Leitner, 2011; Moore & Hackett, 2016; Pasqualetti & Schwartz, 2011; Phadke, 2010). Yenneti, Day, and Golubchikov (2016) write of a case in India of communal land dispossession due to a large-scale solar plant that is similar to the Moroccan case and argue that a lack of attention is given to the social justice aspects of renewables. Additionally, Valentine (2015) discusses the social, economic, technological, and political aspects of wind power siting in the United States, Canada, Europe, China, and Japan and focuses on the "gradualist politics" and "pragmatic planning" that underpin large-scale wind development initiatives in China. This research sheds light on the reasons for the public's resistance to energy facilities sited in their backyards, including the instrumental issues of noise, reduction of property values, visual impacts, and traffic, as well as the sociocultural issues of sense of place and place attachment, values for endangered species and wildlands protection, urban-rural inequity, and opposition to centralized power systems. While the majority of the literature focuses on developed countries, challenges of place-making, place attachment, and livelihoods related to land use are not limited to developing countries but exist across the globe. Moreover, the issues at stake extend beyond place-making to include sustainable development and livelihood concerns, procedural justice aspects, food-energy-water nexus considerations, and the effects of centralized versus distributed solar system configurations.

Research is lacking that evaluates whether transnational renewable energy export poses new considerations related to the social sustainability of centralized renewables. This chapter makes speculations about the potential social sustainability implications at the local level of the export of renewable electricity across nation-state borders. While no concrete plans exist to export the electricity from Moroccan CSP plants, this case offers a thought experiment for considering the connection of justice outcomes across scales (from local to global) in a future transnational renewable electricity system. In Chapter 1, I recounted sitting along the Friedrichstraße in Berlin wondering what the ramifications would be of powering this shopping district with renewable energy generated from power plants adjacent to small villages in Morocco that lack clean drinking water. I asked: What would the outcomes be for social and ecological justice if the power system were globalized to bring together disparate places, such as the commercial center of East Berlin with its storied past and a small village at the door of the Sahara that lacks clean drinking water? Will "premium networked spaces" be expanded, or will locals shoulder the burdens of energy development without gaining any benefits?

While Chapter 5 considered citizens who have been excluded from national sustainable development plans, this chapter focuses on citizens who could by harmed by policies intended to promote sustainable development. I made six trips to Ouarzazate to conduct this research: two independently (in 2012 and 2013), four as a co-facilitator of a study-abroad experience for students (in 2013–2016), and one as a participant in a portion of a research project headed by Boris Schinke and a team of German, Egyptian, and Moroccan researchers (in 2014). This project sought to understand how the siting, construction, and operation of Noor I would affect communities living in the region. The report—*Social CSP*—details the existing and projected social impacts of Noor I (Terrapon-Pfaff et al., 2015), and Terrapon-Pfaff, Fink, Viebahn, and Jamea (2017) discuss the technocratic methodology used for social impact assessment of various criteria as well as the participatory methodology with input from focus groups. In this chapter, I use some data collected while working with the *Social CSP* team (including some data that was not in the report), as clearly noted throughout, as well as data that I collected during my independent research.

This chapter begins with a comparison between Morocco's dam-building policy under King Hassan II and the CSP-building policy under King Mohammed IV. Then, it overviews the existing inequities in the Ouarzazate region and discusses the siting process for CSP plants in Ouarzazate, Morocco. People wanted a say in development projects sited in their backyards. However, the community's participation was impeded by its lack of capacity due to preexisting inequities, the structure of the process, and the lack of legitimacy of community representatives. Even so, I found that citizens opposed the process and its effects on land use rather than the technology itself. Next, this chapter discusses energy-water nexus challenges associated with CSP plants. This chapter concludes by comparing the benefits and drawbacks of the Noor-Ouarzazate facility to local development projects initiated by NGOs in Ouarzazate, arguing that both are simultaneously required in order to achieve socially sustainable development.

Doing better than dams: the legacy of large-scale energy development in the Atlas Mountains

The first CSP facility under the Moroccan Solar Plan is under construction in Ouarzazate, Morocco. This facility, called Noor or Noor-Ouarzazate, is expected to be a 580 MW power plant for proof-of-concept of CSP trough and tower technologies, although the last phase will deploy large-scale solar PV panels. The Noor facility is mostly funded by concessionary financing, including loans from eight development banks and governments, and is expected to cost roughly $2.6 billion.[1] The leading project developer is a subsidiary of ACWA Power, a Saudi company described in Chapter 5. The first phase, Noor I, is a 160 MW wet-cooled solar trough plant with three hours of molten salt storage, started in 2013 and commissioned in February 2016. According to a MASEN interviewee, the electricity from Noor I costs roughly 18 US cents per kWh. Noor II is a 200 MW CSP dry cooled trough plant with seven hours of molten salt storage, expected to be completed in 2018. Its electricity price is estimated to be 14 cents/kWh. Noor III is a 150 MW dry cooled CSP tower plant with seven hours of molten salt storage, which is under construction and expected to be completed in 2018. Electricity from Noor III is expected to cost 15 cents/kWh, which is higher than

Noor II because a less proven CSP technology will be developed. Finally, Noor IV is a 72 MW utility-scale demonstration of solar PV panels, which ACWA Power will complete in 2018. With the costs of utility-scale solar PV panels plummeting worldwide, electricity from Noor IV is expected to cost only 5 cents per kWh. It also appears that the CSP technology came down the cost curve because of experience, as expected. In September 2017, ACWA Power won a bid to develop a CSP project with molten salt storage in Dubai at record low costs ($73/MWh) ("ACWA Power scales up tower-trough design to set record-low CSP price," 2017). Saudi Arabia plans to construct 200 GW of solar by 2030 for domestic use and export, and it recently switched its plan from CSP to PV technology (St. John, 2018).

In contrast to the European imaginary of Saharan technological systems grafted onto an empty desert landscape (described in Chapter 2), Noor is intentionally being built in populated areas in order to include the region's inhabitants in socioeconomic development. It is important to note that while the overall system cost estimates for Desertec and similar visions highly depend on solar insolation levels, in reality choosing sites is a social process. Solar insolation is just one of many factors that shape siting decisions. Ouarzazate is one of the "doors" to the Sahara, which Bisson (2003) described as populated areas on the outskirts of the Sahara Desert that indelibly shape and are shaped by the Sahara. The population of the city of Ouarzazate in 2014 was 113,752, and the population of the Province of Ouarzazate was 297,502 (Haut-Commissariat au Plan, 2014). Ouarzazate is a relatively poor region with a largely agrarian economy except for its long-standing movie business that has produced films such as Lawrence of Arabia and Gladiator. The main economic activities outside the movie business include subsistence agriculture, tourism, a few cash crops such as dates, and small handicrafts (e.g., textiles, rugs, ornamental daggers).[2]

An interviewee from MASEN explained that Morocco's power plants have been historically built in coastal areas, and the "veins of the grid inject energy into the mountainous areas." He said that the Moroccan Solar Plan would reverse this situation, and the mountainous and desert regions would become the major energy producers for coastal population centers. The areas of the country with the most sunlight are also the poorest areas, and the Moroccan Solar Plan could change the geography of energy and the distribution of winners and losers from energy development. Thermal power plants indeed have been historically built near Morocco's coasts, dating back to the first coal-fired power plant, Roches-Noires, which was built in the port city of Casablanca. However, mountainous areas do not provide the bulk of Morocco's power capacity, but they have historically been sites of renewable energy production from hydroelectric facilities built under Morocco's dam-building policy, *la politique des barrages*. The French Protectorate initiated this policy, and King Hassan II continued it after Morocco gained independence (Swearingen, 1998). The policy has interwoven Morocco's agricultural policy with its energy policy through the concomitant construction of irrigation and hydropower in the Atlas Mountains.

Advocates of CSP have framed it as a farming and irrigation technology since the 1800s.[3] Today, CSP plants are metaphorical farms integrated into Morocco's electricity and water system. The Mansour Eddahbi Reservoir provides cooling water for Noor I. It was built under *la politique des barrages* in 1972 along the Draa River, with a height of 70 meters and an initial reservoir capacity of 560 million cubic meters (Swearingen, 1988). It was designed to irrigate 19,000 hectares, to contain flooding, and to provide electricity (ibid.).[4] Noor I is a solar farm irrigated by the *politique des barrages* legacy.

While *la politique des barrages* sought to provide for Morocco's nutritional needs (ibid.), CSP policy seeks to provide for Morocco's increasing electricity needs. Through dam building, the Kingdom of Morocco promoted regional development to raise the rural standard of living and to stem rural exodus (ibid.). The same goals apply to CSP. Just as King Hassan II hoped that agricultural projects made possible by intensive irrigation would provide cash crops for export (ibid.), the current government views green electrons from CSP plants as a potential export commodity. Interviewees from Ouarzazate spoke of electricity exports as a means of "earning foreign currency," which was also a goal of *la politique des barrages*. As Morocco transitions from *la politique des barrages* to the Moroccan Solar Plan, the king is balancing Morocco's past identity as an agrarian society with its planned future as a knowledge economy with a robust national innovation system. CSP, therefore, straddles the gap between an agrarian and industrialized society, and CSP is framed as a solar farming technology that contributes to the development of a domestic supply chain and a green commodity for export. In Chapter 4, I illustrated that the European promoters of CSP view innovation as accelerating the adoption of a proven technology rather than as developing new technologies. In contrast, Moroccan CSP policy hybridizes the past, present, and future, reflecting a balancing act that has long been important to Morocco's development. Shale oil or gas could be exported but would not offer the historical continuity that CSP provides.

In addition to the similarities between *la politique des barrages* and the Moroccan Solar Plan, dam building can be used as a heuristic benchmark for comparison to solar power. The stipulation that people must benefit locally from sustainable development from solar power demands a higher standard than dam building. Under *la politique des barrages*, villagers from Idelsane were displaced when their land was flooded by the creation of the Mansour Eddahbi Reservoir, and they were excluded from socioeconomic development relating to the project. Being disadvantaged by energy development has lasting effects, which I witnessed when I visited the village of Idelsane across the highway from the Noor facility. Villagers with whom I spoke perceived themselves as still being disadvantaged by their displacement in the 1970s. They knew little about the Noor facility, and they were mostly concerned about their precarious water availability situation. Despite living near the reservoir, they must walk one kilometer up the highway to reach drinkable water because the reservoir's water quality is poor. They have dug wells only to reach brine. After a Moroccan colleague and I interviewed the villagers, we walked to the reservoir. There was no path, and it was a challenging and dirty trek. The water in the reservoir was low because of years of drought. The reservoir's outskirts resembled an estuary: The ground was coated with white powder and cracked, salty mud. The water was green from algal blooms, which reduce water quality. According to data from Swearingen (1988), the reservoir was not designed to provide drinking water, which Ouarzazate badly needs today. Interviewees said that people sometimes drink water from the reservoir even though it is dirty because of their lack of access to clean drinking water. This anecdote illustrates systemic problems associated with dam building (Khagram, 2004; World Commission on Dams, 2000) and emphasizes the importance of avoiding past problems caused by centralized energy infrastructure.

Diverging from past dam building processes, MASEN sought to engage the local community in Noor's development and to provide them with benefits through an integrated project plan. As previously discussed, Morocco seeks to align technological development end goals (i.e., the provision of electrons) with social development goals (e.g., increasing jobs, developing rural, areas improving education). While Dii saw itself as a project accelerator, MASEN viewed itself as an "animator" for local development. A MASEN interviewee stated that MASEN's projects should have an impact from kilometer zero to kilometer 1,000, or the entire country. He continued, "Since there is a negative correlation in Morocco between solar radiation and development, we are working in the poorest areas of the country." MASEN's main development requirement for Noor I was to locally source 30% of the power plant's materials in order to develop supply chain industries and to foster small enterprises that provide services such as maintenance and security. By September 2014, ACWA Power had hired 1,200 to 1,600 construction workers for Noor I. About 80% of the jobs went to low-skilled laborers, 10% to semi-skilled, and 10% to skilled (Terrapon-Pfaff et al., 2015). However, Hanger et al. (2016) found that more than half of the public from the Ouarzazate area wanted to work at the Noor facility, and they thought that there would be more available jobs. Moreover, Noor I's workforce needs did not align with the skills of people from neighboring villages.

In addition to discussions about the distribution of benefits and drawbacks at the local and national scales, there are discussions of exporting electricity from localities such as Ouarzazate across nation-state lines. There are similar visions to export electricity from dams. For example, electricity from the proposed Grand Inga Dam in Democratic Republic of the Congo is envisioned for export as far as Europe and the Middle East (Green, Sovacool, & Hancock, 2015). If solar electricity exports are pursued in Morocco, how would people living near power plant sites expect to benefit, and how much benefit would they view as sufficient? I observed two prevailing perspectives at Dii conferences and in European media related to electricity exports from North Africa to Europe. One was that electricity exports will drain electricity out of North Africa and, therefore, will be neocolonial. Surprisingly, interviews conducted by the *Social CSP* researchers in the Province of Ouarzazate found nearly unanimous general public support of exporting electricity from Moroccan CSP plants to Europe, provided that people living in the province benefit. A second perspective was that

North African countries want to export electricity and that socioeconomic benefits will automatically trickle down to the local population through macroeconomic growth. In contrast, the case study of Noor-Ouarzazate illustrates that while there could be socioeconomic benefits from electricity exports, they are not automatic but instead are shaped by existing inequities in Morocco and the Mediterranean region. The case study also shows that people wanted a say in how they would be affected by the Noor facility, which suggests that they would also want a say in how they would be affected by the export of electricity. Moreover, local benefit must be ensured in renewable energy development projects built for export in order to avoid, for example, tying siting drawbacks in Morocco to benefits people gain from energy wealth in Germany.

Social inequity in the Ouarzazate region

The Ouarzazate region reflects broader development trends and inequities in the Mediterranean region. Therefore, it provides a useful case study not only for understanding the implications of the development of CSP in Morocco but also for anticipating these implications in other rural areas in North Africa. Roux and Guerraoui (1997) described the Mediterranean as a region in which wealthy coastal areas with higher rates of economic and population growth receive the majority of the fruits of development, while the depressed mountainous zones are left with weak economic growth and dilapidated infrastructure. (The French Protectorate derisively called these areas "*le Maroc inutile,*" or unuseful Morocco.) Roux and Guerraoui argued that the inequities between these coastal and mountainous areas are more significant than the inequities between the north and south of the Mediterranean. The previous chapters of this book demonstrated that there are indeed significant inequities and political disagreements between North and South. However, Roux and Guerraoui's argument highlights the layers of inequity within and across the Mediterranean that would affect the energy justice outcomes for a Mediterranean-wide electricity system.

Inequity and inequality in Morocco impede the capacity of citizens to participate in energy governance. It is through participation that citizens can ensure that they will not be the losers of renewable energy development. The government views electricity development and provision as boons to the equalization of Morocco's inequalities, but I see them as insufficient to the task. Rida Lamrini's 2006 book—*Is There a Future for Morocco, Yasmina Asked Me*—discussed the effects of social inequity on Morocco's development. The first time Yasmina asked Lamrini whether there is a future for Morocco, Lamrini considered the improvements in development statistics made since the 1990s. He considered whether nearly universal electrification rates might be indicative of a promising future for Morocco stating, "The countryside has rejected the mantle of darkness that enveloped the evening to offer an electrified spectacle of starry constellations" (p. 10).[5] In the end, he was cynical about Morocco's future because of a combination of state dysfunctions (e.g., corruption, lack of competitiveness compared to other emerging economies, lack of political will) and indifference of Moroccan citizens (e.g., Moroccan youth who favor emigrating rather than staying to improve the country). A Moroccan interviewee blamed Morocco's problems on social structures stating,

> The first danger for men is men. There are people who think there are first- and second-class people—that's a problem. It's not a lack of resources [Morocco suffers from] but a lack of respect.. . . Morocco must valorize people before natural resources in order to develop renewable energy.

These so-called second-class people are prevented from participating in renewable energy development.

The situation in the Ouarzazate region reflects these problems related to the marginalizing effect of social structures and to the inadequacy of statistics to the task of measuring human development. Social inequality between the rural Ouarzazate region and major cities, such as Marrakesh, is not determined by—but is shaped by—geographic isolation. To borrow a metaphor from Chapter 3, it is "islanded," meaning that it has the opposite of the characteristics of a "premium networked space" because it is remote, cut off, and lacks economic diversification. Ouarzazate is situated on a plateau south of the High Atlas Mountains. Other than a commercial flight from Casablanca that serves tourists and business people, Ouarzazate is reached by a winding road from Marrakesh through the High Atlas Mountains. It is only 120 miles from Marrakesh to Ouarzazate, but the trip takes about five and a half hours by bus. It requires crossing the Tichka Pass, which is among the most dangerous stretches of road in Morocco (see Figure 7.1). Moroccan friends with whom I traveled always called their family to let them know when they had safely crossed the pass. During the winter, the pass is sometimes snowed in, and buses from Marrakesh must turn back.

Ouarzazate's relative isolation impedes socioeconomic development. For example, when I visited Noor I during construction, the workers told me that the first shipment of mirrors had arrived broken because of bad roads. Many people in Ouarzazate viewed this isolation as the causal mechanism for their disadvantages. For example, I talked to small-scale producers of olive oil in the Province of Ouarzazate who emphasized how difficult it is to reach the market in Marrakesh. Inequality between rural areas and major urban centers results in rural exodus, which is often driven by desperation rather than a genuine desire to relocate. I visited the town of Agdz, which is an hour and a half drive from Ouarzazate, with researchers from the *Social CSP* group. I asked one of the Moroccan researchers about economic opportunities in Agdz. She explained that since Agdz does not have large businesses or factories that employ people, many young men migrate to the cities of Casablanca or Rabat looking for construction jobs. Many rural migrants end up living in slums in the outskirts of major cities. They pay 2,000 dirhams (roughly $210) per month to share a crowded apartment. If they get married, they marry someone from Agdz and only see their family on several holidays per year, which threatens social cohesion.

Urban-rural inequalities are significant, but focusing only on this category overlooks numerous layers of inequity that shape the capacity of the community

Figure 7.1 The Tizi n'Tichka Road
Photo by author

to participate in decision making. First, there are inequities even within cities. For example, in the outskirts and poor neighborhoods of Ouarzazate city people lack water and sanitation services, while the rest of the city has access (Terrapon-Pfaff et al., 2015). Second, there are inequalities in access to and affordability of health care. People in Ouarzazate who can afford it travel five and a half hours to Marrakesh for health care because specialists and hospitals are lacking in the Province of Ouarzazate. Third, Moroccan women lack equal rights and access to education. For example, in the Province of Ouarzazate, 84% of women are illiterate compared to 67% of men (Terrapon-Pfaff et al., 2015).[6] According to the head of a local NGO I interviewed, girls often cannot attend school because their family needs them to do domestic chores; families see it as unsafe for them to travel long distances to school; and school sanitation services are sometimes lacking, making attendance difficult after girls reach puberty. The NGO supports a boarding school for girls in Ouarzazate so that they can access a high school education.

Access to electricity also affects development in Ouarzazate. Almost 96% of households in the province have access to electricity (Terrapon-Pfaff et al., 2015). People in Ouarzazate primarily use electricity for lighting but also for television; cooking; and, in some cases, heating and cooling (ibid.). While electricity access ought to improve human capabilities, the director of a development NGO headquartered in Marrakesh—who is American—argued that electricity has harmed development in Morocco because it is used for television, which threatens social cohesion because people spend less time participating in community activities and searching for employment and more time on entertainment. Furthermore, he said that villagers have trouble paying the bills, and the electricity is not reliable. Extensive interview research in the Province of Ouarzazate found that the majority of people can afford their electricity bills, although they complained about price increases (Terrapon-Pfaff et al., 2015).[7] It is possible that TV usage threatens social cohesion in the province, but I did not study this issue.

While energy systems are seemingly invisible and taken for granted in Western countries, energy usage is visible in the Province of Ouarzazate. When a group of US students and I stopped for lunch outside Ouarzazate in 2013, a shopkeeper charged the students a high price for water. When they questioned the price, the shopkeeper—who did not speak English—gestured toward the bottle of butane running the refrigerator. Our drink was not cold for free. Butane improves human capabilities: It heats water for showers, fuels portable stovetops and refrigerators, and pumps water for irrigation. It also avoids the negative health implications caused by inhaling particulate matter from burning charcoal, which disproportionately affects women. However, there is a lack of access to and affordability of butane, even with subsidies. Villagers trek to Ouarzazate city and haul butane bottles back over treacherous roads with the help of donkeys. Some traditional fuels are also used in the region; I often saw hunchbacked Amazigh women plodding along the highway with large bushels of firewood slung over their backs. The firewood is used for cooking and heating; it is a cheaper fuel source but increases risk of lung disease. An interviewee who works in local development said that the most demanded renewable energy technology in rural areas is a subsidized solar water pump. These pumps allow people to avoid the journey to obtain butane bottles.

Ouarzazate's inequities impede the capacity of people from rural areas to become full citizens in energy development. The Noor facility is being built within the context of these socioeconomic inequities. It is under construction next to rural villages within the Rural Commune of Ghassate. According to the commune's president in an interview, the commune includes 44 villages and 10,000 inhabitants spread over 1,034 square kilometers (399 square miles) of land. Socioeconomic development in the commune has improved since 2002. The president said that reaching a 99% electrification rate was particularly important to the commune's development.

I observed that the villages adjacent to the power plant suffer from lower levels of development compared to the rest of the commune, and they lack clean drinking water and equal access to education. Tasselmante is the closest of a grouping of villages to Noor and is only yards from where construction is underway. It has a population of 326 people with 40 households (as of 2008) (Terrapon-Pfaff et al., 2015). The wells in the village provide salty water appropriate for washing but not fit for human consumption. When I visited Tasselmante with a group of US students in 2014 and 2015, we were asked to bring our own bottled water, even for boiled tea, so that no one from our group would become ill. Additionally, some interviewees reported that several households in the villages adjacent to Noor lack electricity, and these households will now live in the shadow of a major energy facility. (We could not confirm whether these houses do lack electricity). These villages urgently require clean water and energy access to meet sustainable development goals.

Citizenship and representation: the Noor-Ouarzazate public engagement process

In this section, I briefly overview the Noor siting process, which is discussed more extensively in Rignall (2016) and Terrapon-Pfaff et al. (2015). In Morocco, legal changes have given citizens the opportunity to become engaged in sustainable development projects. Framework law 99–12, *Charte Nationale de l'environnement et du Développement Durable*, requires adequate information provision and participation in sustainable development projects, and Article 12 of Morocco's new constitution allows for the participation of civic groups in public affairs (Terrapon-Pfaff et al., 2015). The siting of Noor-Ouarzazate included a public engagement process that was in compliance with legal regulations but was insufficient in terms of substantive participation. MASEN held three public meetings in a hotel in Ouarzazate in November 2010, March 2010, and November 2012 (ibid.). The first challenge was that Ouarzazate was far for people from Tasselmante and other villages to travel considering their lack of access to adequate transportation, and the hotel was an intimidating venue for them to visit (ibid.). Second, the documents explaining the project were only available in French and FusHa (Classical Arabic), whereas many people from the villages speak local dialects of Arabic and Tamazight and are illiterate (ibid.). Third, some people, especially women, perceived that they were not sufficiently educated to participate in the process (ibid.). One rural woman interviewed in her home by the *Social CSP* team said that she did not know of such things because "these walls are my frontiers." In short, although a public participation process existed, multiple inequities—illiteracy, gender disparity, lack of transportation, and isolation—resulted in the community's lack of capacity to meaningfully participate in the process.

The World Bank and contractors conducted an Environmental and Social Impact Assessment, and MASEN conducted additional studies, which made legible the anticipated impacts of the Noor-Ouarzazate power plants. While the Moroccan government framed the 3,043 hectares of land on which the Noor facility is being constructed as barren and unproductive, the land provided essential services for nearby villagers. Ghassate commune collectively owned the land, and villagers used it for grazing, gathering rocks for building materials, and gathering sticks for firewood (ibid.). (Rignall (2016) illustrates that land dispossession in Morocco and collective ownership of land has colonial roots.) Proceeds from the sale of the land were deposited into a compensatory fund for the commune to support the Social Development Plan (ibid.).

When I visited the village in early 2014, villagers complained that they had to sell their sheep that they used to graze on the land and that they could no longer gather firewood and rocks from the site. Additionally, construction encroached on the perimeter of the village, much closer than villagers understood would be the case. While MASEN constructed a new paved road for villagers to reach the highway to Ouarzazate, the power plant obstructed their typical footpath to Ouarzazate. Furthermore, the proceeds from the land sale largely benefited the center of the commune some 10 kilometers away, instead of those who had directly lost capabilities provided by the land. The villages closest to the power plant site lack access to clean water but will now live in the shadow of a major energy facility. As such, CSP is resulting in uneven development in the Ouarzazate region as those villages most affected were afforded the fewest benefits.

The process did not allow for direct citizen engagement at all stages. In those stages, representatives from the community were called on that the community did not necessarily consider to be legitimate. Representation substitutes for direct democracy when citizens agree to a social contract and then select representatives to speak for their needs (Brown, 2009). In the Noor case, both initial consent to a social contract and legitimate representation were lacking. Six representatives from the commune and nine governmental authorities from ONEE and MASEN approved the land sale for 1 Moroccan dirham per square meter, or 30.5 million dirhams (roughly $3.2 million). In a focus group conducted by Germanwatch, participants argued that their representatives in the sale did not consult the local population and that they did not truly represent the population (Terrapon-Pfaff et al., 2015).

Systemic issues exist in Morocco with lack of legitimate representation for villagers and elites capturing participatory processes (Catusse, Vairel, & Bono, 2010). The representatives do not necessarily share the interests of the local population—a fact that is often masked by the assumption that the key disparities in Morocco exist between urban and rural areas rather than within rural areas. The problems with representation and procedural injustice in Ouarzazate resulted in conflict. Villagers from Tasselmante and Tidgheste protested several times against what they perceived to be a lack of fair compensation for the land, and several protestors were even jailed (Terrapon-Pfaff et al., 2015).[8]

MASEN recognized that it was building power plants in an economically disadvantaged and isolated area, and it did, in fact, seek to improve the capacity of the community. In addition to the industrialization and employment goals previously mentioned, MASEN undertook "charity work" in the region to improve the region's resilience and guarantee MASEN's social license to operate. For example, MASEN funded entrepreneurship classes, eyeglasses, dentistry, Internet, and buses and bicycles for children to ride to school. It also supported a dormitory that allows rural girls and young women to attend school in Ouarzazate. MASEN's charity framing was well intentioned and some aspects of it went beyond preparing the region for economic investment to addressing the community's lack of capacity. However, under the "charity" framing, the government did not view villagers as true participants in deciding what projects should be supported under the Social Development Plan from the land sale.

In summary, there were engaged citizens who wanted to participate in the solar siting process who had important ideas, opinions, and knowledge, but they did not have the ability to participate because inequities and issues with representation and procedure impeded their capacity. At the national level, an interviewee stated, "What is important is to create the conditions for [everyone affected by new energy development] to become citizens." The Noor siting process did not enable this. Just as I argued in previous chapters that capacity building is needed at the national level to put Morocco on a level playing field in international energy negotiations and R&D with Europe, capacity-building is needed for local citizens in order to improve their ability to participate in decision making. This lack of capacity is a systemic issue that is not limited to a single power plant siting case.

The Solar Hajj: local pride and national legitimacy

While villagers were dissatisfied with the siting process and the compensatory mechanisms, local conflict had surprisingly little to do with the solar technology itself. The Moroccan government's decision to pursue a particular technology, instead of letting the free market or local population decide, was not a source of conflict. In Chapter 5, I argued that solar power has become part of national pride and nation-building in Morocco. In Ouarzazate, I observed that this sociotechnical imaginary had permeated the local level. Participants in the *Social CSP* interviews and focus groups expressed a sense of patriotism related to the project, which they were proud to have in their backyards. Survey research conducted by Hanger et al. (2016) found remarkably high levels of public support for Noor-Ouarzazate, with 91% of the local population in support.

In 2013, I gazed out of the window of a car going from Zagora to Agdz, several hours southeast of Ouarzazate. We passed a faded ONEE sign on which someone had scrawled *allahu akbar* (God is the greatest). We drove through winding roads with canyons covered by rocks and shrubs, free of visible technological infrastructure other than the road ahead and high-voltage transmission lines. The transmission lines took sharper turns and dives than the winding road, draping across the road and down canyons. They crisscrossed over hills on which the Arabic words for "God, Country, and King" were printed in white. I was reminded of the ONEE logo, in which a transmission line resembles the star on the Moroccan flag. The transmission lines, therefore, stood out to me as symbols of national pride etched onto a landscape free of other infrastructure. This scene reflected both the region's remoteness and its unusually high electrification rate compared to rural areas in sub-Saharan Africa. These lines are symbols that rural Moroccans see every day: technological feats that have been imprinted on the local conscience since Morocco's Rural Electrification Program began in 1996. This program laid the groundwork for the rural electricity system and fostered local pride in a centralized electrical power system that reaches most Moroccan villages.

Not only did villagers embrace national pride related to CSP development, they also afforded strong legitimacy to the king and largely blamed the problems that they had experienced on local representatives. One participant in the *Social CSP* Tasselmante focus group in February 2013 stated,

> The sale procedure should have been more transparent and clear. If the local authorities had told us clearly that they would want the land for the solar project, and especially that his majesty is involved, we would not have sold the land, we would have given it for free.

An interviewee for the *Social CSP* study said, "All in all, the CSP project is good for the region, it gives a certain honor to our area, and we would say that it is not only a local project, it is a national project." Interviewees were proud of Noor I and thought that it improved the reputation of the Ouarzazate region. The President of the Ghassate commune hoped that the Noor project would turn Ouarzazate into the site of a Hajj for solar energy.[9] He emphasized that, unlike a movie, Noor provided the region with a long-term project, attracting foreign companies and collaborators. People would make the Solar Hajj to Ouarzazate to work, study, and see large-scale solar power. As I made additional trips to Ouarzazate, I heard the term solar Hajji used to bestow respect upon outsiders who traveled to Ouarzazate to visit Noor and to learn about CSP.

A brief comparison-contrast of the CSP siting process in Morocco to CSP siting processes in the United States explains why I find it notable that people in Morocco did not oppose the CSP technology itself. I found that the siting process for the Ivanpah Solar Electric Generating System in California had procedural justice deficiencies (Moore & Hackett, 2016), as did Pasqualetti and Schwartz (2011) in CSP plant siting cases in Arizona. In Arizona, communities of Mexican immigrants who lived near the site lacked the necessary capacity to participate in the siting process, while community groups with more capacity opposed the CSP technology itself, partly because of its water usage (ibid.). In both California and Arizona, members of the public who wanted to participate in solar siting were not consulted early in the process, and they later faced a deluge of information, including lengthy reports written in technoscientific prose (e.g., the Solar Energy Development Programmatic Environmental Impact Statement for the deserts of the US Southwest totaled 11,000 pages). As in Ouarzazate, project developers framed the land slated for CSP development as unproductive, barren, and "previously disturbed" (Moore & Hackett, 2016). In California, some of the involved citizens and groups opposed the CSP technology itself for being too expensive and large, for disrupting viewsheds and "pristine" desert landscapes, and for killing or displacing threatened desert tortoises (ibid.). Many of these groups, such as Solar Done Right, support solar photovoltaic panels sited on rooftops not CSP plants sited in desert areas.

Such opposition to the CSP technology itself was distinctly absent in Morocco, where citizens had faith in the country's broader energy policy but wanted a voice in plans for local socioeconomic development projects and in decisions made about the use of their land and water. CSP is differently perceived and interpreted in various sociopolitical contexts. However, shortcomings in CSP siting processes applied in both the Kingdom of Morocco and the liberal democracy of the United States. These shortcomings underscore an urgent need to improve community engagement in order to achieve sustainable development outcomes. Effective citizen input would improve the procedural justice aspects of large-scale energy planning and improve the trade-off calculus of the project.

That said, CSP offers advantages over large-scale hydroelectric projects, which are still being pursued worldwide. For example, Noor did not displace communities; it did not lead to increased waterborne illnesses; and it cannot collapse and flood communities, which is a risk in the cases of the Kariba Dam in between Zambia and Zimbabwe and the Mosul Dam in Iraq. According to research I have conducted on transnational grid projects worldwide, most electricity export projects focus on hydroelectricity. While exporting electricity from CSP plants has trade-offs, it offers advantages over exporting from hydroelectric facilities. It is also important to note for CSP siting cases in other developing countries that opposition would have been greater if the local population had been displaced from their homes, compensatory mechanisms had not been offered, and national legitimacy did not exist. While Rignall (2016) is right to criticize Noor for its imposition on people's sovereignty over communal land, my focus overall in this chapter is to consider the difficulty of making energy policy choices related to "wicked problems" in a carbon-constrained world. I further consider these difficult choices in the remainder of this chapter.

CSP's water usage: trade-offs and win-win discourses

Win-win discourses have been prevalent on large-scale solar power plants and on visions for transnational renewable energy grids. McShane et al. (2011) criticize sustainable development discourses that set unrealistic expectations for wins across the triple bottom line (people, planet, and profits). Instead, the authors recommend evaluating a project's difficult decisions, value conflicts, conflicts between social goods, strategic choices, and problem frames. An interviewee expressed that, even at the local level, people hesitated to discuss the trade-offs of the Noor facility. He said that discussing the negative aspects of the Noor facility could bring *hshuma*—a Moroccan concept of shame—on the elites in the area. He drew a comparison to a marriage contract that does not mention divorce. However, it is important to discuss trade-offs in order to avoid negative outcomes where possible. The process of considering trade-offs unmasks the challenges associated with achieving sustainable development outcomes from large-scale solar projects and from the export of green electricity. If they are not addressed, these challenges can become intractable over time.

Building CSP plants requires trade-offs between energy and water usage. This is important in Morocco, which has capricious and unevenly distributed water resources and long periods of drought. The future of water resources in the Province of Ouarzazate is important, uncertain, and under researched. The province is water-scarce, with 360 cubic meters of available water per capita per year versus 700 cubic meters nationally: less than 1,000 cubic meters per capita per year is defined as the threshold for water scarcity (Terrapon-Pfaff et al., 2015). Around 98% of the population in the province has access to water, but there are large disparities in the quality and quantity of the water available within Ouarzazate city and across the broader province (ibid.). Water in the region has naturally high salinity, but salinity is increasing because of groundwater overdraft. Irrigation from the Mansour Eddahbi Reservoir remains essential for supporting subsistence and other types of agriculture in the region. Due to this water stress, new demands on water, including Noor, concern people living in the region. Moroccans suffered from significant water stress in late 2017, and people in Zagora held protests against the lack of drinking water (Mekouar, 2017).

Addressing the energy-water nexus problems of CSP technology is more complex than reducing greenhouse gas emissions alone. The Environmental and Social Impact Assessment projected that Noor I would use 1.75 million cubic meters of water per year for cooling and for washing mirrors, which the study classified as a minor to moderate impact (5 Capitals, 2012). It is unclear whether Noor I's water use will worsen water stress in the region because this has been insufficiently studied. Noor II and III will use dry cooling technology, which will significantly reduce water usage: reductions of 90% are typically expected from use of dry cooling. Diekkruger et al. (2010) illustrate that data on water resources in the Draa Valley in the Province of Ouarzazate are limited and crude where available. They also argue that water availability in the future is contingent on uncertain social and governance choices as well as on climate change. Reactions to Noor-Ouarzazate thus relate to decision making under uncertainty as well as to the perceptions of people coping with a precarious water situation. Additionally, if the reservoir were to run out of water, Noor I would be unable to function.

This situation also illustrates that, depending on what technology is used, exporting electricity could de facto export water, and renewable electricity exports could deplete a country's natural capital. Dasgupta (2010) refers to hidden resource uses as "implicit subsidies" on exports. In a multi-scalar power system, integrated challenges would connect the wealth of the commercial streets of Berlin to the water scarcity of Moroccan villages. Alternatively, water-scarce North Africa could de facto export water to countries in sub-Saharan Africa with more abundant water resources although with less electricity availability, resulting in a conflict between social goods across scales. Future water management at the Moroccan local and national scales would affect future electricity policy at the regional scale in complex and possibly unpredictable ways, particularly as climate change worsens. As described in Chapter 4, Desertec Design I included integrated solutions, combining water desalination with electricity generation. Design II excluded this focus on integrated solutions at the energy-water nexus; the Noor I case study illustrates the drawbacks of this change. As construction began on CSP plants in Morocco, the energy-water nexus gained attention at the local level. The Desertec Foundation posted a picture on Facebook of Noor I under construction with the caption, "The future has begun.. . . Help to spread the word: Energy Transition is possible!". Several commenters quipped that the word should be spread that water stress is possible in the MENA region. A just energy future has not begun for Morocco if it results in people at the local scale lacking access to a sufficient quantity of clean water. The Noor siting process illustrates that win-wins are difficult across scales. Elites in Europe and Morocco may benefit from a transnational renewable energy system, while Moroccans in Tasselmante may be negatively affected or bypassed by benefits. The project may contribute to global climate change mitigation, but construction may damage the local environment. There may be instances in which a trade-off does not become apparent until construction begins and in which value judgments concealed in a discourse of technological and market determinism mask trade-offs.

The de facto export of water also applies to agricultural and other exports, although this fact is often overlooked. Lack of water makes an agricultural export economy problematic for Morocco, especially because climate change forecasts indicate that water scarcity will increase. As an interviewee from the Moroccan government stated, "People see Morocco as an agricultural society, yet we have no water and we're well positioned within the region to excel at other exports." An interviewee from ONEE said,

> The problem is that today water, it's a resource that's rarer and rarer. And so, the dam policy was really successful in the past, but today we cannot continue to count on the dams for responding to water needs. It is necessary to go toward different procedures . . . like water desalination . . . but these procedures are . . . consumers of electricity.. . . When we think of the future, when we reflect on the availability of water, we must think about the availability of energy for responding to the availability of water.[10]

In 2016, I considered the difference between exporting water through agricultural exports versus through electricity exports when I visited a village in the Ouarzazate region that grows saffron. Saffron is not water intensive and is a cash crop that yields a higher rate of return than the village's other economic activities: carpet making and olive oil. (Saffron is not labor intensive during the growing season, although it is very labor intensive during the harvest. It takes 130 flowers to make 1 gram of saffron, yielding 4 to 5 kilograms of saffron per hectare.) A 20 kW solar PV system, installed in cooperation with Spanish researchers, pumps water from a 20 meter well for irrigation. Averaged over a year, saffron consumes about 100 cubic meters of water per day for a field of 60 hectares.[11] When I visited Noor I in 2016, managers reported that it used an average of 340 cubic meters of water per day for wet-cooled electricity production. However, MASEN has taken criticism of Noor I's water usage seriously. In a one-hour meeting with a MASEN official in 2016, he mentioned saving water at least 15 times. While dry cooling reduces the power plant's efficiency, MASEN and ACWA Power successfully lowered the cost of CSP because of experience with Noor I, which will make dry cooling more economical in future plants. With 90% water usage reductions from dry cooling technology, dry-cooled CSP plants will likely use less water than traditional crops that are not water intensive such as saffron. Therefore, the implicit subsidy on exported electricity from dry-cooled CSP likely would be lower than for most agricultural exports, depending on the quantity exported.[12] While dry cooling alone does not solve the nexus issues—particularly the perceptions of people who suffer from water scarcity in the region—it is a pronounced improvement that should be emulated in CSP development elsewhere in the region, especially where exports are planned.

Comparing centralized to distributed energy development in Morocco

The goal of electricity development from an energy justice and socially redistributive perspective is to redress the sort of inequities described in this chapter and to improve people's capabilities to live out rich and fulfilling lives (Sovacool & Dworkin, 2014). Therefore I ask: Do the shortcomings of the Noor siting process suggest that an energy transition that replaces centralized fossil fuels with centralized renewable energy would result in fewer improvements to human capabilities than one that replaces a centralized fossil fuel system with a distributed renewable electricity system? Distributed systems underpin a popular vision of a just energy system. E. F. Schumacher was an early advocate of small-scale technologies such as solar photovoltaics. His writing informed the appropriate technology movement, which emphasized that a democratic and open society can only be built using decentralized systems because centralized technologies, such as nuclear power plants, subject society to an insular, technocratic, and undemocratic form of governance (Sclove, 1995; Winner, 1986). In this section, I compare the Noor facility to distributed integrated development initiatives led by local NGOs.

Morocco's renewable energy strategy focuses on large-scale solar, wind, and hydroelectric power plants and illustrates that national policymakers conceive of a modern electricity system for Morocco as a centralized system (see Chapter 5). Distributed solar technologies have received less attention in Morocco than CSP plants and utility-scale solar PV panels. Morocco mostly lacks the necessary policies to develop a distributed electricity system (Mahmoud, Myrsalieva, El Forgani, Bouadane, Sarant, 2016). For example, rooftop and small-scale PV systems cannot be connected to the grid. Morocco developed a net metering policy that would allow owners of solar PV systems to receive credit for surplus electricity fed into the grid, but it has not been implemented at the household level (ibid.). Independent power producers can sell electricity directly to large industrial consumers, but they cannot yet access medium and high-voltage lines (ibid.). Some focus has been placed on developing a distributed solar hot water market to reduce demand for butane, but solar hot water has mostly been used in hotels that require large quantities of reliable hot water for tourists (Samborsky, Myrsalieva, & Mahmoud, 2013).

Unlike in the United States where solar PV panels are often seen as a sign of wealth, in Morocco solar PV panels have been stigmatized as appropriate only for the poorest, most remote dwellings that were not connected to the grid during the rural electrification program of the 1990s. When talking with citizens in Ouarzazate, I noted that they thought that having centralized grid access was more desirable than having access to solar PV panels. Similarly, over lunch at Dii's 2013 conference, Solar Impulse Captain Bertrand Piccard (see Chapter 3) recounted that the villagers he met in Ouarzazate were convinced that "the cable" was the only desirable method of electricity access. A partnership formed in 2017 between Masdar Clean Energy, a Emirati company that builds utility-scale solar and wind facilities, and ONEE to deploy solar home systems to Moroccans who lack grid access, including 290 Watt solar panels with three days of battery storage, a small refrigerator, and LED lamps (Masdar, 2018). So far, they have deployed over 19,000 of these systems (ibid.).

While I discussed earlier that there is no social movement that opposes CSP in Morocco, a few interviewees criticized the centralized nature of Morocco's solar plan and argued that exports from CSP could result in a rentier model in which government elites will benefit from rents from solar electricity exports, and the economy will depend on a single natural resource. One interviewee stated, "CSP poses a governance risk by keeping the current power structure intact by purchasing steam turbines . . . with funding provided by a rentier state [Saudi Arabia]." The interviewee believed that all centralized technologies have authoritarian and anti-democratic properties. Instead, energy policies ought to improve energy citizenship, meaning that they ought to promote citizen engagement in and ownership of energy development and to distribute power among private and public actors.

The decentralized integrated development initiatives I studied at the local level sought to address citizens' lack of capacity and to redress inequities. An NGO called Association Tichka, headquartered in Ouarzazate, provides local community development services to the Province of Ouarzazate through a model that conceives of citizens as sovereign. Tichka coordinated the saffron farm discussed in the previous section, and it runs boarding schools that allow rural boys and girls to pursue their education in the city. The villagers come to Tichka with a proposal for a technology. Often they request a hospital on wheels for their village. Association Tichka asks them to critically examine the root of the health problems in their village, which often objectively relate to water and sanitation services. They engage the population early and often rather than waiting until the project execution stage. Villagers ultimately decide for themselves whether they want a water filtration system or the project that they initially requested. Then Tichka and the villagers prepare a contract that each household must read (or have read to them) and sign. The local population provides the workforce for the project, and they commit to a portion of the work in the contract. Overall, Tichka works to directly engage citizens and seeks to avoid paternalism. Their bottom-up model of development focuses on public engagement rather than charity. It seeks to improve quality of life and capacity at the local scale. Generally, the development literature suggests that citizen-led development efforts are more successful than foreign-led ones, and Association Tichka offers the benefit of being run by Moroccans for Moroccans.

To observe Tichka's local version of integrated project development, I visited the village of Zaouit Sidi Ahmed in 2013 and 2016. The local population in Zaouit Sidi Ahmed had limited success with previous development projects, and many young people were leaving the village for urban areas, embodying Morocco's urbanization trend. Latent capacity existed in the village—ideas, people willing to volunteer, a cooperative, and an association—but they had insufficient resources. They planted olive trees, but they experienced mechanical problems with their butane pump used for pumping groundwater, and the trees died from lack of water. Tichka funded an 8.5 kW solar PV system to run a 7.5 kW water pump. The villagers grew new olive trees using this solar irrigation system. (On my return visit, the villagers complained that the panels had lost efficiency, but the system remained operational.) The village purchased an olive press to independently produce a higher quality oil. Association Tichka purchased a solar hot water heater for the olive press. The village rents out the press to olive producers from other villages who would have otherwise taken their olives to Marrakesh to be pressed. The villagers also purchased livestock for a continual source of income. They use chickens to produce eggs, and they can sell these chickens quickly if they require cash. They use milk from cows to make butter and cheese. Additionally, they grow spirulina, a type of algae that is a popular health food in France. They export about half of it to France and sell the rest in Ouarzazate, Marrakesh, and Rabat for a higher price than olive oil. This local-scale integrated development model enables bottom-up participation, and over time it is likely to improve the community's resilience.

There are other NGOs that work in the region. Just as MASEN called itself a "animator" for the region, NGOs send "animators" into villages who lead activities and songs to engage children in learning. Additionally, several Moroccan researchers who conducted interviews for the *Social CSP* study started an NGO called The Human Touch to alleviate suffering in Ouarzazate province. The leaders of these organizations are part of Morocco's educated but unemployed youth demographic. Their actions reflect the conception of the role of citizens as volunteers but with a focus on improving the capacity of citizens to lead sustainable lives rather than on cleaning up the environment (see Chapter 5). These NGOs offer hope that there is, in fact, a bright future for Morocco.

Given the community's lack of capacity to benefit from Noor, local capacity-building initiatives ought to be undertaken alongside large-scale projects focused on national capacity building. A 2015 working group convened by the environmental NGO Germanwatch in Rabat found that representatives from government, industry, and academia all wanted more opportunities for citizens and NGOs to participate in renewable energy development. They also found that development initiatives in addition to CSP are necessary to reduce the economic marginalization of Ouarzazate (Schinke, 2015). I agree that decentralized integrated development projects are needed to meet immediate needs and improve capabilities at the local level. Examples could include increasing solar irrigation and ensuring clean water; however, local communities ought to choose development projects themselves through model similar to Tichka's model.

Considering the benefits of and the need for decentralized local development projects, it would be easy to assume that these projects are more just than centralized projects. However, after studying centrally-planned, large-scale renewable energy systems as well as local development initiatives, I see benefits and drawbacks to each and potential synergies stemming from the simultaneous pursuit of both. For example, Noor-Ouarzazate's greatest drawback is that actors failed to address the needs of the people living next to the power plant site who lost their grazing land. One drawback of local development is that it is not of sufficient scale to address Morocco's national goals to become a knowledge economy with reduced dependence on water-intensive agriculture and carbon-intensive energy systems. Tichka's development project effects are immediate and urgently needed to address basic needs, whereas the goals for CSP development are centered around a broader long-term macroeconomic vision for Morocco. Additionally, there are advantages stemming from being connected to centralized, large-scale power sources, such as powering factories that create jobs.

In addition to facilitating macroeconomic growth, large, dense sources of power will likely be necessary for water desalination in the MENA region in the future. CSP plants could theoretically provide electricity for the desalination of water. However, there are drawbacks associated with siting CSP plants near the ocean because particulates from evaporating ocean water interfere with solar insolation, and coastal communities often protest the siting of new infrastructure near scenic shorelines that generate tourist revenue. One cannot assume, as de Souza, Bosco, Cavalcante, and da Costa Ferreira (2018) do, that CSP plants can desalinate water to make up for the amount of water that CSP plants use for cooling to produce electricity because the spatial arrangements matter. For example, Morocco would have to pipe water from desalination plants on the coast into or across the High Atlas Mountains to reach some of the existing and planned CSP sites. It is too simplistic to assume that water scarcity in one region of a country can be made up for with additional water production in an area that is hundreds of miles away. However, it is the case that a large quantity of energy will be required for providing drinking water in the MENA region in the future.

Brown and Purcell (2005) criticize the privileging of the local scale, calling it "the local trap" because there is nothing "intrinsically more desirable [about the local scale] than other scales" in terms of outcomes for sustainability, justice, and democracy (p. 607). The local trap entails a faulty assumption that localized decision making is intrinsically more democratic (Purcell, 2006). Similarly, localism has long been viewed as central to the democratization of technology. In his book on technology and democracy, Sclove (1995) recommended that technologies that promote local self-reliance should be developed. While Sclove argues that local autonomy and decision making will result in more just outcomes than national top-down decision making, people living in the Ouarzazate region see themselves as disadvantaged because of their isolation and their lack of interconnections with the rest of Morocco as well as the Mediterranean region. Similarly, the Desertec vision's proponents viewed areas disconnected from premium networked spaces as isolated islands that do not benefit from the reliability and power density of a centralized grid. Falling into the local trap risks overlooking the potential benefits of reducing isolation and building interdependencies. Each region has comparative advantages and complementary attributes that could provide benefits through integration provided that just political integration processes are developed. Moreover, the recommendation of Sclove (1995) to build small-scale, decentralized technologies was underpinned by an assumption that this development would start with a blank slate and, therefore, his recommendation did not address how existing infrastructure, with its sunk costs, would be refurbished and transformed.

A rooftop solar PV panel is better suited than a CSP plant to provide light and small-scale household uses. However, just as a centralized energy facility can be poorly planned so too can distributed energy generation. As an illustrative anecdote, I visited a nomad family living in the Sahara to whom foreign tourists had donated solar panels. The family had no use for the panels because they had no light bulbs or appliances and no battery storage for nighttime. Additionally, development banks have used solar PV panels to offer rural communities a small amount of electricity that, while beneficial, could further their marginalization by keeping them isolated from the grid for the foreseeable future (Byrne, Toly, & Glover, 2006). Once a household gains access to a small solar panel, it is often classified as electrified and may not benefit from increased energy access in the near to medium term future, especially in a country that is struggling to meet increasing electricity demand. It is also problematic for development officials from outside the community to define what counts as an appropriate technology for citizens from that community. Comparative, or relative, justice in our globalized, high-tech world requires more than simply meeting basic needs.

Discussion: achieving a balance

In this chapter, I described existing inequities in Ouarzazate and then explained how these inequities related to the community's lack of capacity to participate in the Noor siting process and to benefit from the Noor solar power plants. The drawbacks associated with the Noor CSP plants related to the deficient public engagement process as well as to water use in the water-scarce Province of Ouarzazate. Several NGOs have developed standards for CSP siting and local development, including the "DESERTEC Criteria" developed by the Desertec Foundation and the *Social CSP* study that categorized every project impact and recommended reducing water usage, ensuring socioeconomic benefits, and avoiding reductions in biodiversity. I recommend more substantive and earlier citizen engagement to improve the procedural justice aspects and thus the outcomes of large-scale energy planning and policymaking. Additionally, long-term efforts must be made to address systemic local-scale inequities. This could be achieved through distributed development projects over which affected publics have ownership.

Despite these drawbacks, well-planned CSP plants offer vast improvements over hydroelectric facilities. For example, Noor did not displace communities; it did not lead to increased waterborne illnesses; and it cannot collapse and flood communities, which is a risk in the cases of the Kariba Dam in between Zambia and Zimbabwe and the Mosul Dam in Iraq. It is also important to note for CSP siting cases in other countries that opposition would have been greater if the local population had been displaced from their homes, if compensatory mechanisms had not been offered, and if national legitimacy did not exist. According to in-progress research I have conducted on transnational grid projects worldwide, most electricity export projects focus on hydroelectricity. While exporting electricity from CSP plants has trade-offs associated with water usage, it offers advantages over exporting electricity from hydroelectric facilities. In the case of exports, considerations of justice cross scales and nation-state borders. Chapter 8 will aid in considering just processes and outcomes in complex, multi-scalar energy systems.

While CSP has relative benefits compared to hydroelectricity, it is unreasonable to expect that it will result in win-wins and will meet all of a country's socioeconomic development needs in addition to meeting increasing electricity demand. Rignall (2016) reasonably criticizes Noor for its imposition on people's sovereignty over communal land. It is inappropriate for people suffering from inequity to accept a disproportionate share of the drawbacks of energy development. That said, studies that evaluate specific siting cases almost invariably find numerous faults with the individual case, but these studies often do not make an energy policy argument that grapples with the systems-level challenges. My focus in this chapter has extended beyond the individual Noor case study to consider the difficult energy policy choices related to the "wicked problems" of achieving sustainable development in a carbon-constrained world. While I have argued that we should not uncritically accept win-win discourses, energy policy decisions must address the broader system challenges of building electrical power systems, which entail difficult trade-offs and require a balancing act among multiple goals at the local and national scales. I have recommended that win-win discourses be critically examined and that the best practices for siting CSP be paired with local development to meet immediate needs as defined by the affected citizens. It is important to achieve a balance between both national and local needs and conflicting values.

Notes

1 Funders of Noor-Ouarzazate include the World Bank, the International Bank for Reconstruction and Development, the African Development Bank, the French Development Agency, the European Investment Bank, the KfW development bank, the European Neighborhood Investment Facility, and the German Ministries of Environment and Development Cooperation.

2 There are also manganese, cobalt, and copper mines. Increasing international competition had harmed the movie sector, but business has recently improved (Terrapon-Pfaff et al., 2015).

3 See Chapter 2 for this history.

4 The latest available bathymetry in 1998 indicated a capacity of 433 million cubic meters (Diekkruger et al., 2010).

5 Translated from French.

6 I worked with and met many articulate, intelligent, and assertive Moroccan women, and statistics do not adequately reflect the situation of women in Morocco.

7 Electricity in the area is purchased through prepaid charge cards. Depending on their usage levels, consumers in the Province of Ouarzazate pay between 0.9 Moroccan dirhams per kWh and 1.44 dirhams per kWh (Terrapon-Pfaff et al., 2015).

8 The World Bank halted construction for several weeks while it investigated the conflict. The villagers were released from jail after community members paid fines on their behalf.

9 The Hajj is the pilgrimage to Mecca that is one of the five pillars of Islam.

10 Translated from French.

11 I heard several different figures for saffron's water usage while visiting the village. I chose this figure because it corresponds with numbers referenced for saffron irrigation in the scholarly literature (Lichtfouse, Navarrete, Debaeke, Véronique, & Alberola, 2009).

12 A technical study that precisely compares figures for water used for electricity to water used for other exports would be a valuable contribution to practitioner and academic knowledge.

Works cited

5 Capitals Environmental and Management Consulting. (2012). Ouarzazate solar power complex, phase 1 Morocco. Specific Environmental and Social Impact Assessment Volume 1. Retrieved from http://www.masen.ma/media/uploads/documents/Masen_NOORoI_SESIA_Volume1.pdf

ACWA Power scales up tower-trough design to set record-low CSP price. (2017, September 20). *New Energy Update: CSP.*

Apostol, D., Palmer, J., Pasqualetti, M., Smardon, R., & Sullivan, R. (Eds.). (2016). *The renewable energy landscape: Preserving scenic values in our sustainable future* (1st ed.). London: Routledge.

Bisson, J. (2003). *Mythes et réalités d'un désert convoité: le Sahara.* Paris, France: L'Harmattan.

Brown, C. J., & Purcell, M. (2005). There's nothing inherent about scale: Political ecology, the local trap, and the politics of development in the Brazilian Amazon. *Geoforum, 36*(5), 607–624. Retrieved from https://doi.org/10.1016/j.geoforum.2004.09.001

Brown, M. (2009). *Science in democracy: Expertise, institutions, and representation.* Cambridge, MA: MIT Press.

Byrne, J., Toly, N., & Glover, L. (2006). *Transforming power: Energy, environment, and society in conflict.* Piscataway, NJ: Transaction Publishers.

Carlisle, J. E., Kane, S. L., Solan, D., & Joe, J. C. (2014). Support for solar energy: Examining sense of place and utility-scale development in California. *Energy Research & Social Science, 3,* 124–130. Retrieved from https://doi.org/10.1016/j.erss.2014.07.006

Catusse, M., Vairel, F., & Bono, I. (2010). *Le Maroc de Mohammed VI: Mobilisations et action publique.* Paris, France: Karthala.

Dasgupta, P. (2010). Nature's role in sustaining economic development. *Philosophical Transactions of the Royal Society B: Biological Sciences, 365*(1537), 5–11. Retrieved from https://doi.org/10.1098/rstb.2009.0231

de Souza, L. E. V., Bosco, E. M. G. R. L., Cavalcante, A. G., & da Costa Ferreira, L. (2018). Postcolonial theories meet energy studies: "Institutional orientalism" as a barrier for renewable electricity trade in the Mediterranean region. *Energy Research & Social Science, 40,* 91–100.

Devine-Wright, P. (2011). *Renewable energy and the public: From NIMBY to participation.* Oxford: Earthscan.

Diekkrüger, B., Busche, H., Klose, A., Klose, S., Rademacher, C., & Schulz, O. (2010). Impact of global change on hydrology and soil degradation- Scenario analysis for the semi-arid Drâa catchment (South Morocco). In Conference Proceedings of *The Global Dimensions of Change in River Basins: Threats, Linkages and Adaptation.* Bonn, Germany.

Freudenburg, W. R. (1981). Women and men in an energy boomtown: Adjustment, alienation, and adaptation. *Rural Sociology, 46*(2), 220–244.

Green, N., Sovacool, B. K., & Hancock, K. (2015). Grand designs: Assessing the African energy security implications of the Grand Inga Dam. *African Studies Review, 58*(1), 133–158. Retrieved from https://doi.org/10.1017/asr.2015.7

Hanger, S., Komendantova, N., Schinke, B., Zejli, D., Ihlal, A., & Patt, A. (2016). Community acceptance of large-scale solar energy installations in developing countries: Evidence from Morocco. *Energy Research & Social Science, 14,* 80–89. Retrieved from https://doi.org/10.1016/j.erss.2016.01.010

Haut-Commissariat au Plan. (2014). Recensement général de la population et de l'habitat de 2014, Region Draa-Tafilalet. Retrieved from https://www.hcp.ma/draa-tafilalet/attachment/884248/

Hunold, C., & Leitner, S. (2011). 'Hasta la vista, baby!' The Solar Grand Plan, environmentalism, and social constructions of the Mojave Desert. *Environmental Politics, 20*(5), 687–704. Retrieved from https://doi.org/10.1080/09644016.2011.608534

Khagram, S. (2004). *Dams and development: Transnational struggles for water and power*. Ithaca, NY: Cornell University Press.

Lamrini, R. (2006). *Y a-t-il un avenir au Maroc me demanda Yasmina*. Paris, France: Marsam Editions.

Lichtfouse, E., Navarrete, M., Debaeke, P., Véronique, S., & Alberola, C. (Eds.). (2009). *Sustainable agriculture* (2009 ed.). New York, NY: Springer.

Mahmoud, M., Myrsalieva, N., El Forgani, M., Bouadane, H., & Sarant, L. (2016). *Arab future energy index (AFEX) 2016: Renewable energy*. Cairo, Egypt: RCREEE. Retrieved from www.rcreee.org/sites/default/files/final_afex_re_2016.pdf

Masdar. (2018). Masdar celebrates installation of 19,438 solar home systems in rural Morocco. Retrieved June 1, 2018 from http://www.masdar.ae/en/media/detail/masdar-celebrates-installation-of-19438-solar-home-systems-in-rural-morocco

McShane, T. O., Hirsch, P. D., Trung, T. C., Songorwa, A. N., Kinzig, A., Monteferri, B., . . . O'Connor, S. (2011). Hard choices: Making trade-offs between biodiversity conservation and human well-being. *Biological Conservation*, *144*(3), 966–972. Retrieved from https://doi.org/16/j.biocon.2010.04.038

Mekouar, H. (2017, October 15). "Thirsty protests" hit Morocco over water shortages. *Phys.Org*. Retrieved October 31, 2017, from https://phys.org/news/2017-10-thirsty-protests-morocco-shortages.html

Moore, S., & Hackett, E. J. (2016). The construction of technology and place: Concentrating solar power conflicts in the United States. *Energy Research & Social Science*, *11*, 67–78. Retrieved from https://doi.org/10.1016/j.erss.2015.08.003

Pasqualetti, M. J., & Schwartz, C. (2011). Siting solar power in Arizona: A public values failure? In P. Devine-Wright (Ed.), *Renewable energy and the public: From NIMBY to participation* (pp. 167–186). Oxford: Earthscan.

Phadke, R. (2010). Defending place in the Google Earth age. *Ethics, Place & Environment*, *13*(3), 267–281. Retrieved from https://doi.org/10.1080/1366879X.2010.516495

Purcell, M. (2006). Urban democracy and the local trap. *Urban Studies*, *43*(11), 1921–1941. Retrieved from https://doi.org/10.1080/00420980600897826

Rignall, K. E. (2016). Solar power, state power, and the politics of energy transition in pre-Saharan Morocco. *Environment and Planning A: Economy and Space*, *48*(3), 540–557. Retrieved from https://doi.org/10.1177/0308518X15619176

Roux, B., Guerraoui, D. (1997). *Les zones défavorisées Méditerranéennes*. Paris, France: Editions L'Harmattan.

Samborsky, B., Myrsalieva, N., & Mahmoud, M. (2013). *Arab future energy index AFEX 2013*. Cairo, Egypt: RCREEE.

Schinke, B. (2015, February 20). *Civil society perspectives on renewable energy policy making in Morocco*. Germanwatch, Heinrich Boll Stiftung. Retrieved from https://germanwatch.org/en/download/10931.pdf

Sclove, R. (1995). *Democracy and technology*. New York, NY: Guilford Press.

Sovacool, B. K., & Dworkin, M. H. (2014). *Global energy justice: Problems, principles, and practices*. Cambridge: Cambridge University Press.

St. John, J. Saudi Arabia and SoftBank's 200-Gigwatt Solar Plan offers 'so many questions, so few details.' *Green Tech Media*. Retrieved June 1, 2018 from https://www.greentechmedia.com/articles/read/saudi-arabia-200-gigawatt-solar-plan#gs.0v1GDtE

Swearingen, W. D. (1988). *Moroccan mirages: Agrarian dreams and deceptions, 1912–1986*. London: I.B.Tauris.

Terrapon-Pfaff, J., Borbonus, S., Schinke, B., Veibahn, P., Fink, T., & Brand, B. (2015). *Social CSP energy and development: Exploring the local livelihood dimension of the*

Noor I CSP project in Southern Morocco. Bonn, Germany: Wuppertal Institute; Germanwatch. Retrieved from www.wupperinst.org/en/projects/details/wi/p/s/pd/449/

Terrapon-Pfaff, J., Fink, T., Viebahn, P., & Jamea, E. M. (2017). Determining significance in social impact assessments (SIA) by applying both technical and participatory approaches: Methodology development and application in a case study of the concentrated solar power plant NOOR I in Morocco. *Environmental Impact Assessment Review*, *66*, 138–150. Retrieved from https://doi.org/10.1016/j.eiar.2017.06.008

Valentine, S. V. (2015). *Wind power politics and policy*. Oxford: Oxford University Press.

Winner, L. (1986). *The whale and the reactor: A search for limits in an age of high technology*. Chicago, IL: University of Chicago Press.

World Commission on Dams. (2000). *Dams and development: A new framework for decision-making*. London: Routledge.

Yenneti, K., Day, R., & Golubchikov, O. (2016). Spatial justice and the land politics of renewables: Dispossessing vulnerable communities through solar energy mega-projects. *Geoforum*, *76*, 90–99.

8 Conclusion

Energy justice and security in visions of multi-scalar systems

Visions for energy transformations: how long and what would it take?

This book discussed a vision for the transformation of the electricity systems in the Mediterranean region—a vision to provide for a 90% clean energy future, either through the export of electricity from clean, dispatchable power sources from North Africa to Europe or through a regional supergrid that allows electricity to be traded at different times of day between North and South. Power systems momentum in the Mediterranean is being disrupted in multiple ways, and the system imaginers sought at least to guide the system's direction, if not to affect its mass and velocity. Their lack of success was not due to cost alone. Sovacool (2016) questions what and how long it will take to achieve a transition to low-carbon electricity. Both the techno-economic and the socioinstitutional aspects of major changes in energy systems should be taken into account (Sovacool & Geels, 2016).

Because regional power systems are proposed for other regions of the globe, it is useful to consider what it would have taken to achieve the Desertec vision. A variety of complex socioinstitutional changes would be needed. First, a socio-technical vision must fit within states' imaginaries of what should be done, along with the anticipated social, economic, and political benefits of doing it, so that decision makers possess a certain degree of faith that the vision is feasible and desirable. Sociotechnical imaginaries are "collectively held, institutionally stabilized, and publicly performed visions of desirable futures animated by shared understandings of forms of social life and social order attainable through, and supportive of, advances in science and technology" (Jasanoff & Kim, 2015, p. 4). Techno-economic feasibility studies are useful in motivating latent desires that align with imaginaries already supportive of the vision's process and outcomes, but feasibility studies themselves cannot produce imaginaries. Second, a common policy and regulatory organization would be needed to govern the process of electricity integration, founded on a just procedure between North and South. Third, a trusted transmission system operator would be needed, or at least substantial collaboration among the system operators, as described in Chapter 4.

Changes in numerous techno-economic factors would need to be effected as well. First, cost estimates for a Desertec system were almost incomprehensibly high—at least half a trillion dollars. However, it was difficult to predict the costs of such a large system, and to account for new, unique costs compared to the cost of replacing aging infrastructure. Different modeling assumptions also substantially changed the system price (Backhaus, Gausling, & Hildebrand, 2015). Decision makers may need to acquiesce that the true system cost is somewhat unknowable, which would require transformations to be motivated at least partly by social and environmental policy goals, not just by economic rationality. Second, the costs of CSP would need to be reduced. However, it is worth noting that Moroccan policymakers chose to build CSP plants for industrial policy reasons in spite of their high cost, and that first mover commitment has helped to bring CSP technology down the cost curve. Third, the cost of HVDC would need to come down. However, without sufficient values-based policy goals to motivate the construction of HVDC across the Mediterranean, it became easy for stakeholders to dismiss Desertec as too expensive in spite of predictions that the investment would pay off in the long term. Moreover, building transmission lines across the Mediterranean is at least as difficult for socioinstitutional reasons as it is for techno-economic reasons. Fourth, new economic mechanisms for pricing dispatchability or storage would be needed to facilitate overall system change. Market liberalization may be needed in addition, but not all states agree to this, as discussed in Chapter 6.

It is difficult, and even ill-advised, to fully separate the techno-economic from the socioinstitutional factors, as the techno-economic factors are perceived and framed through socioinstitutional lenses. I have illustrated that a vision for an energy system goes well beyond the technical components, encompassing framings of feasibility and technological maturity versus innovativeness; geopolitical construction of the space for which a system or technology is planned; system properties, such as dispatchability or baseload power; a depiction of what a modern energy system looks like; cost and economics; managing organizations; and the characterization of societal benefits and risks (see Table 4.1). Multiple actors and forces shaped the vision, including the system imaginers and stakeholders; facets of the broader sociopolitical context; wildcards, or unexpected events; media framings; and ideology. Technological feasibility was not an objective reality, but rather was socially constructed.

Within the energy transitions discussion, CSP is sometimes framed as a techno-fix, or a plug-and-play clean energy option that could be integrated with relative techno-economic ease into the broader grid without changing the grid's overall socioinstitutional configuration (see Chapter 2). However, my inquiry into a vision that sought to leave the overall centralized system configuration intact has illustrated that even so-called techno-fixes require substantial social, political, and economic reorganization. For example, the overall system cost estimates for visions of low-carbon regional energy systems greatly depend on solar irradiation levels (Backhaus et al., 2015), but site selection is a sociotechnical rather than a techno-economic process, on which the social sustainability of CSP hinges. Failing to acknowledge this social complexity through the techno-fix framing could mask potential challenges and injustices. Yet instead of using history as evidence that transitions are too difficult today from a techno-economic standpoint, historical lessons could be brought to bear to avoid past mistakes and to improve sociopolitical feasibility. Numerous missteps could have been avoided if the actors had operated with greater knowledge of the history of injustice related to large-scale engineering projects and solar power for the Sahara Desert.

Temporal complexity: Because the sociotechnical visioning phase plays an important role in the innovation system and in shaping the electrical power systems of the future, it is important to improve it. In contrast to the system builders of the late 1800s and early 1900s who worked to construct a system around their newly invented technology (Hughes, 1983), today's system imaginers act as project accelerators and catalysts working to situate technologies viewed as "already ready" in an inevitable future. System imaginers used temporal tactics to frame and control the future in an attempt to speed up their desired outcomes. A framing that focuses on acceleration, rather than on reflexivity, complexity, and procedural justice, is problematic. This focus on acceleration constrains the visioning phase by reducing the time spent on invention, exploration, and stakeholder participation. Thus, the visioning phase serves less as a period of time to imagine than an impediment to imagination and a set of constraints placed on imagination. These constraints end up closing down the very pathways to sustainability and sustainable development the visioning phase sought to open up (Stirling, 2014).

As I discussed in Chapter 4, some stakeholders came to see a building blocks approach to grid integration as more feasible than an overall system plan. The German *Energiewende* offers a cautionary tale about a lack of system planning. Sturm (2017) argues that in spite of Germany's success increasing renewable installed capacity, customers have borne high costs. She points out that Germany's scale-up of renewable energy has depended on dirty backup generation run in economically inefficient and environmentally harmful ways. Emissions reductions have been much smaller than anticipated because of increased reliance on coal and a concomitant phaseout of nuclear power, which is zero-carbon in the generation phase (ibid.). While I would afford some leeway to Germany as an early mover, system imaginers ought to consider how targets set for renewable generation will alter the broader system, rather than to assume that economic incentives to spur the scale-up of renewable energy, such as the feed-in tariff, will automatically reshape systems, improve societal outcomes, and achieve greenhouse gas emissions reduction targets. As difficult as it is to achieve, systems-level planning is necessary for just and effective energy transformations.

Systems planning and innovation will require an overall change in energy stakeholders' mindsets, as well as the development of institutions that play an anticipatory role in energy governance. Adam (2003) pointed out that there are no organizations that manage the governance of time. Institutions that enact temporal governance are crucial for anticipating problems, representing stakeholders, apportioning the future trade-offs of energy systems, and ensuring that injustices made visible in the design phase are not cemented for decades to come. In this process of temporal governance, changes and innovation in both the system configuration and the technologies embedded in the system are needed—changes that go beyond prioritizing the typical standards of low cost and reliability. The remainder of this chapter explores these principles.

Energy justice and social sustainability in multi-scalar systems

Energy justice theoretical overview: While one section of the energy transitions literature focuses on what it would take and how long it would take to accelerate transitions along pathways that reduce greenhouse gas emissions, another focuses on developing theories and principles for achieving a transition that is overall just and sustainable. Sovacool and Dworkin (2014) conceptualize a just energy system or technology as one that conforms to the principles of availability, affordability, due process, good governance, sustainability, intergenerational equity, intragenerational equity, and responsibility. Sovacool, Burke, Baker, Kotikalapudi, and Wlokas (2017) supplement these principles with the principles of resistance to energy injustice and of intersectionality, or attention given to the multi-layered power dynamics and systemic discrimination related to intersecting identities, such as race, gender, citizenship, class, and sexual orientation.

In subtle contrast to the approach of developing principles for a just energy system, Jenkins, McCauley, Heffron, Stephan, and Rehner (2016) focus on theories of justice, arguing for a "three-pronged approach." The first prong, distributive justice, relates to how the burdens and benefits of the supply and demand sides

Table 8.1 Theories of Justice

Justice theory	*Brief description*	*Examples from empirical material*
Recognition justice	Marginalized social groups should be recognized and afforded political rights.	– Colonists imagined the Sahara Desert as *terra nullius* – CSP sites in North Africa were framed as empty
Procedural justice	There should be a just and representative process for making energy decisions.	– Process for choosing CSP sites and compensating people for losses lacked buy-in from most affected stakeholders – Weaker countries could be coerced in the development of a regional energy system – Visioning process was not inclusive and lacked appropriate stakeholder representation
Distributive justice	The burdens and benefits of energy systems and technologies should be equitably distributed among individuals and social groups.	– Greater initial focus in the vision on the EU importing green electrons from North Africa than on meeting growing demand in Morocco – Vulnerable populations shoulder a disproportionate burden of the land and water usage of CSP – Imazighen people in the Rif Mountains were excluded from national sustainable development plans

Justice theory	*Brief description*	*Examples from empirical material*
Cosmopolitan justice	Justice should be afforded to all citizens of the world rather than priority given to those with whom one shares national citizenship.	– Certain social groups are treated as if they do not matter, or as if they are second-class people. – Western lives are often prioritized, as critiqued by the *Je Ne Suis Pas Charlie* hashtag
Capabilities justice	Just institutions are insufficient. Development practitioners must examine an individual's capacity to live a good life. Energy services could enhance human capabilities.	– Capabilities and needs of the local communities were ignored – Energy services for populations in North Africa were not consistently prioritized by system imaginers
Intergenerational equality, equity	The current generation's failure to mitigate climate change harms the future generation.	– Certain social groups benefit from clean energy development today, but other social groups do not. The descendants of these social groups will not likely be better off in the future even if natural capital is maintained overall. – The Al Hoceima case illustrates how intergenerational equity differs from intergenerational equality
Ecological justice	All organisms have non-anthropocentric value. Distributive justice should be extended to non-humans. (In contrast, environmental pragmatists argue for a just procedure that bases environmental decision making on pluralistic values).	– CSP plant sites affect habitat – Climate change will have deleterious effects on the natural environment – Fishing communities harm the marine ecosystem by overfishing or fishing out of season

of energy are distributed among individuals and social groups. An often-cited example of distributive injustice is the lack of access to and affordability of electricity and primary fuels for populations that often bear a disproportionate share of the drawbacks of dirty energy generation and extractive industries (Reames, 2016). The second prong, procedural justice, refers to the fairness, inclusivity, and representativeness of the process and practice of energy decision making. The third prong, recognition justice, calls attention to the importance of recognizing disadvantaged and marginalized social groups and providing them with full political rights. Intersectionality could be viewed as an articulation of recognition justice that affords nuanced attention to the power dynamics of intersecting identities. Table 8.1 describes theories of justice used in this empirical study.

Instead of imposing scholarly frameworks and principles on communities, Smith and High (2017) call for anthropological research that examines how social groups and communities themselves define what counts as a desirable energy system, an approach the authors term "energy ethics." What counts as a desirable and just transition pathway indeed differs among localities and nation-states, which is one reason why transnational cooperation on energy transformations is difficult. The examination of actors' imaginaries, visions, priorities, and goals provides rich empirical data for understanding energy injustice within particular cultures and contexts. However, scholars must afford attention to the power dynamics underlying these empirical conceptualizations of energy justice. For example, Newell and Mulvaney (2013) question "who defines what is just, and for whom" across global power and wealth divides (p. 138). Theoretical conceptualizations of justice are useful because actors sometimes lack a vocabulary for identifying and diagnosing energy injustice. Thus, both empirical analysis of communities' and individuals' conceptions of energy justice and a theory for diagnosing and characterizing energy injustice are needed. Moreover, empirical and theoretical conceptualizations of energy justice could aid in describing and assessing social sustainability in energy systems.

Layering urgent questions of "climate justice" onto energy justice further complicates the task of achieving just energy transformations (Newell & Mulvaney, 2013). For example, it is necessary to mitigate climate change to avoid its negative distributive effects while extending electricity access to those currently living in energy poverty. Additionally, the international community must develop governance mechanisms for mitigating climate change, while addressing distributive, recognition, and procedural injustices stemming from the performance of power across gaping divides in economic equality and geopolitical clout (ibid.). Energy justice concerns are sometimes obscured in the public discourse by the urgency of the threats of global climate change, reflecting what Davies and Selin (2012) term "climate change as trump card arguments" (p. 128). Yet redressing energy injustice is also urgent, as numerous people and social groups live without access to energy and its associated services and capabilities and have their rights and livelihoods impeded by energy systems—for example, through displacement from their land for energy development and through the negative health repercussions of extracting and burning fossil fuels. The urgency of addressing energy injustice is amplified by the reality that actions taken in the energy sector today will have significant effects for decades to come. Power plants last for 30 to 40 years and are integrated into entrenched and obdurate system configurations that distribute the benefits and drawbacks of both energy supply and demand.

Addressing energy injustice and effecting socially sustainable development is complex and multifaceted and requires a higher standard than low-carbon energy. We ought not to let the urgency of climate change excuse the construction of systems that will perpetuate energy injustice. Instead, we ought to build energy systems that both power and empower people and perform the difficult work of balancing societal and environmental goals. Difficult trade-offs will be required in the transformation process: to suggest otherwise would be naïve. Islar, Brogaard, and Lemberg-Pedersen (2017) argue that geopolitics, biophysics, and scarce resources pose "feasibility constraints" to energy justice theory. Theories of justice propose an ideal that in reality must be negotiated across multiple stakeholder perspectives and scales of power, although I view geopolitics and scarce resources as intrinsic to the justice challenge rather than as a set of extrinsic feasibility constraints. A scholarly critique that pushes decision makers toward what is ideal and optimal across multiple dimensions, unmasking unseen and undebated trade-offs, is useful even though real-world decisions are never perfect.

Conceptualizing winners and losers in multi-scalar technologies and systems: The win-win discourse prevalent in the Desertec case study is observed frequently in sustainable development projects (McShane et al., 2011). Discourses on winners and losers typically do not afford attention to whether the wins and losses are absolute or relative, who defines winners and losers and at what scale and over what time period, whether winners and losers can self-identify, and on what criteria the wins and losses are based (O'Brien & Leichenko, 2003). O'Brien and Leichenko (2003) examine the epistemological underpinnings of the winners and losers discourses in two processes of global change—climate change and economic globalization. Win-win discourses related to economic globalization are underpinned by a paradigm from neoclassical economics that assumes that everyone will win from increased economic efficiency and opportunity (O'Brien & Leichenko, 2003). Given enough time, temporary losses will be equalized through global economic efficiency that improves living standards across the globe (ibid.). Social Darwinist perspectives view winners and losers as natural and inevitable, whereas a political economy perspective views winners and losers as "socially and politically generated" (ibid.).

Discourses on the regionalization and globalization of electrical power systems often reflect the neoclassical economics paradigm, assuming that reducing state control over electrical power systems and expanding free markets will increase competition and result in an invisible hand distributing efficient win-win outcomes. The win-win discourse consistently overpromises and fails to account for the multidimensional complexity of defining what counts as a win versus a loss and for whom. Energy systems intersect with water availability and quality, poverty alleviation, and the provision of food, but actors sometimes draw narrow boundaries around the system, examining only a single dimension or scale to justify a win-win. Desertec's win-win framing generally referred to wins at a regional and abstract scale, in broad categories, such as peace and security, regional stability, and development. Such categories for diagnosing wins are "ideographs," or words that evoke emotion but are too broad to contain specific meaning and substance (Van Lente, 2000). Win-win discourses could mask distributive injustice by excluding losers and those whom technological progress bypasses from the problem-framing. For example, Morocco may benefit as a nation-state from reducing electricity imports by building CSP plants, or by increasing the export of green electrons, but CSP siting negatively affected the water-impoverished village adjacent to the first power plant. Moreover, the fruits of energy development bypassed Imazighen living along the Mediterranean coast whose inequitable lack of access to livelihood opportunities resulted in unsustainable fishing practices.

Thus, there are no simplistic, win-win solutions to providing electricity and fuels in sustainable and equitable ways across multi-scalar sociotechnical systems. Complex global supply chains and transnational energy projects, such as pipelines and transmission lines, interweave spatial scales, linking low-income communities with locales that enjoy profound energy wealth. Sociotechnical systems effect different outcomes and trade-offs at each temporal stage, from imagined to constructed to decommissioned. They also interweave past, current, and future injustice. Furthermore, to transition these systems, system builders must retrofit existing path-dependent energy infrastructure rather than starting anew, layering systems onto systems.

Designing just energy systems in an energy interdependent world will require mapping out the complex ways in which energy injustice crosses scales. Because of this, Jenkins et al. (2016) recommend a "whole energy systems" approach as a tool for considering energy justice across the life cycle, from mining to disposal. Similarly, Bickerstaff, Walker, and Bulkeley (2013) use the concept of the "sociotechnical assemblage" to describe networks that transcend different scales with messiness and variety, across which agency is diffused. An actor's location within the assemblage affects the actor's access to power and level of energy vulnerability (ibid.). Diagnosing energy injustice will require difficult and intentional choices that delineate system boundaries. Without processes that encourage reflexivity in the process of drawing system boundaries, these choices are likely to reflect power dynamics. This problem is made even more difficult because multi-scalar technologies are embedded within broader, multi-scalar systems.

One aspect of multi-scalar energy justice is to consider the effects of individual energy technologies across scales. Sovacool et al. (2017) emphasize the importance of cross-scalar justice across global supply chains, considering the materials embedded in solar panels, power plants, and transmission lines. In this study, I examined the multi-scalar aspects of CSP, although I did not even trace out the source of the raw materials. In Chapter 2, I discussed critiques of CSP for its effects on desert landscapes in California and for its role in the centralization of energy systems. In Chapter 3, I discussed how the export of green electrons from CSP could include local "implicit subsidies" related to the use of land and water resources. In Chapter 7, I discussed critiques of CSP for consuming water and land resources in Morocco, with disproportionate effects on nearby water-scarce villages in need of improved capabilities. Even with individual technologies, the effects are not contained within a single scale.

The challenge of energy justice is that it is not only a matter of examining the multi-scalar life cycles of individual technologies but it also requires building just multi-scalar sociotechnical systems. Critiques of individual technologies often do not pay attention to difficult trade-offs related to constructing a system that is efficient and sustainable overall. The benefits and drawbacks of a particular technology across its life cycle may change when it is situated within a system context. For example, the siting effects of CSP, provided they are locally mitigated and fairly distributed, could be more bearable if CSP reduces the need for dispatchable natural gas plants or baseload coal-fired power plants, thereby mitigating climate change. At tension with this is the critique of CSP by some activist groups for enforcing a centralized energy system that they see as intrinsically inefficient and inequitable (Moore & Hackett, 2016). The processes available for stakeholder engagement in energy systems design often focus on individual technologies alone rather than on systems, contributing to procedural injustice in which the process for evaluating a technology is incompatible with critiquing the broader system framing (ibid.).

While Desertec seems to encompass an impossibly large scale for enacting a deliberative procedure, perhaps it is not so different from other energy systems and technologies, which in reality all cross scales of time and space. The main difference lies in the intentional construction of a material system across nation-state borders. Next, I more concretely describe and characterize the energy injustices revealed in this empirical study of a vision for a multi-scalar energy system, as well as how they intersect with issues of energy security. This discussion offers lessons for the numerous electricity integration projects underway worldwide, as overviewed in Chapter 1.

Nation-state and transnational scale: justice in North-South collaboration

Neocolonialism and injustice: Chapter 2 described the similarities between colonial-era discourses related to large-scale engineering projects for the Sahara Desert and contemporary regional energy integration discourses. This illustrates a lack of sensitivity to and appreciation for the historical continuity of injustice in the region. However, the empirical discourse about neocolonialism often served as a cryptic substitute for a substantive critique. Just as colloquially calling something unsustainable or sustainable is often a mild way of calling it bad or good (Thompson, 2007), the neocolonial critique was, at times, utilized to express uneasiness with a vision that connected dozens of countries across North-South power and wealth disparities. Recognition, distributive, and procedural justice theories aid in fleshing out the neocolonial critique to characterize problems and act on them.

In energy visions for the Mediterranean, indigenous people and people living in desert landscapes often were not recognized. They were written out of the landscape, and desert landscapes were cartographically redrawn to match the vision for a technological system rather than the sociocultural terroir of the places identified as suitable for CSP development. Chapter 3 illustrated how visualizations and metaphors associated with the Sahara Desert continue to evoke Orientalist tropes and to frame the Sahara as an interstitial zone to be bridged. Those who tried to counter what de Souza, Bosco, Cavalcante, and da Costa Ferreira (2018) term an "institutional orientalist" perspective, one that blames North African nations for regulatory and institutional deficiencies impeding energy integration, reinforced Orientalist stereotypes by framing the solar terroir as devoid of anyone who could disrupt it.

The neocolonial critique relates also to distributive injustice, in which lower- and middle-income countries would accept a disproportionate share of the burdens of energy generation while citizens in wealthy countries would reap the benefits. Mitchell (2011) describes how a global framework was established in which the colonial powers put in place protectorates over other countries to control their resources, sending the benefits to citizens in the metropole, not to the local community. In the worst-case scenario, green electricity production in North Africa would reflect the unjust distribution of benefits and drawbacks observed in colonial and neocolonial extractive industries, such as uranium mining in Africa (Hecht, 2012). Concern also arose over whether in a transnational supergrid the most powerful countries with more ability to pay would maintain reliable electricity in the case of a failure. These distributive injustices could be acted on by developing institutional arrangements that ensure benefits are more fairly distributed, such as compensating people for land and water, ensuring that electricity access is provided for everyone, and meeting demand in North Africa.

Developing a just process across North and South is also essential for avoiding these distributive injustices. Not all collaboration between North and South Mediterranean is intrinsically neocolonial, and assuming as much may impede opportunities for fruitful collaboration. Some Moroccan interviewees emphasized that the outcomes of the Desertec vision could be ethical and desirable, but that the procedure available to participate in the visioning phase included very few voices from North Africa. As one interviewee put it, “I decide, you provide” is indicative of an unjust decision-making procedure, even if both sides of the Mediterranean could, in theory or practice, “win” from the distributive outcomes. A substantive and inclusive process for electricity integration is needed, with a diversity of perspectives—including those of nation-states from the South, affected publics, and marginalized populations, such as the Amazigh people.

While some well-intentioned European actors assumed that North Africans would view the Desertec vision as intrinsically neocolonial, many Moroccan stakeholders supported exporting green electricity to Europe and viewed Morocco as unfairly cut out of the European electricity market. Perceiving Morocco only through the lens of the past injustices it has suffered—essentially stereotyping it as an abused former colony—rather than seeing it as a state with agency, expertise, and social capital, perpetuates injustice. This perception itself is a performance of power that affords 100% of the agency to Western European countries and 0% to Morocco. In reality, Morocco is intentionally addressing the possible distributive injustices at the nation-state scale through local content requirements and other policies that seek to retain benefits for Morocco. These policies aim to bolster national capacity and prevent exploitation and coercion.

Energy sovereignty, global energy security, and freedom from coercion: Issues of neocolonialism relate to the concept of nation-state sovereignty because former colonies and protectorates lost their sovereignty for decades to European colonial powers. The concept of sovereignty is important in realist and neo-realist international relations theory that views the international community as being in a state of anarchy in which sovereign states pursue their own narrow self-interests, partly to secure scarce, geopolitical resources (Dannreuther, 2010). Energy sovereignty refers to the right of communities and nation-states to have the ultimate authority over how their energy systems are designed and used. Holleman (2012) argues that countries in the Global South will not be energy sovereign as long as they serve as "supply depots" for the North. Smith and Frehner (2010) define energy sovereignty for indigenous communities, which have endured a long history of violence from settler colonialism, as developing a "sustainable and self-sufficient community" that allows for self-determination and more equitable control of science, technology, and geopolitical space.

Some energy policymakers view energy security as synonymous with "energy autarky," or extreme energy autonomy, viewing any dependence or interdependence as insecure (Nance & Boettcher, 2017). In contrast, the idea of "global energy security" was advanced by the Desertec Foundation and by other actor discourses, such as the Energy Charter Treaty (Trombetta & Dyer, 2013). Global energy security may imply a collaborative rather than a combative view of energy security. It counters the energy autarkic discourse and focuses instead on securitizing energy interdependence. Goldthau and Witte (2010) argue that instead of viewing energy security as a zero-sum game, energy governance actors ought to focus on developing institutions and "rules of the game" that could result in win-win solutions and interdependencies that improve global energy security. However, I argued earlier that this win-win discourse is often incomplete. The idea of global energy security emphasizes how security and environmental problems cross nation-state boundaries and could be addressed through building better governance institutions, thus reflecting a liberal IR paradigm. Similarly, Desertec actors argued that solving climate change, which transcends borders in a globalized world, should be prioritized over nationalism and sovereignty.

Could a country be energy sovereign if its grid were shallowly or deeply integrated with other countries, or is this contradictory to the definition of energy sovereignty? Would global energy security impede energy sovereignty as defined above? Is global energy security preferable to autarky for achieving energy justice? On one hand, a non-autarkic view is preferable because if energy autarky were necessary for achieving energy justice, the trend toward the transnationalization, or even globalization, of electricity would inevitably result in injustices, as would supply chains for individual energy technologies. System complexity also makes energy autarky infeasible in reality as energy must be sourced from supply chains, and interdependencies are needed to maintain reliable energy systems (Nance & Boettcher, 2017). Energy autarky is already infringed on by multiscalar supply chain dependencies. Those dependencies are more difficult to see than are connections in a system intentionally built to materially cross borders. On the other hand, the idea of global energy security could wrest energy from its relevant historical context of injustice and from states' sociotechnical imaginaries that define what a desirable energy system is within a local context. The liberal IR framing of global energy security may overlook the history of why energy sovereignty might ideologically matter to states that were colonized.

The idea of the sociotechnical imaginary for energy, and the goals embedded within, provides a vocabulary for discerning nation-states' sovereign goals for energy. Sustainable development and conceptions of "the good life" and the modern life are differentially defined in distinct nation-state imaginaries. These imaginaries include tacit or explicit visions of what the "social value of energy" is (Miller, Altamirano-Allende, Johnson, & Agyemang, 2015), as well as how a particular state within a historical and cultural context imagines a secure energy future. The Moroccan imaginary contains a vision for Morocco's future tied to certain congruencies with its past. It is situated within a problem-framing that Morocco lacks domestic fossil fuel wealth and must bolster its independence and overall socioeconomic capacity to benefit from energy systems. However, Morocco's conception of energy security within this imaginary is non-autarkic. Rather, it emphasizes the potential security and sustainability of regional interdependence and the socioeconomic benefits of green energy exports, while emphasizing that the procedure for establishing these interdependencies must be just and must not reproduce colonial energy-extractive relationships.

I have questioned whether a regional imaginary that views a deeply interconnected energy system as advantageous to achieving regional prosperity would need to be shared across all nation-states in order for countries to buy into the deeply interconnected supergrid envisioned in Desertec Design II. I identified in Chapter 6 motivations shared between North and South to ensure the security of supply, but I also identified barriers to collaboration, such as different views of the role of the state versus the market, different end goals for climate change mitigation, and a lack of trust. Additionally, Spain's imaginary did not contain a vision for a regional energy system or see integration as advantageous to its economic security. The Desertec actors lacked a vocabulary for describing these differences, and even incongruences, and, instead, viewed solar power as apolitical while conceiving of deep technological integration as something that could occur separately from, or even determine, political integration. It remains unclear whether shared imaginaries, or substantial synergy and overlap among imaginaries, would need to exist for deep integration to take hold, or whether making visible the different goals and histories behind these imaginaries and developing a just process for designing social and technological infrastructure that accounts for these different goals could be sufficient for achieving integration. It is also important to consider whether shared imaginaries, or at least shared definitions of security, would be necessary for achieving global energy security.

The principle of freedom from coercion is one possible means of negotiating the tensions between sovereignty and interdependence without resorting to autarky. Sen (2009) argues that freedom of choice matters for achieving justice; someone who chooses to stay at home for the day has more freedom than a person coerced into staying at home by a gang of thugs, even though the end result is the same. Morocco is currently free to motivate national low-carbon energy development as part of a socioeconomic development strategy rather than an EU-focused climate change mitigation strategy, but both framings ideally will result in mitigation. Morocco thus has the freedom to imagine and to set its own goals for its energy systems but is still open to collaboration on outcomes desired by multiple countries. In contrast, in Chapter 5, I discussed how European pressure on Morocco to follow environmental regulations on sword fishing resulted in conflict because it reflected the EU's environmental sustainability goals rather than Morocco's sustainable development needs, suggesting a coercive relationship in which differing goals have not been recognized and negotiated. Freedom from coercion would be important in establishing situations in which mutual interdependence is perceived on the part of two or more nation-states, rather than lopsided dependence. Capacity building of the kind described in Morocco in Chapters 5, 6, and 7 is likely a precursor to avoiding coercion so that Morocco can improve its negotiating skills and capacity to benefit from energy integration.

Regional procedural justice for nation-states and citizen stakeholders: The principle of freedom from coercion leads to a more difficult, practical question of how to achieve a just, noncoercive process for regional energy integration. The transnational space in which global energy security would be defined and enacted requires an examination of the intersection between energy security and energy justice. Further complicating the goal of energy sovereignty is that the framing of justice as administered only by the nation-state is no longer sufficient in today's globalized world in which injustices are neither contained by nation-state borders nor legislated by the state alone (Fraser, 2009). If a sovereign state were needed to apply the principles of justice, and if geographical proximity were required for negotiating the principles of justice, achieving energy justice would be impossible in complex systems that cross borders where non-state actors have significant power in energy governance. Fraser (2009) terms this the space of "flows" rather than "places." In a regional electricity system, both transnational and domestic inequities relate to energy injustice. The scale of justice is therefore no longer at the level of state alone, but rather at the level of the region in which energy flows.

Fraser's theory of "post-Westphalian democratic justice" recommends the institutionalization of a "transnational meta-democracy" that is representative and that enacts fair procedures, as well as decides who counts as a subject of justice (2009). In Chapter 3, I questioned whether a global framing of energy security should aim to achieve equal distribution among states or equitable distribution among citizens and social groups within and across states. A focus on energy equality emphasizes only maldistribution, not misrecognition and lack of representation in a transnational process. Within a framing of global energy security as equality, a developing country could be derided for "resource nationalism" for refusing to export energy resources that would increase global supply. However, this country likely lacks access to a just transnational procedure that would facilitate achieving equitable distribution of the benefits stemming from energy extraction, and therefore it may be justified in focusing on domestic supply.

Fraser recommends an institution that would decide "who should count with respect to what in a post-Westphalian world" (p. 6). Who counts as a subject of justice in a system of regional energy flows? Can all stakeholders be adequately represented in practice? In the Mediterranean, representation only by a private industry consortium such as Dii, or other transnational corporations, is insufficient because Mediterranean publics have not elected these corporations to represent them. Transnational elites and nation-states are not the only relevant "who" of justice—citizens are as well, which requires opening up the technocratic domain of energy decision making. Since direct participation at the regional scale is not feasible, citizen representatives should adequately speak for the populations they represent rather than communicating only the views of elites, in contrast to the case of the rural commune outside Ouarzazate described in Chapter 7. Achieving regional or global energy security would require a just procedure for envisioning and designing energy systems, one that seeks to establish mutual interdependence. In regions seeking to integrate their electricity systems, an energy policymaking and regulatory body should be established that is representative, with an anticipatory outlook. While this is admittedly difficult, electricity regionalism projects largely have failed even to recognize and frame this as a challenge. Such a process likely would be impossible for a global grid, such as the GEIDCO vision discussed in Chapters 1 and 4.

Citizen scale: a cosmopolitan critique of energy systems envisioned and built as if certain lives do not matter

On February 20, 2018, the well-reputed US public TV news program *PBS NewsHour* reported on the Moroccan Solar Plan in a 6.5-minute segment entitled "Morocco Turns the Sahara Desert into a Solar Energy Oasis." The segment asserts that the Spanish company SENER is building the Noor-Ouarzazate CSP plant in Morocco. It does not acknowledge that the consortium is led by the Saudi company ACWA Power, in spite of including an interview clip with ACWA Power's CEO, Paddy Padmanathan. *NewsHour* states that Noor-Ouarzazate is in the Sahara Desert rather than on the edge of the Sahara. It frames Ouarzazate as a "tiny" "tourist stop" that produces movies. In reality, the Province of Ouarzazate has a population of 300,000 people and is much more than a tourist stop, as discussed in Chapter 7. The segment then moves without transition to discuss Desertec. Francisco Stepo, president emeritus of the Club of Rome, argues that "Desertec set the guidelines on how to produce renewable energy in the desert to supply electricity to Europe." Stepo states that Noor is a pilot project of the Desertec idea, an assertion that I realized was a fallacy on my first day of fieldwork in Morocco in 2012. The segment then frames the Arab Spring and the attack on the Algerian natural gas facility (discussed in Chapter 5) as the demise of Desertec. Stepo states that "geopolitical instability is always the main obstacle for the implementation of these kinds of projects." *NewsHour* uses a clip from the movie *Casablanca* to discuss water scarcity in the desert. Edwards (2005) states that "*Casablanca* is the paradigmatic example of 'American Orientalism,'" which was filmed in California and relegates Moroccans to the background (Edwards, 2005, p. 71). He explains that the film "remaps North Africa as [it] plots [its] vision of Americanness onto the cleared space" (p. 61). The clip *NewsHour* uses shows the character Captain Louis Renault telling Rick Blaine that there is no water because Casablanca is in the desert. But Casablanca is not in the desert; it is a coastal city (ibid.).

NewsHour's assertion that Noor is a European-led project overwrites South-South collaboration and agency. The Moroccan Solar Plan, MASEN, and the other domestic institutions and policies crucial to implementing solar power as part of a socioeconomic development strategy for Morocco are not discussed. Rather, *NewsHour* essentially frames Noor as part of a European imaginary grafted onto the Sahara Desert. The segment has much more to do with the American imaginary of Morocco than the Moroccan Solar Plan. It is remarkable that such inconsistencies and mistruths about Desertec and the Moroccan Solar Plan still widely circulate in spite of years of scrutiny and criticism focused on them. The energy policy goal ought be to build a resilient, clean electricity system, but what instead can be viewed as resilient is the Orientalist framing of North Africa that Edward Said has long critiqued (for example, Said, 1979).

Resilient forms of discrimination also relate to energy systems within countries. Writing of the Black Lives Matter movement in the United States, a social movement that has launched a critique that Black people are treated as if they do not matter, Lennon (2017) identifies a need to "radically decolonize energy" rather than to democratize it. He states, "The colonial apparatus transformed energy—the ability to change matter—into a commoditized form that made certain lives not matter" (p. 19). The grid long has functioned as a colonizing sociotechnical system in its uneven distribution of pollution and negative effects in communities of color in the United States. Black Lives Matter activists have given up on the grid as a social institution since it has resulted in misrecognition and maldistribution.

Lennon's argument can be applied to sociotechnical systems in the Mediterranean through numerous examples discussed in this book in which certain populations remain undervalued, and at times unseen, in systems and visions that cross broad power and wealth disparities. For example, I discussed how colonial visions sought to reengineer the Sahara into a landscape of energy production and mobility interconnection and into a water and food source for Europe, while viewing some populations as undeserving of reliable electricity. The colonial electricity system in Morocco served extractive industries rather than local populations (Saul, 2002). A Moroccan wind developer argued that wind resources in the Atlantic have been used historically to marginalize and exploit rather than to power and empower. A Moroccan professor criticized a structural social system that creates different classes of people who have inequitable access to opportunities and resources, as well as an energy policy that perceives Morocco's sunlight, rather than its citizens, as Morocco's greatest resource to be valorized. A nomad woman living in the Sahara Desert struggled to meet her basic needs, illustrating a lack of equity of opportunity for citizens to access sustainable development. Opportunities were constrained for the citizens living closest to the Noor solar power facility to participate in the siting process and to define appropriate development projects for their community. Citizens protesting in Al Hoceima, Morocco wanted to be included in the nation-state imaginary of socioeconomic development, and protested a police force that treated human beings as garbage.

A resilient energy system is one that responds to perturbations and shocks without failure. Currently, Orientalism and racism serve as resilient social infrastructures that result in systemic discrimination. From a technological standpoint, a regional grid could respond elegantly to disruption, but if it is built as if certain social groups do not matter, it will not be socially resilient. Sovacool et al. (2017) argue that deliberate resistance to energy injustice should be a principle of energy justice, but I argue that what must be resisted is more fundamental and foundational. Building a just energy system requires not only building a resilient sociotechnical system, but also resisting and dismantling resilient social systems that legislate inequity. Otherwise, we are left with energy systems that have potential equalizing benefits but end up failing to serve, and even harming, certain social groups. In seeking to build a resilient grid, we must also avoid building juggernauts that are inflexible to change. How, then, can the challenges of mitigating climate change, generating clean energy, and decolonizing energy be addressed? As a starting point, I discuss intergenerational equity rather than equality, human capabilities, and "deep energy security."

Intergenerational justice and social groups: Sovacool and Dworkin (2014) identify "intergenerational equity" as one principle of energy justice, defined as: "Future generations [having] a right to enjoy a good life undisturbed by the damage our energy systems inflict on the world today" (p. 367). The Brundtland Report also implicitly references intergenerational justice within its definition of sustainable development as "development that meets the needs of the present without compromising the ability of future generations to meet their own needs" (Brundtland, 1987). The full environmental costs of energy consumption to both present and future generations are rarely taken into account (Sovacool & Dworkin, 2014). In addition, the econometric practice of using a discount rate in climate change modeling prioritizes the needs of present over future generations (Matson & Carasso, 1999).

The conditions for achieving intergenerational equity can partly be met through procedural justice, as future generations could be spoken for during planning and decision-making processes. It would be a conceptual mistake, however, to see the recognition of future generations as a form of recognition justice because recognition justice is identity-based, focusing on social groups, not generations. While Sovacool and Dworkin (2014) also include intragenerational equity in their framework, or "all people [having] a right to fairly access energy services," the definition of intergenerational equity they present could more accurately be labeled intergenerational equality (p. 367). Without recognition and distributive justice for present generations of social groups who lack equity of opportunity, the natural capital left to future generations will continue to be inequitably distributed among social groups, even if equivalent amounts are sustained between generations.

Attention should be given to power dynamics and social inequity in intergenerational justice theory. Powerful actors make decisions in the present about what is left to the future, and these decisions sometimes call on social groups in the present generation to make sacrifices unlikely to benefit their social group in future generations. Instead, generations are often discussed as if they are homogenous, with clear beginning and end points that allow for comparison.

Norton (2005) critiques the "ignorance problem," that philosophers cannot predict the needs of future generations, as nonsensical since the basic needs of future humans can safely be assumed. Similarly, it can be assumed that there are systemic and contingent causes of poverty and social exclusion that have continuity from a particular social group in one generation to the next. We know how long socio-energy systems last and that design choices made today will result in maldistribution for particular social groups during that time period. Norton's "convergence hypothesis" suggests that environmental policy goals based on long-sighted anthropocentric values would converge in the long term with those based on non-anthropocentric values (Minteer, 2009). I argued in Chapter 5 that the interests of disadvantaged social groups will not eventually converge with other human and nonhuman interests without equity of opportunity for these social groups to chart new paths for their progeny. Additionally, "weak sustainability," which focuses on the overall preservation of natural capital from one generation to the next through technological change, overlooks the fact that technological advancements are fostered within sociotechnical imaginaries that encompass particular views of what sustainable development ought to look like while excluding particular social groups.

Cosmopolitanism, capabilities, and "deep energy security": The previous sections have illustrated that wins cannot be assumed for all individuals and groups under a global energy security framework. Energy security relates not only to physical resources but also to broader claims about what a good life entails, the power and the framing of multiple actors, the collective imagination of a society, and the contestation of social groups clamoring for access to the sociotechnical imaginary or for their own space to imagine. Thus, security has expanded from the bounds of sovereignty over resources, and intersects with claims about justice and fairness among states, regions, groups, and also individuals. A commitment to a global conceptualization of energy security that is secure for individuals may require a concomitant commitment to cosmopolitanism in which the needs of citizens in all nations are valued universally.

Cosmopolitan justice focuses on individuals rather than nation-states and universally applies justice (Sovacool, Heffron, McCauley, & Goldthau, 2016). A cosmopolitan approach confers equal rights to all individuals, regardless of their nationality or citizenship (Mandle, 2006). Activists on social media who questioned the *Je Suis Charlie* hashtag, as described in Chapter 3, made visible the priority given to French lives over the lives of people living in North Africa and the Middle East. Cosmopolitanism conceives of people as citizens of the world rather than only of nation-states (Heffron, McCauley, & Sovacool, 2015). It applies procedural and distributive justice globally, arguing that each person has an intrinsic worth and should have access to energy (Sovacool & Dworkin, 2014). Rather than focusing on increasing oil supply for all, a cosmopolitan framing of global energy security would also value the human rights of people living in sites of oil extraction.

Under the cosmopolitan capabilities approach to justice, it is insufficient to develop just institutions in society; the capabilities of individuals to secure their rights and to live out fulfilling and enriching lives should also be improved (Sen, 2009). Aggregate measures of human development, such as GDP, are insufficient metrics for measuring individual capabilities because they do not account for the distribution of benefits from economic growth (ibid.). It can be inferred that energy's worth ought to be measured not simply using electrification rates and energy usage per capita, but by assessing the services that electricity provides to improve human capabilities, such as lighting a house so that children can do homework at night. Through this lens, renewable energy should not be considered as a good in and of itself but as a tool for achieving human development goals (Goldemberg, Johansson, Reddy, & Williams, 1988). Nussbaum (2007) views capabilities as the foundations that afford people the capacity to secure their human rights. For example, better education can improve a person's articulateness, aiding in securing his or her right to free speech (ibid.). Additionally, improved capabilities for individuals within disadvantaged social groups would be necessary for achieving intergenerational equity.

Sen's approach to justice allows people to chart their own definitions of freedom and development (2009). Rather than viewing the role of citizens in sustainable development as volunteers who clean up the environment, or as people in need of discipline to follow rules that impose the needs of the collective over those of the individual, citizens are afforded the agency to imagine what a good life would look like. Tidwell and Tidwell (2018) emphasize the normative importance of examining citizens' sociotechnical energy imaginaries even if these imaginaries are not collectively held. This methodology could aid in building energy systems that improve people's capabilities on their own terms. To improve human capabilities and human security, capacity building is needed not just at the nation-state level, but also at the scale of social groups that have been excluded from the national sociotechnical imaginary or whose visions of the good life go unseen. At the local level, people living near Noor-Ouarzazate are among the most vulnerable in the nation in terms of human security, but are the least equipped to participate in the process of energy development and to benefit from it. Therefore, capacity building is needed at the local level, in addition to the nation-state level, so that marginalized communities can more easily secure benefits.

The concept of "deep energy security" adds a cosmopolitan sensibility to energy security by prioritizing the needs of individuals rather those of nation-states alone. "Deep energy security" is "energy security that contributes to human security over space (from the local to the global) and time (that is, now and for future generations)" (Karlsson-Vinkhuyzen & Jollands, 2013, pp. 513–514). This normative perspective calls attention to the possibility that a win at the nation-state scale could be a loss for individuals' human security. Rather than viewing energy as a prize to enhance nation-state security, clean energy should be viewed as a public good for global citizens (Karlsson-Vinkhuyzen, Jollands, & Staudt, 2012). "Deep energy security" offers a useful additive to global energy security.

I do not believe that cosmopolitanism, and a focus on the individual, necessarily demands energy decentralization. I recommended in Chapter 7 that a mix of decentralized and centralized energy systems could benefit both the local and nation-state scales. A centralized grid offers potential to provide electricity as a social service that improves human capabilities. The centralized approach in

North Africa has provided greater access to energy services than what is available in many sub-Saharan African countries today. The scalability of a fully distributed energy system is questionable, as such a system would require individuals and communities to raise significant capital, secure individual access to resources, and bear risks individually rather than collectively. A secure and just grid ought to meet the justice and security criteria presented in this chapter to the greatest extent possible. If it is well designed, a grid can provide the service of reliable, resilient, and low-cost energy for all, and if the energy is clean, then it could provide a public good under the "deep energy security" approach.

Improving human security and human capabilities and reducing energy inequity across the globe should be the first priorities of a global framing of energy security. An energy system in which millions live in energy poverty and outcomes are inequitably distributed fails to ensure human security and, therefore, is difficult to conceive of as globally secure. Whether the energy infrastructures we build are centralized, decentralized, or both, we must build them as if security and justice for all humans and all social groups matter.

Works cited

Adam, B. (2003). Reflexive modernization temporalized. *Theory, Culture & Society*, *20*(2), 59–78. Retrieved from https://doi.org/10.1177/0263276403020002004

Backhaus, K., Gausling, P., & Hildebrand, L. (2015). Comparing the incomparable: Lessons to be learned from models evaluating the feasibility of Desertec. *Energy*, *82*, 905–913. Retrieved from https://doi.org/10.1016/j.energy.2015.01.100

Bickerstaff, K., Walker, G., & Bulkeley, H. (2013). *Energy justice in a changing climate: Social equity and low-carbon energy*. London and New York, NY: Zed Books.

Brundtland, G. H. (1987). *Report of the World Commission on Environment and Development: Our common future*. United Nations.

Dannreuther, R. (2010). *International relations theories: Energy, minerals and conflict* (Polinares Working Paper no. 8). Polinares. Retrieved from www.polinares.eu/docs/d1-1/polinares_wp1_ir_theories.pdf

Davies, S. R., & Selin, C. (2012). Energy futures: Five dilemmas of the practice of anticipatory governance. *Environmental Communication*, *6*(1), 119–136. Retrieved from https://doi.org/10.1080/17524032.2011.644632

de Souza, L. E. V., Bosco, E. M. G. R. L., Cavalcante, A. G., & da Costa Ferreira, L. (2018). Postcolonial theories meet energy studies: "Institutional orientalism" as a barrier for renewable electricity trade in the Mediterranean region. *Energy Research & Social Science*, *40*, 91–100. Retrieved from https://doi.org/10.1016/j.erss.2017.12.001

Edwards, B. T. (2005). *Morocco bound: Disorienting America's Maghreb, from Casablanca to the Marrakech Express*. Durham, NC: Duke University Press.

Fraser, N. (2009). *Scales of justice: Reimagining political space in a globalizing world*. New York, NY: Columbia University Press.

Goldemberg, J., Johansson, T. B., Reddy, A. K. N., & Williams, R. (1988). *Energy for a sustainable world* (1st ed.). New York, NY: John Wiley & Sons Inc.

Goldthau, A., & Witte, J. M. (Eds.). (2010). *Global energy governance: The new rules of the game*. Washington, DC: Brookings Institution Press.

Hecht, G. (2012). *Being nuclear: Africans and the global uranium trade*. Cambridge, MA: MIT Press.

Heffron, R. J., McCauley, D., & Sovacool, B. K. (2015). Resolving society's energy trilemma through the Energy Justice Metric. *Energy Policy*, *87*, 168–176. Retrieved from https://doi.org/10.1016/j.enpol.2015.08.033

Holleman, H. (2012). *Energy justice and foundations for a sustainable sociology of energy*. University of Oregon (Doctoral dissertation). Retrieved from https://scholarsbank.uoregon.edu/xmlui/handle/1794/12419

Hughes, T. (1983). *Networks of power*. Baltimore, MD: Johns Hopkins University Press.

Islar, M., Brogaard, S., & Lemberg-Pedersen, M. (2017). Feasibility of energy justice: Exploring national and local efforts for energy development in Nepal. *Energy Policy*, *105*, 668–676. Retrieved from https://doi.org/10.1016/j.enpol.2017.03.004

Jasanoff, S., & Kim, S-H. (2015). *Dreamscapes of modernity: Sociotechnical imaginaries and the fabrication of power*. Chicago, IL: University of Chicago Press.

Jenkins, K., McCauley, D., Heffron, R., Stephan, H., & Rehner, R. (2016). Energy justice: A conceptual review. *Energy Research & Social Science*, *11*, 174–182. Retrieved from https://doi.org/10.1016/j.erss.2015.10.004

Karlsson-Vinkhuyzen, S. I., & Jollands, N. (2013). Human security and energy security: A sustainable energy system as a public good. In M. J. Trombetta & H. Dyer (Eds.), *International handbook of energy security* (pp. 507–525). Cheltenham, UK: Edward Elgar Publishing.

Karlsson-Vinkhuyzen, S. I., Jollands, N., & Staudt, L. (2012). Global governance for sustainable energy: The contribution of a global public goods approach. *Ecological Economics*, *83*, 11–18. Retrieved from https://doi.org/10.1016/j.ecolecon.2012.08.009

Lennon, M. (2017). Decolonizing energy: Black Lives Matter and technoscientific expertise amid solar transitions. *Energy Research & Social Science*, *30*, 18–27. Retrieved from https://doi.org/10.1016/j.erss.2017.06.002

Mandle, J. (2006). *Global justice* (1st ed.). Oxford: Polity Press.

Matson, R. J., & Carasso, M. (1999). Sustainability, energy technologies, and ethics. *Renewable Energy*, *16*(1–4), 1200–1203. Retrieved from https://doi.org/10.1016/S0960-1481(98)00482-0

McShane, T. O., Hirsch, P. D., Trung, T. C., Songorwa, A. N., Kinzig, A., Monteferri, B., . . . O'Connor, S. (2011). Hard choices: Making trade-offs between biodiversity conservation and human well-being. *Biological Conservation*, *144*(3), 966–972. Retrieved from https://doi.org/16/j.biocon.2010.04.038

Miller, C. A., Altamirano-Allende, C., Johnson, N., & Agyemang, M. (2015). The social value of mid-scale energy in Africa: Redefining value and redesigning energy to reduce poverty. *Energy Research & Social Science*, *5*, 67–69. Retrieved from https://doi.org/10.1016/j.erss.2014.12.013

Minteer, B. (Ed.) (2009). *Nature in common? Environmental ethics and the contested foundations of environmental policy*. Philadelphia, PA: Templeton University Press.

Mitchell, T. (2011). *Carbon democracy: Political power in the age of oil*. London: Verso.

Moore, S., & Hackett, E. J. (2016). The construction of technology and place: Concentrating solar power conflicts in the United States. *Energy Research & Social Science*, *11*, 67–78. Retrieved from https://doi.org/10.1016/j.erss.2015.08.003

Nance, M. T., & Boettcher III, W. A. (2017). Conflict, cooperation, and change in the politics of energy interdependence: An introduction. *Energy Research & Social Science*, *24*, 1–5. Retrieved from https://doi.org/10.1016/j.erss.2016.12.020

Newell, P., & Mulvaney, D. (2013). The political economy of the "just transition". *The Geographical Journal*, *179*(2), 132–140. Retrieved from https://doi.org/10.1111/geoj.12008

Norton, B. (2005). *Sustainability: A philosophy of adaptive ecosystem management*. Chicago, IL: University of Chicago Press.

Nussbaum, M. C. (2007). *Frontiers of justice: Disability, nationality, species membership (Tanner Lectures on Human Values)*. Cambridge, MA: Belknap Press of Harvard University Press.

O'Brien, K. L., & Leichenko, R. (2003). Winners and losers in the context of global change. *Annals of the Association of American Geographers, 93*, 89–103. Retrieved from https://doi.org/10.1016/S0959-3780(00)00021-2

Reames, T. G. (2016). Targeting energy justice: Exploring spatial, racial/ethnic and socio-economic disparities in urban residential heating energy efficiency. *Energy Policy, 97*, 549–558. Retrieved from https://doi.org/10.1016/j.enpol.2016.07.048

Said, E. W. (1979). *Orientalism*. New York, NY: Pantheon Books.

Saul, S. (2002). L'électrification du Maroc à l'époque du protectorat. *Outre-Mers, 89*(334–335), 491–512.

Sen, A. (2009). *The idea of justice*. Cambridge, MA: Harvard University Press.

Smith, J., & High, M. M. (2017). Exploring the anthropology of energy: Ethnography, energy and ethics. *Energy Research & Social Science, 30*, 1–6. Retrieved from https://doi.org/10.1016/j.erss.2017.06.027

Smith, S. L., & Frehner, B. (2010). *Indians & energy: Exploitation and opportunity in the American Southwest*. Santa Fe, NM: SAR Press.

Sovacool, B. K. (2016). How long will it take? Conceptualizing the temporal dynamics of energy transitions. *Energy Research & Social Science, 13*, 202–215. Retrieved from https://doi.org/10.1016/j.erss.2015.12.020

Sovacool, B. K., Burke, M., Baker, L., Kotikalapudi, C. K., & Wlokas, H. (2017). New frontiers and conceptual frameworks for energy justice. *Energy Policy, 105*, 677–691. Retrieved from https://doi.org/10.1016/j.enpol.2017.03.005

Sovacool, B. K., & Dworkin, M. H. (2014). *Global energy justice: Problems, principles, and practices*. Cambridge: Cambridge University Press.

Sovacool, B. K., & Geels, F. W. (2016). Further reflections on the temporality of energy transitions: A response to critics. *Energy Research & Social Science, 22*, 232–237. Retrieved from https://doi.org/10.1016/j.erss.2016.08.013

Sovacool, B. K., Heffron, R. J., McCauley, D., & Goldthau, A. (2016). Energy decisions reframed as justice and ethical concerns. *Nature Energy, 1*(5), 16024.

Stirling, A. (2014). Transforming power: Social science and the politics of energy choices. *Energy Research & Social Science, 1*, 83–95. Retrieved from https://doi.org/10.1016/j.erss.2014.02.001

Sturm, C. (2017, January 9). Inside the Energiewende: Policy and complexity in the German utility industry. *Issues in Science and Technology*. Retrieved February 12, 2018, from http://issues.org/33-2/inside-the-energiewende-policy-and-complexity-in-the-german-utility-industry/

Thompson, P. (2007). Agricultural sustainability: What it is and what it is not. *International Journal of Agricultural Sustainability, 5*(1), 5–16.

Tidwell, J. H., & Tidwell, A. S. D. (2018). Energy ideals, visions, narratives, and rhetoric: Examining sociotechnical imaginaries theory and methodology in energy research. *Energy Research & Social Science, 39*, 103–107. Retrieved from https://doi.org/10.1016/j.erss.2017.11.005

Trombetta, M. J., & Dyer, H. (2013). *International handbook of energy security*. Cheltenham, UK: Edward Elgar Publishing.

Van Lente, H. (2000). Forceful futures: From promise to requirement. In N. Brown, B. Rappert, & A. Webster (Eds.), *Contested futures: A sociology of prospective techno-science* (pp. 43–64). Burlington, VT: Ashgate.

Index

Page numbers in *italics* indicate figures and in **bold** indicate tables on the corresponding pages.